Studies in Systems, Decision and Control

Volume 392

The series "Studies in Systems, Decision and Control" (SSDC) covers both new developments and advances, as well as the state of the art, in the various areas of broadly perceived systems, decision making and control–quickly, up to date and with a high quality. The intent is to cover the theory, applications, and perspectives on the state of the art and future developments relevant to systems, decision making, control, complex processes and related areas, as embedded in the fields of engineering, computer science, physics, economics, social and life sciences, as well as the paradigms and methodologies behind them. The series contains monographs, textbooks, lecture notes and edited volumes in systems, decision making and control spanning the areas of Cyber-Physical Systems, Autonomous Systems, Sensor Networks, Control Systems, Energy Systems, Automotive Systems, Biological Systems, Vehicular Networking and Connected Vehicles, Aerospace Systems, Automation, Manufacturing, Smart Grids, Nonlinear Systems, Power Systems, Robotics, Social Systems, Economic Systems and other. Of particular value to both the contributors and the readership are the short publication timeframe and the world-wide distribution and exposure which enable both a wide and rapid dissemination of research output.

Indexed by SCOPUS, DBLP, WTI Frankfurt eG, zbMATH, SCImago.

All books published in the series are submitted for consideration in Web of Science.

More information about this series at http://www.springer.com/series/13304

Tomasz Szapiro · Janusz Kacprzyk
Editors

Collective Decisions: Theory, Algorithms And Decision Support Systems

Springer

Editors
Tomasz Szapiro
Collegium of Economic Analysis
SGH Warsaw School of Economics
Warszawa, Poland

Janusz Kacprzyk
Polish Academy of Sciences
Systems Research Institute
Warsaw, Poland

ISSN 2198-4182 ISSN 2198-4190 (electronic)
Studies in Systems, Decision and Control
ISBN 978-3-030-84996-2 ISBN 978-3-030-84997-9 (eBook)
https://doi.org/10.1007/978-3-030-84997-9

This Springer imprint is published by the registered company Springer Nature Switzerland AG
The registered company address is: Gewerbestrasse 11, 6330 Cham, Switzerland

Professor Gregory E. Kersten (1949–2020) in memoriam

Preface

Very many decisions, in particular in important and relevant contexts, are made jointly, that is by a group of individuals or agents who are usually organized in various forms exemplified by teams, committees, juries, panels, partnerships, etc. In many cases, such a mode of decision-making is advantageous, preferable to individual decision-making, from a general point of view, even if decisions reached in such a way can be for many members of the group different than decisions individually made. Of course, this advantage of a collective decision that so often occurs is at the expense of more time needed to reach such a decision because there is a need for a proper deliberation, discussion and dialogue, which can be time consuming, and then a need to develop a proper procedure for making decisions in a collective way, for instance consensus decision-making which is often followed.

In our context, of special relevance is the negotiation which basically boils down to a dialogue between two or mode parties, individuals, small or larger human groups but more and more often agents (usually software) and their groups. Negotiations are usually needed when a conflict emerges with respect to one or more aspects, for instance, salaries, and the parties involved need to find an agreement by resolving points of difference between the parties so that various interests be satisfied, may be not of all parties but many, and well in general. This involves many aspects, for instance proposal and deal and making concession to attain an agreement.

It is obvious that negotiations are omnipresent in virtually all non-trivial human activities and enter more and more into group decisions in multi-agent systems. Evidently, this omnipresence and crucial importance of negotiations has triggered an urgent need for developing formal models to support various kinds of negotiation processes. These models involve various models of group decisions, team decisions, social choice and voting, game theoretic models, etc. This all is augmented, notably in recent years, by the use of tools and techniques from the huge field of social media, data analytics, cognitive and psychological models, etc. Moreover, to make the models developed useful in practice, they are implemented as software packages for supporting negotiations and various collective decision models.

This volume is a token of appreciation of our colleague and friend, the late Professor Gregory E, Kersten, who unexpectedly passed away some months ago. He was one of the most prominent and active researchers and scholars in the broadly perceived field of collective decisions, notably negotiations, the author of numerous influential papers, books and edited volumes. He was a great scientist and also a loyal friend and colleague, a mentor for many younger people in the field. Therefore, we—on behalf of the research community—have put together this volume to commemorate Gregory and his great scientific achievements. The papers included present a wide perspective of various topics, approaches and tools and techniques that are of relevance for the broadly perceived collective decision-making, notably in its negotiation related aspects.

Part I "foundational aspects, and tools and techniques" is devoted to the presentation of some basic aspects of collective decisions, notably voting, auctioneering, group decision-making, etc. The contributions will span the whole area in the sense of covering all types of models, from those in roots in economics, psychology and cognitive sciences, to those based on formal, mathematical analyses and computational algorithms.

Hannu Nurmi, Janusz Kacprzyk and Słąwomir Zadrożny (Voting Systems in Theory and Practice) are concerned with various aspects of voting which is a very important way of arriving at a joint opinion. The papers focus mainly on the presentation and analysis of some important voting paradoxes, which have been known for many years, even centuries. Then, some remarks on main theoretical results stating incompatibilities between various choice desiderata are given. Moreover, a discussion of possible reasons for omitting these negative results and their dangers in the design of practical voting procedures is outlined. A brief evaluation of the newly proposed voting procedure is provided with emphasis on the importance of a comprehensive comparative analysis of several voting rules and criteria of performance in the rule selection process continuing the previous works by the authors.

Piotr Faliszewski, Stanislaw Szufa and Nimrod Talmon (Optimization-Based Voting Rule Design: The Closer to Utopia the Better) provide a comprehensive and full-fledged presentation of novel results related to the specification of requirements for the popular and powerful multiwinner voting rule so that it provides committees that correspond to some desirable intuitive notions exemplified by individual excellence of particular committee members or their diversity. Several such desired requirements, referred to as "utopias", are described. With such utopias available, an optimization-based mechanism for constructing committee scoring rules is developed that provide results as close to these utopias as possible by testing the results on the weakly separable and OWA (ordered weighted averaging)-based rules. This new approach provides some interesting insights as well as makes it possible to recover some believed connections between known multiwinner voting rules and certain applications and—moreover—makes it possible to find other interesting properties and insights.

Gregory E. Kersten, Tomasz Szapiro and Shikui Wu (A Procedure for Multiattribute Reverse Auctions with two Strategic Parameters) deal with the reverse auctions which are often used in the procurement of goods and services, with the single-attribute, typically price-only, auctions being the most widely employed. Multiattribute auction mechanisms have also been proposed and implemented, which are obviously more complex, and a dilemma that designers of multiattribute auction mechanisms face is the trade-off between the buyer's preference revelation and the ability of the sellers (bidders) to construct effective offer, that is those which are preferred by the buyer over offers made earlier. Full transparency, i.e. the requirement that the buyer gives a complete preference information to the bidders, is often rejected by the buyers because of an obvious possible future problems when competitors obtain this information. Zero transparency can make the auctions ineffective because bidders lack the information necessary for bid construction. In the paper, a novel procedure is developed in which preference information is transformed into the information on reservation levels which are perturbed so that the bidders cannot compute the buyer's preferences. The perturbation is controlled by two parameters which have strategic character for the auction mechanism and jointly represent the owner's trade-off between the maintaining of the preference secrecy and the getting of an efficient winning bid.

Miłosz Kadziński, Grzegorz Miebs, Dariusz Grynia and Roman Słowiński (Aggregation of Stochastic Rankings in Group Decision Making) propose a novel method for group decision-making which is applicable for multiple criteria ranking problems in which alternatives need to be ordered from the best to the worst by multiple decision-makers. In the first stage, incomplete preference information of each decision-maker is analysed via the Monte Carlo simulation which is applied for exploiting the space of preference model parameters compatible with each preferences of the decision-maker. The values of stochastic acceptability indices are estimated that quantify the support given to the preference, indifference, and incomparability relations for each pair of alternatives. Then, the stochastic rankings obtained are aggregated into a group compromise recommendation that minimizes either an average or a maximal distance from the input of each decision-maker. In addition to the accounting of the utilitarian and egalitarian perspectives, the dedicated mathematical programming models developed deal with the processing and constructing of complete or partial rankings. The proposed method is coupled with the robust variants of the well-known and widely used PROMETHEE I and II methods; however, it can be combined with any method from the broad family of stochastic multicriteria acceptability analysis (SMAA) techniques. An application for the ranking of project proposals at a research funding agency is shown.

Bilyana Martinovski (Quantitative Measures for Recognition of Negotiation Style and Activity) is concerned with some important aspects of group decision and negotiation support systems which are meant to provide support for different types of communication activities. The main purpose of the study is to identify and test cross-linguistic measures for recognition of communication activities in order to provide support for various forms of group decision and negotiation support systems, including e-negotiation support systems. Modern and advanced tools and techniques of the quantitative cross-activity and cross-linguistic analysis are employed in order to test if complex linguistic measures, such as caution, liveliness, stereotypicality and other ones, can be used to recognize spoken language negotiation activity and style. The results obtained show that the modern automated linguistic analysis is a useful tool for the recognition of activity type in spontaneous speech and that the activity factor influences the interaction in a more profound way than the language factor or the national culture factor. This may have a considerable impact of the development of group decision and negotiation support systems.

Gregory E. Kersten, Ewa Roszkowska and Tomasz Wachowicz (How Well Agents Represent their Principals' Preferences: The Effect of Information Processing, Value Orientation, and Goals) are concerned with an important problem related to the situation that some agents negotiate on behalf of their principals with an aim of setting the state for negotiations. The focus of this study is related to the negotiation preparation phase when the agents are given information in both text and graphical forms which unambiguously express their principal's preferences which should clearly be followed. The online bilateral negotiation experiments show that the majority of the agents make both ordinal and cardinal errors in the reconstruction of the principal's preferences. It shows that the agents' intrinsic motivation contributes to the errors and has a negative effect on their preparation. To determine how different types of motivation contribute to the agents' inaccuracy, a hierarchical framework of the epistemic, social and identity motivation is proposed which is based on a structural equation model (SEM). The new model developed shows relations among the motivation types and between the motivation and accuracy. The results indicate that the rational motivation system has a significant direct effect on social motivation (accommodating, competing and collaborating intentions) and also on the identity motivation (learning and substantive goals). The model also shows that only the identity motivation has a direct effect on the accuracy of the preferences formulated by the agents in comparison with their principal's preferences.

Part II, "decision and negotiation systems", comprises papers focusing on the implementation of various approaches, algorithms and tools and techniques through computer-based systems, notably of a decision support type.

Sabine T. Koeszegi (Automated Decision Systems: Why Human Autonomy is at Stake?) is concerned with the issues related to the model- and data-based support systems for supporting decision-making that has been used for over a century, hoping that decisions derived via decision support systems will be better, more objective and fairer, that is, more efficient and less biased. With data-based artificial intelligence (AI) systems, this hope has experience a revival. The AI systems are meant to support, for instance, physicians in diagnosing illnesses, assist managers when recruiting personnel, to just mention a few. Furthermore, the AI systems could even replace humans when deciding whether or not applicants receive a bank loan, an insurance policy or can a lease of a car. Even governments and public institutions employ such automated decision systems including such diverse applications as setting bail in legal proceedings, risk assessment in youth welfare services or determining the eligibility for state assistance to the unemployed and the access to health services. The results of the current applications of automated decision systems are, however, problematic. In this work, a crucial question will be examined of how the automated decision systems impact the autonomy of the humans and what requirements are to be placed on the automated decision systems in order to protect individuals and society.

Bo Yu, Rustam Vahidov and Gregory E. Kersten (Supporting Users' Utilitarian Needs for Systems in Online Negotiations) deal with a very general aspect of automated systems for online negotiations. Basically, from users' perspective, their use of an e-negotiation system is meant for the achievement of their objectives (e.g. an agreement) to the highest possible extent. This perspective on the use of e-negotiation systems aligns well with the view of the use of an utilitarian system that conceptualizes the use of a system as embedded inside a process within which a user attempts to attain an exogenous goal. Given this conceptualization, the users' assessment of a system use can take place in three tiers, i.e. their assessment of goal achievement, technology interaction and the utilitarian value of the system. These theoretical aspects are tested by conducting an experiment involving e-negotiations. The results support the proposed new theoretical model and also show that social psychological impacts on the users, while using a system, can influence their assessment of the use of the system.

Simon Alfano, Bo Yu, Gregory E. Kersten, Dirk Neumann and Nil-Jana Akpinar (Setting the Right Tone: The Role of Language Sentiment in E-negotiations) deal with the important problem of various aspects of social interaction in e-negotiations including the linguistic features of messages exchanged. More specifically, the analysis of the importance of language dynamics, which is not often undertaken, is considered. The role of the language tone (i.e. sentiment) in messages on outcomes in bilateral e-negotiations is considered. The study builds on the messages extracted from 1,092 bilateral e-negotiations of the inspire experiment. The results obtained suggest that positive language is helpful for attaining an agreement. The avoidance of negative language is a stronger driver of agreement value than an increased levels of positive language. Intriguingly, successful e-negotiations exhibit a more positive sentiment during the opening, relationship building stage, and a reduced usage of both positive and negative words during the core e-negotiation phase. It is proposed that the interpersonal skills, such as a careful language choice, remain crucial despite the more transactional nature of e-negotiations.

In Part III, (Applications) two very interesting and convincing real-world applications are presented.

José Craveirinha, João Clímaco and Rita Girão-Silva (Mathematical Based Models for Group Decision Support in Telecommunication Network Design and Management - Challenges and Trends) are concerned with telecommunication which is an extremely important area as the extremely rapid evolutions of telecommunication network technologies and services and their interactions with complex socio-economic environments, calls for the application, in certain areas of network planning, design and management, some group decision-making approaches. In this area, there is a significant number of decision problems focused on various issues of network planning and design in which more than one decision-maker is involved or in which it is possible to develop mathematical formulations of the problems considering multiple decision-makers. Moreover, the evolution of these telecommunication networks and their related industries leads to a great variety of multifaceted and complex problems, usually involving multiple dimensions, very frequently of a conflicting nature. These factors justify the interest in the applications of mathematical models for group decision support. Many mathematical models have been developed within operations research, systems science, game theory, etc., and they are an essential part of group decision and negotiation support systems. However, this work is limited to multicriteria models and game theoretic models as they seem to be adequate to discuss the challenges and trends in the telecommunication applications considered. An outline of more relevant evolutions of telecommunication network technologies and services is first presented followed by a brief overview of major concepts concerning the multi-criteria group decision (MCGD) and game theory (GT) approaches and methods, relevant to these areas, and then with an overview of representative contributions in these areas, based on the MCGD and GT methodologies. Finally, an analysis and discussion of current and future research trends and challenges concerning the use of the MCGD and GT approaches in this broad area of decision support, also focusing on some relevant methodological issues, are discussed.

Danielle C. Morais, Andre M. Araújo, Eduarda A. Frej and Adiel T. de Almeida (Group Decision Process for Evaluating a Mango Variety to be Planted in New Agricultural Farms) are concerned with fruticulture which is one of the most important sectors of the Brazilian agribusiness being a strategic segment for the country's socio-economic development, with mango having a significant participation. One of the most complex problems faced by this sector is to make decision which is variety of mango to grow in new farms, given the long period of time to have the first production and only then, to verify the result of the cultivation decision. Furthermore, this kind of choice may consider different technical aspects and stakeholders' viewpoints. The paper presents a case study of an agribusiness organization which is one of the greatest exporting Brazilian companies of mango that needs to make the above decision which is variety of mango to plant in new farms intending to double its cultivable area in the next five years. A group decision process is proposed, appropriate to the company's organizational structure, with four phases: 1) definition of the actors, criteria and identification of alternatives, 2) individual assessment by each decision-maker, 3) application of the framework for choosing a voting procedure and 4) collective result. Based on the results obtained, in addition to the recommendation of the mango variety to be planted, it is also possible to extend the analysis to the expansion of planted areas for supporting strategic planning for the company's growth in a sustainable way.

We strongly believe that the valuable and informative contributions included in this volume will constitute, first of all, a token of a great appreciation of the group decision-making and negotiation community for our colleague, friend and peer, Professor Gregory E. Kersten whose contribution and inspiration to the area and to all of us cannot be overestimated. Second, the papers included will provide all the readers with a comprehensive account, new perspectives and visions about the fields and applications.

We wish to thank the contributors for their great works. Special thanks are due to anonymous referees whose deep and constructive remarks and suggestions have helped to greatly improve the quality and clarity of the contributions.

And last but not least, we wish to thank Dr. Tom Ditzinger, Dr. Leontina di Cecco and Mr. Holger Schaepe for their dedication and help to implement and finish this important publication project on time, while maintaining the highest publication standards.

Spring 2021

Tomasz Szapiro
Janusz Kacprzyk

Contents

Foundational Aspects, and Tools and Techniques

Voting Systems in Theory and Practice

Hannu Nurmi, Janusz Kacprzyk, and Slawomir Zadrożny

Abstract Voting procedures are typically motivated by social choice desiderata or criteria of performance. These are, however, largely omitted in the design and evaluation of voting rules in practice. We present some important voting paradoxes and theoretical results stating incompatibilities between various choice desiderata. We also discuss possible reasons for omitting these results in institution design. While most rule innovations are based on other than social choice results, one relative recent one takes its cues from the theory. We briefly evaluate the proposed voting procedure and stress the importance of a comprehensive comparative analysis of several voting rules and criteria of performance in the rule selection process.

Keywords Voting rule · Condorcet extension · No-show paradox · Plurality runoff · Monotonicity

1 Introduction

Gregory Kersten's scholarly work spans over many fields with the primarily focus undoubtedly in group decision and negotiation. Of particular interest to him were electronic negotiation support systems, an area where he made pioneering and seminal contributions. His vast research experience convinced him of the importance of decision and negotiation context in the design of the appropriate support systems. We share this conviction and try to apply this insight in a field that stands next to

H. Nurmi (✉)
Department of Philosophy, Contemporary History and Political Science, University of Turku, Turun Yliopisto, 20014 Turku, Finland
e-mail: hnurmi@utu.fi

J. Kacprzyk · S. Zadrożny
Systems Research Institute, Polish Academy of Sciences, Newelska 6, 01-447 Warsaw, Poland
e-mail: kacprzyk@ibspan.waw.pl

S. Zadrożny
e-mail: zadrozny@ibspan.waw.pl

WIT – Warsaw School of Information Technology, Newelska 6, 01-447 Warsaw, Poland

T. Szapiro and J. Kacprzyk (eds.), *Collective Decisions: Theory, Algorithms And Decision Support Systems*, Studies in Systems, Decision and Control 392,
https://doi.org/10.1007/978-3-030-84997-9_1

negotiation in human interaction systems: group decision methods. Our main motivation is in the strange interaction—or rather lack thereof—between the theory of voting as a subfield of abstract social choice theory and the development of voting systems for various purposes in group decision making in committees, parliaments, board rooms, civil associations and other assemblies including informal gatherings of people searching for the 'will of the electorate'. On several occasions we had the privilege to talk about this lack of theory-practice interaction with Gregory Kersten. It is an honor for us to dedicate this chapter to the memory of our esteemed colleague, Gregory Kersten.

The theory of voting in general and its formal background, social choice theory, in particular, is notorious for the incompatibility results often phrased as paradoxes (see, e.g. Felsenthal 2012; Kelly 1978; Nurmi 1999). Many of these pertain to procedures that are widely used in practice in electing presidents, governors, chairpersons, etc. And yet the results seem to play an insignificant role in choosing and designing election procedures. One is thus naturally led to ask if there are plausible reasons for ignoring those results. This then suggests the more general question of which are the desiderata of reasonable voting systems and whether the scholarly community is largely in agreement about these.

Our chapter proceeds as follows. We first demonstrate the main motivation for the study of voting rules, viz. different rules may end up with different outcomes when the opinions of voters are kept fixed. In other words, the opinions of voters are only one determinant of voting results, equally important role is played by the voting procedures. We then touch upon some classic voting paradoxes before moving to some seminal results in the modern social choice theory. Thereafter we address the puzzling observation that most debates on voting procedures ignore even the most well-known social choice results. Are there plausible reasons for this observation? We single out some. Finally we discuss a relatively recent proposal for the electoral reform which—in contrast to many others—not only does take a social choice perspective into account, but is based on it. Our own suggestions for a proper balance between theory and practice conclude the chapter.

2 Rules Matter

The main reason for studying voting procedures is straight-forward enough: over the past centuries up to the present times numerous procedures have been invented for apparently the same purpose, viz. to find out the collective opinion about alternative actions, policies, candidates or applicants for various positions. That the sameness of purpose does not necessarily translate into identical outcomes of voting can illustrated with the following example (Table 1) involving five alternatives and nine voters. Keeping the voter opinions fixed and varying the procedure we get five different winners so that each procedure produces a different winner. All procedures included here are in some sense democratic, i.e. reflect some aspect of the opinion distribution and do not discriminate for or against any alternative or voter.

The procedures are[1]:

- plurality voting (a.k.a. first-past-the-post): each voter can vote for one and only one alternative. The alternative with the largest number of votes wins.
- plurality runoff: same as the plurality voting if the winner gets more that 50% of the votes. Otherwise, the two alternatives with the largest support face off and the winner is the one preferred to the other by more voters than the other way around.
- the Borda count: if there are k candidates, each voter gives $k - 1$ (Borda) points to his/her first ranked alternative, $k - 2$ points to his/her second ranked one and so on, 0 points to his/her last ranked one. The alternative with the largest point sum is the winner.
- the Copeland rule: each alternative is confronted with all other alternatives in pairwise comparisons. In each comparison, the alternative gets 1 point if it is supported by a majority of voters against the opponent. Otherwise it gets 0 points. The alternative with the largest point sum is the winner.
- the approval voting: each voter can give each alternative either 1 or 0 points. The alternative with the largest point sum is the winner.

Table 1 represents an decision making body of nine voters partitioned into three groups of identically minded voters with respect to five candidates A, B, C, D and E. In the 4-voter group the preference ranking is ABCDE, in the 3-voter one EDBCA and in the 2-voter group DCBEA.

Assuming sincere voting, the plurality winner is A with the largest number of first ranks. The plurality runoff winner is E since in the ensuing A vs. E runoff, E is preferred to A by 5 votes against A's 4 votes. The Copeland winner is D since it defeats A, B and C with a $5 - 4$ margin and E with a $6 - 3$ margin. The Borda count results in B which narrowly beats D (22 vs. 21). To determine the approval voting winner one has to make additional assumptions. Assuming—purely *ad hoc* for the sake of argument—that the 4-voter group approves of three of their highest ranked alternatives, while the other 5 voters approve of their two highest ranked ones, the winner is C with 6 approvals. So, clearly the procedures are important determinants of voting outcomes.

Table 1 5 candidates, 5 winners

4 voters	3 voters	2 voters
A	E	D
B	D	C
C	B	B
D	C	E
E	A	A

[1] For more detailed definitions of these and quite a few other procedures, see Felsenthal and Nurmi (2019).

The reader familiar with the basic concepts of voting theory observes that the preference profile of Table 1 has an alternative that in pairwise contests would beat all others by a majority of votes, a Condorcet winner. This is D. Indeed, all procedures that always end up with a Condorcet winner when one exists—the so-called Condorcet extensions—result in D in this profile. In profiles where no such winner exists, these procedures can lead to different outcomes. Some of the discrepancies between Condorcet extensions are exhibited in Table 2. There three Condorcet extensions are applied to a profile of 29 voters and four alternatives. Copeland's method, defined above, ends up with a three-way tie: A, B, C because each of these beats two others by a majority in pairwise contests. D is ranked last by an absolute majority of voters and is thus an absolute loser. This is a special case of the Condorcet loser, i.e. an alternative that is defeated by all others in pairwise majority contests.

Two other Condorcet extensions, the max-min and Dodgson's rule, result in D. The former rule determines, for each alternative, the minimum support it receives in all pairwise comparisons. The alternative with the largest minimum support is declared the winner. Dodgson's rule, in turn, determines, for each alternative, the minimum number of binary preference switches in voters' rankings needed to make that particular alternative the Condorcet winner. Obviously, if the profile has a Condorcet winner at the outset, it is *ipso facto* the Dodgson winner. D's minimum support is 14, while the other alternatives have a strictly smaller minimum support. Moreover, D requires only 3 pairwise switches in voter preferences to become the Condorcet winner, while the other alternatives require strictly more such preference inversions.

It is worth noticing that in Table 2 two Condorcet extensions yield the Condorcet loser as the winner. Another Condorcet extension, Copeland's method, on the other hand, results in all alternatives but the one elected by Dodgson's and max-min rules.

Within the class of positional procedures, i.e. those determining the winner on the basis of the positions that the alternatives occupy in the voters' rankings, the possibilities for different choices by different procedures in fixed electorate are downright unlimited. Consider the profile of Table 3. Here the plurality winner is A. If all voters are allowed to vote for two alternatives the winner is B. If they are allowed to vote for three alternatives, the winner is C, and the Borda winner is D. This observation is generalized in Donald G. Saari's theorem (Saari 2001):

Table 2 Discrepancy among some Condorcet extensions

10 voters	7 voters	1 voter	7 voters	4 voters
D	B	B	C	D
A	C	A	A	C
B	A	C	B	A
C	D	D	D	B

Table 3 Discrepancy among positional procedures

2 voters	2 voters	2 voters	3 voters
A	A	C	D
B	D	B	B
C	C	D	C
D	B	A	A

Theorem 1 *(Saari 2001, p. 36) Consider the alternative set* $c_1, \ldots, c_K$ *of at least three elements. Then such a profile exists that alternative* c_j *wins when the voting rule is vote-for-*j *and this holds for* $j = 1, \ldots, K - 1$. *Moreover,* c_K *is the Borda winner.*

The above examples are probably sufficient to support the claim that voting procedures do indeed play an important role in collective decision making. Substituting criteria for voters in these examples extends the same claim to multiple criteria decision settings. But what does the social choice theory tell us about the properties of various voting rules? We now turn to some well-known results that provide answers to this question. We start with two classic results.

3 Two Classics

In 1770 Jean-Charles de Borda addressed the French Royal Academy of Sciences. In his talk he provided a devastating criticism against the plurality rule then—as now—widely in use in various types of elections, including in electing new members to the Academy (English translations of Borda's memoir can be found in De Grazia (1953), McLean and Urken (1995)). The argument he presented used as an illustration a hypothetical electorate whose opinions of three candidates A, B and C are those presented in Table 4.

In plurality voting, A wins narrowly beating B and C. What caught Borda' attention was, however, the fact that A is ranked last by an absolute majority of voters. In modern terminology this makes A the absolute loser, a strong variant of the Condorcet loser. Thus, one of the first choice theoretic argument for an institutional reform saw the light of day. Not to much avail, though: the plurality voting remains even today as one of the most widespread voting procedures. The reason for ignoring Borda's misgivings about the plurality voting is not difficult to conjecture: one typically doesn't know the preference profiles in sufficient detail to find out whether an eventual absolute loser has been elected under it. This is in fact the main reason why the negative theoretical results are most often dismissed in the institution reform debates.

Slightly later than Borda, Marquis de Condorcet published a major treatise on social choice (1785). Its lasting findings were the concepts of a winner and of cyclic

Table 4 Borda's paradox

1 voter	7 voters	7 voters	6 voters
A	A	B	C
B	C	C	B
C	B	A	A

Table 5 Condorcet's paradox

Voter I	Voter II	Voter III
A	B	C
B	C	A
C	A	B

majorities. Today these are called the Condorcet winner and the Condorcet cycle, respectively. A special case of the latter is also known as Condorcet's paradox. An example is in Table 5. The crux of the paradox consists of the observation that no matter which of the three candidates one declares the winner, there is a majority (in fact, a majority of the same size, here $2/3$) that prefers another candidate to the declared one. This is because all candidates beat one other candidate by a majority and lose to another one by a majority.

This paradox might thus conceivably be encountered in all voting bodies that resort to pairwise majority voting. A case in point is the amendment procedure widely used in contemporary parliaments and municipal councils. Yet no major protests have been raised because of this possibility. Why? It can conjectured that the reason is similar to the one mentioned in the context of Borda's paradox, viz. one does't know enough of the preference profile to be able to ascertain whether the winner was elected just due to the agenda of pairwise comparisons or whether he/she would have been a Condorcet winner. An additional feature of the amendment procedure should also be mentioned: of k alternatives only $k - 1$ pairwise comparisons are conducted whereby at each stage the winner of the pairwise comparison proceeds to the next comparison so that the winner of the final comparison in declared the overall winner. We thus do not know how the declared winner would have fared in comparisons with the losers of the preceding comparisons. The amendment procedure seems to be based on *the assumption* that the collective preference ranking obtained is transitive: whatever beats the winner, beats necessarily the loser as well. This is precisely what Condorcet's paradox proves to be occasionally false.

4 Some Modern Classical Results

The classics on the 18'th century were able to point out specific problems related to some voting rules. With the advent of the modern social choice theory in the mid-1950's the problem setting took a somewhat more general—or at least different—form. To wit, the focus changed towards the properties of voting rules and, in particular, to the compatibilities between those properties. The best known result of the genre is Arrow's theorem (Arrow 1963).

Theorem 2 *(Arrow 1963) No social welfare function satisfies the following conditions:*

1. *unrestricted domain (U)*
2. *independence of irrelevant alternatives (IIA)*
3. *Pareto (P)*
4. *non-dictatorship (D).*

It will be recalled that a social welfare function assigns to each n-tuple of connected and transitive individual preference relations a (collective) connected and transitive preference relation. Hence, it is a different rule than the one focused upon in the standard social choice theory. In the latter the rules are mappings from the individual preference rankings to the set of subsets of candidates. Yet, Arrow's theorem is certainly relevant for the choice rules as well. It is fundamentally an incompatibility result involving several apparently plausible desiderata of choice rules. The most dramatic interpretations of the result have seen it as a demonstrating that 'collectively rational decisions cannot be reached by democratic voting procedures (if one adopts a few minimalist assumptions)' (Amadae 2005, p. 737). This and similar views certainly exaggerate the issue. For example, the IIA has little if anything to do with democracy. It definitely makes things easier to handle: procedures incompatible with it—e.g. the Borda count or Copeland's rule—call for a holistic computation of winners and social rankings, while IIA procedures can be implemented via pairwise comparisons. Still, IIA is a technical condition, not a normative one. If it were included in the class of necessary conditions of democratic rules, it would exclude nearly all known voting procedures. This may well be the main reason for the fact that Arrow's theorem plays an insignificant role in institutional reform debates.[2]

Another celebrated result in the social choice theory is known as the Gibbard-Satterthwaite theorem (Gibbard 1973; Satterthwaite 1975). It pertains to the strategic aspects of voting and poses the question: what if the voters deem voting as a game of strategy and use their reported preferences as tools for achieving the best possible voting outcomes (instead of merely revealing their sincere opinions)? Would this be a possible scenario under voting systems? The Gibbard-Satterthwaite theory gives a partial answer to these questions. Before stating it, we present two definitions.

---

[2] See (Lagerspetz 2016, pp. 247–257) for a thoughtful analysis of the interpretations of Arrow's theorem.

Definition 1 A social choice function is manipulable (by individuals) if and only if there is a situation and an individual so that the latter can bring about a preferable outcome by preference misrepresentation than by truthful revelation of his/her preference ranking, *ceteris paribus*.

Definition 2 A social choice function is non-trivial (non-degenerate) if and only if for each alternative x, there is a preference profile so that x is chosen.

Theorem 3 *(Gibbard 1973, Satterthwaite 1975). Every universal and non-trivial resolute social choice function is either manipulable or dictatorial.*

This theorem certainly looks dramatic. Upon closer inspection some of the effect, however, diminishes. First, the theorem does *not* say that it is always in the voter's interest to misrepresent his/her preferences. Rather, it says that revealing one's preferences in voting (sincere voting) does not always lead to a Nash equilibrium, i.e. an outcome in which no voter would benefit from deviating from his/her sincere voting strategy. Second, the theorem deals with resolute social choice functions, i.e. choice functions precluding ties of outcomes. Most voting rules, on the other hand, may result in a tie between two or more candidates. Third, the practical preconditions for successful preference misrepresentation may be unrealistic when it comes to the information that the voters have about each others' voting strategies. These preconditions vary a great deal between voting rules ranging from plurality voting, where one only needs to know the distribution of the voters' first ranked candidates in their voting strategies, to single transferable vote where a detailed information about the voters' whole preference rankings is required.[3]

In 1988 Moulin published a seminal result on a phenomenon discovered by Fishburn and Brams (1983) and Moulin (1988). As defined by Fishburn and Brams, the no-show paradox occurs whenever the addition of identical ballots with candidate x ranked last may change the winner from another candidate to x. This paradox is also called the downward monotonicity failure in variable electorates or P-BOT paradox (Felsenthal and Tideman 2013). Some authors say that procedures vulnerable to this paradox fail on the bottom property (Kasper et al. 2019).[4] Procedures characterized by the property that it never harms the voter to join the electorate and reveal his/her true opinion of the candidates satisfy the condition of participation.

[3] The single transferable vote (STV) when applied to a single-member constituency uses the voters' entire preference rankings so that one first determines if some candidate is ranked first by more than 50% of the voters. If such a candidate exists, then he/she is the winner. Otherwise, one ignores, in all ballots, the candidate ranked first by the smallest number of voters and makes the same determination as in the first step. The elimination process is continued until among the remaining candidates one is ranked first by an absolute majority of the voters. This candidate is then the winner. In multi-member districts, in order to get elected the required number of voters ranking a given alternative first depends on the number of candidates returned from the constituency and is always smaller than 50% of the electorate.

[4] A more extensive account on the no-show and related paradoxes can be found in Felsenthal and Nurmi (2017).

Theorem 4 *(Moulin 1988). If there are more than three candidates and at least 25 voters, no voting rule satisfies both the Condorcet criterion and the participation condition.*[5]

From the viewpoint of democratic theory and practice this result pertains to the very fundamentals of governance. If the voters—individually or in groups—can be better off by abstaining than voting according to their opinions, then certainly the incentives for participation in elections are undermined. That the result is related only to the Condorcet extensions is, of course, important. It is relatively easy to show that the main positional methods, plurality voting and the Borda count, are not vulnerable to the no show paradox. These are also monotonic in the sense that improving a winner's position in individual rankings, *ceteris paribus*, retains its position as the winner. The monotonicity of runoff rules is the topic of John H. Smith's important, but less known early theorem.

Theorem 5 *(Smith 1973, p. 1036) No point runoff system involving two or more stages and non-trivial point systems is monotonic. More precisely, if such a system determines the first place first, then a change of votes in a candidate's favor can remove him from the first place. If it determines the last place first, such a change can put a candidate in last place who was not previously there.*

A point system is one assigning each candidate a number of points according to his/her position in each voter's ranking. For example, plurality voting is a point system where all first ranks give a candidate 1 point and all other ranks 0 points. The Borda count assigns each first rank $k - 1$ points, each second rank $k - 2$ point and ... each last rank 0 points. Runoff systems apply a point system to the set of candidates, eliminate some candidates with the lowest point sums, compute the point sums of the remaining candidates and continue until the winners are found. Clearly the plurality runoff system is a point runoff system. The same is true of Nanson's method to name just two point runoff systems. Smith's theorem says that all such systems are non-monotonic, i.e. improving a winning candidate's position in some rankings, *ceteris paribus*, may render it non-winning. This is certainly a major weakness of runoff systems. Another significant one is the vulnerability to the no-show paradox, i.e. the lack of the bottom property (Felsenthal and Maoz 1992, pp. 52–53).

The above results—constituting a small but high-profile sample of similar incompatibility findings—point out serious flaws in the existing voting rules. Yet, they are seldom even mentioned in those institution design debates that really matter. In the following we take a look at possible reasons for the cavalier attitude of the powers that be.

[5] This result has more recently been extended so that the lower bound of the number of voters is 12 (Brandt et al. 2017).

5 Why Nobody Seems to Care?

Although the above results deal with intuitively compelling choice desiderata making their incompatibility demonstrations all the more dramatic, it can be conjectured that perhaps those charged with institution design are not primarily interested in the same properties of voting procedures as the social choice theorists. For example, the main function of elections in democratic systems is to produce actionable outcomes, i.e. results that the voters deem unambiguous and legitimate. In social choice parlance this desideratum is known as decisiveness. If the election results are not clear, the very point of voting vanishes. In view of this, all other desiderata may seem secondary to the practitioner.

Another reason for ignoring the incompatibility results may pertain to their very nature. They express dichotomies: some properties are or are not compatible with others. From the practical point of view—and taking into account the fact that no voting rule is flawless—it would be important to have an idea of how often various incompatibilities and other paradoxes are expected to occur. Probability models and simulation studies can give some hints about the relative likelihoods of paradoxes under various 'cultures' (Gehrlein and Lepelley 2017). Empirical studies on the relative frequencies of various types of profile distributions are, however, not readily available and, yet, those would be particularly useful for institution design.

Instead of viewing the incompatibilities as dichotomous variables one could consider them as matters of degree. Strides in this direction have been taken by Kacprzyk et al. (1992, 2020, 2009). These approaches take explicitly into account the trade-offs between various choice theoretic desiderata and look for similarities of various procedures in terms of those desiderata.

Since the procedures are typically based on intuitive notions of goodness—absolute or relative—of alternatives, it is worthwhile to ponder whether the choice of the procedure should be based on the values or desiderata characterizing procedures. For example, the Condorcet extensions emphasize the necessity to elect the Condorcet winner whenever one exists. In looking for an optimal procedure one could start from constructing a hierarchy of desiderata on the basis of the voters' preferences over those desiderata. We would thus end up with a preference aggregation problem of the usual kind whereby the consensus hierarchy depends on the aggregation method. Some ideas towards making the value relatedness of rule choice explicit are discussed in the literature (de Almeida et al. 2019; Nurmi 2015, Ch. 7).

It is perhaps naïve to expect that political decision making bodies were to adopt these kinds of methods in procedural reforms, but it is downright counterproductive to ignore the performance of the adopted procedures with respect to the main social choice desiderata. Hence it is instructive to take a look at a electoral reform proposal of two towering figures in contemporary social choice theory, Eric Maskin and Amartya Sen.

6 Maskin and Sen on Voting Reform

The case for electoral reform in the U.S. as laid out by Maskin and Sen is based on the following argument:

- The plurality rule (in primaries and in elections at large) *may* result in Borda paradoxes, i.e. in winning candidates that would possibly be defeated by *all* other contestants in pairwise majority comparisons.
- If the voters were allowed to express their full preference rankings, the pairwise majority comparisons would sometimes lead to an indisputable winner (Condorcet) winner).
- Should a Condorcet winner not be present, the plurality runoff system might be used as a secondary (tie-breaking) device.

The background of the proposal is developed by these authors in Maskin (2014), (2020), Sen (2020).

This explicitly social choice theory motivated proposal is to be welcomed. At the same time and on similar grounds one may single out aspects of it that are questionable:

1. The Condorcet winner is not necessarily the intuitively most plausible choice.
2. Condorcet extensions are vulnerable to various paradoxes that do not afflict some other procedures.
3. The absence of a Condorcet winner would, according to the proposal, leave us with a system that is not a Condorcet extension, but would suffer from a similar paradox as the Condorcet extensions

The first point leads us to the age-old discrepancy between Condorcet extensions and positional procedures. Consider the following example devised by Fishburn (Table 6).

Here the Borda winner E seems more plausible choice than the Condorcet winner D since the former has equally many first ranks as D, strictly more second and third ranks and no voter ranks it worse than third, whereas D is ranked next to last by one voter and last by one voter.

Another reason for objecting the choice of a Condorcet winner can be seen in Table 7. Here D, the Condorcet winner is the only candidate that is ranked first by no

Table 6 Fishburn's example (Fishburn 1973, p. 147)

1 voter	1 voter	1 voter	1 voter	1 voter
D	E	C	D	E
E	A	D	E	B
A	C	E	B	A
B	B	A	C	D
C	D	B	A	C

Table 7 No voter ranks the Condorcet winner first

1 voter	1 voter	1 voter
A	B	C
D	D	D
C	A	B
B	C	A

Table 8 Plurality runoff is nonmonotonic

6 voters	5 voters	4 voters	2 voters
A	C	B	B
B	A	C	A
C	B	A	C

Table 9 Plurality runoff violates the bottom property

26	47	2	25
A	B	B	C
B	C	C	A
C	A	A	B

one. Obviously, it would thus not make it to the runoff. Thus Table 7 shows that the proposed system has a primary component that is (trivially) a Condorcet extension, but its secondary component (plurality runoff) is not. This notwithstanding, it is likely that some voters would be upset upon noticing that nobody's favorite can beat all the other candidates under the proposed system.

Ergo: it is not always clear that the Condorcet winner is the most plausible choice. It will be recalled that all Condorcet extensions are—by Moulin's theorem cited above—vulnerable to the no-show paradox. The plurality runoff procedure—which according to the proposal is to applied should there be no Condorcet winner—does not do much better.

To wit, it is not monotonic as Table 8 demonstrates.

The outcome here is A, but if A gets some additional support, *ceteris paribus*, so that the 2 BAC voter become ABC voters, A is no longer the winner (C is).

The plurality runoff is also vulnerable the no-show paradox, i.e. a violation of the bottom property. This is shown in Table 9.

Here A wins. Let now those 47 voters with BCA ranking abstain, then C wins.

It should be noted that monotonic Condorcet extensions exist even though the violation of the bottom property is a pervasive feature in the this class of systems. Hence, picking a non-Condorcet extension for contingencies where a Condorcet winner does not exist seems inappropriate. Of course, one could ask why the Condorcet criterion should be invoked in the first place. As pointed out above, the plausibility of the Condorcet winner is not at all obvious.

The proposed system is plausible to the extent it is based on several desiderata instead of just one. However, given the chosen desiderata it does not pass muster.

7 Concluding Remarks

The plethora of incompatibility results in the social choice theory suggests that no ideal procedure exist. So, in choosing a voting rule one is well advised to focus on some most important desiderata and to perform a comparative analysis in the light of them. The practitioners usually pay no attention at all to the theory of social choice. Avoiding voting paradoxes is, after all, not their main focus. The desiderata of practitioners are often related to guarantees of an unambiguous and transparent voting outcome. Yet, voting paradoxes undermine the meaningfulness of the election results and therefore due attention should be directed towards selecting rules that best avoid the most damaging paradoxes. In the preceding we assessed—admittedly in a cursory manner—one relative recent voting rule proposal that is based explicitly on social choice results and concepts. While we do harbour some doubts on a few details of the proposal, we welcome the input from the social choice community to the development of one of the basic instruments of democracy, voting.

References

1. Amadae, S.: Arrow's impossibility theorem and the national security state. Stud. History Philos. Sci. Part A **36**, 734–743 (2005)
2. Arrow, K.J.: Social Choice and Individual Values, 2nd edn. Wiley, New York (1963)
3. Borda, J.-C.: Mémoire sur les élections au scrutin, translated to English and reprinted in [7, 28] (1781)
4. Brandt, F., Geist, C., Peters, D.: Optimal bounds for the no show paradox via SAT solving. Math. Soc. Sci. **90**, 18–27 (2017)
5. Condorcet, Marquis de (Marie Jean Antoine Nicolas de Caritat): Essai sur l'application de l'analyse à la probabilité des décisions rendues à la pluralité des voix. L'Imprimerie Royale, Paris (1785)
6. de Almeida, A.T., Costa Morais, D., Nurmi, H.: Systems, procedures and voting rules in context. A Primer for Voting Rule Selection. Springer, Cham (2019)
7. De Grazia, A.: Mathematical derivation of an election system. Isis **44**, 42–51 (1953)
8. Felsenthal, D.S.: Review of paradoxes afflicting procedures for electing a single candidate. In: Felsenthal, D.S., Machover, M. (eds.) Electoral Systems: Paradoxes, Assumptions, and Procedures, pp. 19–91. Springer, Heidelberg (2012)
9. Felsenthal, D.S., Maoz, Z.: Normative properties of four single-stage multi-winner electoral procedures. Behav. Sci. **37**, 109 – 127 (1992)
10. Felsenthal, D.S., Nurmi, H.: Monotonocity Failures Afflicting Procedures for Electing a Single Candidate. Springer, Cham (2017)

11. Felsenthal, D.S., Nurmi, H.: Voting Procedures Under a Restricted Domain. An Examinination of the (In)Vulnerability of 20 Voting Procedures to Five Main Paradoxes. Springer, Cham (2019)
12. Felsenthal, D.S., Tideman, N.: Varieties of failure of monotonicity and participation under five voting methods. Theor. Decis. **75**, 59–77 (2013)
13. Fishburn, P.C.: The Theory of Social Choice. Princeton University Press, Princeton (1973)
14. Fishburn, P.C., Brams, S.J.: Paradoxes of preferential voting. Math. Mag. **56**, 207–214 (1983)
15. Gehrlein, W.V., Lepelley, D.: Elections, Voting Rules and Paradoxical Outcomes. Springer, Cham (2017)
16. Gibbard, A.: Manipulation of voting schemes. Econometrica **41**, 587–601 (1973)
17. Kacprzyk, J., Fedrizzi, M., Nurmi, H.: Group decision making and consensus under fuzzy preferences and fuzzy majority. Fuzzy Sets Syst. **49**, 21–31 (1992)
18. Kacprzyk, J., Merigó, J., Nurmi, H., Zadrożny, S.: Multi-agent systems and voting: how similar are voting procedures. In: Lesot, M.J., et al. (eds.) IPMU 2020, pp. 172–184. Springer, Cham (2020)
19. Kacprzyk, J., Nurmi, H., Zadrożny, S.: Reason vs.: rationality: from rankings to tournaments in individual choice. In: Transaction on Computational Collective Intelligence. LNCS, vol. 10480. **27**, 28–39 (2017)
20. Kacprzyk, J., Zadrożny, S.: Towards a general and unified characterization of individual and collective choice functions under fuzzy and nonfuzzy preferences and majority via the ordered weighted average operators. Int. J. Intell. Syst. **24**, 4–26 (2009)
21. Kacprzyk, J., Nurmi, H., Zadrozny, S.: Towards a comprehensive similarity analysis of voting procedures using rough sets and similarity measures. In: Skowron, A., Suraj, Z. (eds.) Rough Sets and Intelligent Systems - Professor Zdzislaw Pawlak in Memoriam, vol. 1, pp. 359–380. Springer (2013)
22. Kasper, L., Peters, H., Vermeulen, D.: Condorcet consistency and the strong no show paradoxes. Math. Soc. Sci. **99**, 36–42 (2019)
23. Kelly, J.S.: Arrow Impossibility Theorems. Academic Press, New York (1978)
24. Lagerspetz, E.: Social Choice and Democratic Values. Studies in Choice and Welfare. Springer, Cham (2016)
25. Maskin, E.: The Arrow impossibility theorem: where do we go from here? In: Maskin, E., Sen, S. (eds.) The Arrow Impossibility Theorem, pp. 43–55. Columbia University Press, New York (2014)
26. Maskin, E.: A modified version of Arrow's IIA condition. Soc. Choice Welfare **54**, 203–209 (2020)
27. Maskin, E., Sen, A.: How majority rule might have stopped Donald Trump. The New York Times, 28 April 2016. https://www.nytimes.com/2016/05/01/opinion/sunday/how-majority-rule-might-have-stopped-donald-trump.html
28. McLean, I., Urken, A.B. (eds.): Classics of Social Choice. University of Michigan Press, Ann Arbor (1995)
29. Moulin, H.: Condorcet's principle implies the no-show paradox. J. Econ. Theory **45**, 53–64 (1988)
30. Nurmi, H.: Voting Paradoxes and How to Deal With Them. Springer, Heidelberg (1999)
31. Nurmi, H.: The choice of voting rules based on preferences over criteria. In: Kamiński, B., Kersten, G.E., Szapiro, T. (eds.) Outlooks and Insights on Group Decision and Negotiation, pp. 241–252. Springer, Cham (2015)
32. Nurmi, H., Kacprzyk, J.: On fuzzy tournaments and their solution concepts in group decision making. Eur. J. Oper. Res. **51**(223), 232 (1991)
33. Saari, D.G.: Chaotic Elections! A Mathematician Looks at Voting. American Mathematical Society, Providence (2001)
34. Satterthwaite, M.: Strategyproofness and Arrow's conditions. J. Econ. Theory **10**, 187–217 (1975)
35. Sen, A.: Majority decision and Condorcet winners. Soc. Choice Welfare **54**, 211–217 (2020)
36. Smith, J.H.: Aggregation of preferences with variable electorate. Econometrica **41**, 1027–1041 (1973)

Optimization-Based Voting Rule Design: The Closer to Utopia the Better

Piotr Faliszewski, Stanislaw Szufa, and Nimrod Talmon

Abstract In certain situations, such as elections in a Euclidean domain, it is possible to specify clear requirements for the operation of a multiwinner voting rule, for it to provide committees that correspond to some desirable intuitive notions (such as individual excellence of committee members or their diversity). We formally describe several such requirements, which we refer to as "utopias". Supplied with such utopias, we develop an optimization-based mechanism for constructing committee scoring rules that provide results as close to these utopias as possible; we test our mechanism on weakly separable and OWA-based rules. Using our method we acquire some interesting insights as well as recover some believed connections between known multiwinner voting rules and certain applications and get other interesting insights.

1 Introduction

Multiwinner voting is a formalism for selecting a set of items to share (a committee), based on the preferences of a group of agents (the voters) (Kilgour 2010; Faliszewski et al. 2017; Faliszewski et al. 2017; Lackner and Skowron 2020). For example, a group of judges may need to select a set of finalists of a competition, a hiring committee may need to select a set of people to invite for on-site interviews, and an internet store may need to decide which items to present on its homepage, depending on how the preferences of its customers are perceived. In each of these examples, we need committees with different properties: the judges should select individually best

P. Faliszewski (✉)
AGH University, Krakow, Poland

S. Szufa
Jagiellonian University, Krakow, Poland

N. Talmon
Ben-Gurion University of the Negev, Be'er Sheva, Israel

T. Szapiro and J. Kacprzyk (eds.), *Collective Decisions: Theory, Algorithms And Decision Support Systems*, Studies in Systems, Decision and Control 392,
https://doi.org/10.1007/978-3-030-84997-9_2

candidates, the internet store should select a diverse set of items that cover interests of as many of its customers as possible, and the hiring committee should, in a certain sense, balance these two requirements (on the one hand, we certainly wish to invite as good candidates as possible, but we also wish to maintain some diversity; e.g., a computer science department is likely to interview candidates working in several areas, and not only in the "hottest" one).

More generally, following the overview of Faliszewski et al. (2017), multiwinner elections might be categorized into three classes, based on what their goals are: *Individual Excellence*, for selecting individually best candidates, as in the judges example above; *Proportional Representation*, for accurately and proportionally representing the electorate views; and *Diversity*, for reflecting the wide spectrum of voters' views,as in the internet store above.[1]

So far, to address these varied goals and needs researchers typically analyzed existing multiwinner voting rules, studied their computational complexity (Procaccia et al. 2008; Lu and Boutilier 2011; Betzler et al. 2013; Baumeister et al. 2015; Cygan et al. 2018; Sekar et al. 2017; Skowron and Faliszewski 2016; Aziz et al. 2015), analyzed their axiomatic properties (Elkind et al. 2017; Skowron et al. 2019; Aziz et al. 2017; Sánchez-Fernández et al. 2017; Faliszewski et al. 2019; Peters 2018; Brandl and Peters 2019; Lackner and Skowron 2019; Pierczynski and Skowron 2019), evaluated them experimentally (Diss and Doghmi 2016; Elkind et al. 2017; Faliszewski et al. 2017; Skowron et al. 2017; Caragiannis et al. 2016; Ayadi et al. 2019), and—based on the evidence they accumulated—argued which rules are best for which application.[2] For example, the k-Borda rule (Debord 1992) is seen as appropriate for choosing individually excellent candidates, whereas the Chamberlin–Courant rule (1983) is appropriate for identifying diverse committees that cover a wide spectrum of opinions. In other words, typically, researchers analyzed existing rules and checked which ones behaved appropriately for a given setting. There are also cases where researchers hand-designed multiwinner rules and specially tailored them to achieve their goals (e.g., for situations such as the hiring committee above, Faliszewski et al. (2017) designed a spectrum of rules achieving various levels of compromise between the goals of excellence and diversity; Elkind et al. (2015), Aziz et al. (2017) and Sekar et al. (2017) proposed multiwinner variants of the Condorcet rule, and Aziz et al. (2018) designed polynomial-time rules that provide extended justified representation).

In this paper we take a different approach from those outlined above: Given a specification of the kind of committees that one is interested in, we use an optimization algorithm to automatically design—in a principled way—rules that match this specification.

[1] While useful, this classification seems preliminary. For example, it seems that the idea of proportionality is more complicated and there are many kinds of proportionality that one may aim at (Peters and Skowron 2020; Faliszewski et al. 2019). Further, there are rules for which it is not clear what goal they aim to achieve (such as the Bloc rule).

[2] The references above are meant to present the wide range of results obtained, and are certainly not complete or comprehensive. We point readers interested in a more systematic treatment to the surveys of Faliszewski et al. (2017), Kilgour (2010), and Lackner and Skowron (2020).

Motivation. Our work is driven by two main motivations. The first one, suggested above, is that we wish to develop a methodology for designing voting rules which would satisfy certain desired properties. There are many multiwinner rules (such as k-Borda (Debord 1992), Chamberlin and Courant (1983), Proportional Approval Voting (Thiele 1895), Monroe (1995), and many others) that seem to have good properties for *some* idealized goals, but here we wish to develop a general mechanism that, when supplied with *any* arbitrary goal (specified in an appropriate way), can output a multiwinner rule suitable for this goal. Indeed, designing voting rules which satisfy certain properties is at the heart of social choice. As a proof of concept, in this paper we focus on three subclasses of the class of committee scoring rules (Elkind et al. 2017; Skowron et al. 2019; Faliszewski et al. 2019) and design an algorithm that searches for an appropriate rule among them. We focus on rules from these classes because they are parametrized through sets of numeric parameters that we can tweak to manipulate their properties; this aspect is important for our optimization-based approach.

Our second motivation relates to the richness of the class of committee scoring rules. So far, researchers have analyzed the general structure of committee scoring rules (Skowron et al. 2019), considered a few of their subclasses (Elkind et al. 2017; Faliszewski et al. 2018; Faliszewski et al. 2019), and studied several specific rules (Elkind et al. 2017; Skowron and Faliszewski 2016; Aziz et al. 2017) or spectra of rules (Faliszewski et al. 2017). However, as there are infinitely-many committee scoring rules, it might be that, in spite of this effort, some important rules might have been missed. Our mechanism of designing rules tailored for particular goals explores specified subclasses of committee scoring rules and, for each setting, either finds one of the already-known rules (thus confirming that it is highly appropriate for a given setting) or discovers a new rule. Thus, if applied extensively, our mechanism is able to uncover important uncharted territories.

Methodology. Providing intuitive descriptions of what kinds of committees are appropriate for certain types of multiwinner elections is quite easy and, indeed, in the opening paragraph of this chapter we only needed a few words to provide several examples. Providing formal specifications is far less trivial because there are many ways to understand intuitions and many ways to formalize them. Here we take an approach inspired by a recent work of Elkind et al. (2017) and consider Euclidean elections, in which it is relatively easy to interpret election results.

In the Euclidean election model—popularized by Enelow and Hinich (1984; 1990)—we assume that each candidate and each voter is represented by an ideal point in a given Euclidean space. The closer a candidate's point is to a given voter's point, the more this voter likes this candidate (e.g., in the context of politics, positions on a 2-dimensional plane may indicate the level of a person's belief, say, in personal and economic freedom).

To design a new voting rule, we proceed as follows. First, we assume some distribution of ideal points of the candidates and the voters (we use the uniform distribution on a 1D interval). Second, we specify the desired distribution of committee members; we refer to this distribution as the *utopic distribution* (e.g., as suggested by the experiments of Elkind et al. (2017), for committees with individually excellent candidates we use the distribution that puts all the weight on the single center point, whereas for the case of diverse committees, we use the uniform distribution over the whole interval). Third, for a given multiwinner voting rule we evaluate how closely its results follow the utopic distribution (using a measure inspired by the Earth mover's distance). We perform this evaluation on rules from a given parametric class and seek one whose distance to the utopic distribution is the smallest. Specifically, to identify the rule which is "the closest to a given utopia", we use an optimization algorithm.

Search Space of Voting Rules. We consider the ordinal model of voting, where the voters rank the candidates from those they like most to those they like least. However, occasionally we also mention the approval setting, where each voter lists the candidates he or she finds acceptable. We focus on three subclasses of committee scoring rules, the class of weakly separable rules, the class of OWA/Borda-based rules, and the class of OWA/Approval-based rules. Briefly put, a weakly separable rule is defined by a vector of points that the voters assign to the candidates, depending on how highly they rank them. We add up the points that the candidates receive from the voters and the candidates with the highest scores form the winning committee. This class includes, for example, the well-known k-Borda, SNTV (single non-transferable vote), and Bloc rules. Intuitively, under k-Borda we choose candidates that, on the average, are ranked highest be the voters; under SNTV we choose those that are ranked first most frequently; and under Bloc each voter specifies his or her most favorite committee and we select those candidates that are mentioned most frequently.

OWA-based rules are defined by vectors of points associated with how highly a voter ranks a given candidate (as in the previous case; we focus on the Borda scoring function and on the t-Approval scoring functions) and the ordered-weighted average operators (the OWA vectors (Yager 1988)), which specify the importance of the committee members, depending on their ranking by a voter. OWA-based rules include, e.g., the Chamberlin–Courant rule and the PAV rule (see, e.g., the works of Skowron et al. (2016) and Aziz et al. (2015)). Indeed, together the classes of weakly separable rules and OWA-based rules cover a large fraction of the committee scoring rules analyzed to date.

Our Contribution. We believe that our most important contribution is a proof of concept: We show that it is possible to use optimization algorithms that minimize the distance from utopic distributions as a tool for designing rules with given properties. Further, we achieve the following results:

1. Our search for weakly separable rules that identify committees close to the given utopias is mostly unsuccessful an, we conclude that the class of weakly separable rules is not expressive enough for most of our settings. Nonetheless, our results do have some very meaningful interpretations. For example, we confirm that k-Borda is among the best rules for finding individually excellent candidates (and, in general, weakly separable rules seem to be very good for such tasks),

and that SNTV is among the best weakly separable rules for finding diverse committees (this, however, is a negative result because SNTV is viewed as an unappealing rule; we hoped to find a more appealing diversity-oriented weakly separable rule).

2. For the case of OWA/Borda-based and OWA/Approval-based rules our search turned out to be very successful. In particular, we confirm that several previously studied rules seem to be the best for their respective goals (e.g., the Chamberlin–Courant rule is among the best rules for the diversity goal). We also find several appealing rules that were not studied before. In general, we find that OWA/Borda-based and OWA/Approval-based rules have quite complementary properties—for some utopias OWA/Borda-based rules are significantly better than OWA/Approval-based ones, and for other utopias this reverses.

We believe that our results provide an interesting perspective on the class of committee scoring rules. Our methodology can be useful both for designing new rules for particular settings (as shown in this paper) and for other applications. In particular, in a follow-up work we have used it to find an OWA-based approximation of the STV rule (which is not OWA-based) and used this approximation to compare STV and PAV rules, both of which are viewed as achieving proportional representation of the voters, but in different ways (Faliszewski et al. 2019).[3]

2 Related Work

Let us now discuss some of the related works in more detail. The class of committee scoring rules were introduced by Elkind et al. (2017), who also studied its basic normative properties. Then, Skowron et al. (2019) characterized the class in the space of all possible multiwinner rules, and Faliszewski et al. (2019) discussed its internal structure (yet, the classes of weakly-separable rules and OWA-based rules are due to Elkind et al. (2017) and Skowron et al. (2016), respectively; for the OWA-based rules with approval preferences, we also point the reader to the work of Aziz et al. (2017; 2015)). Many other papers also study the class of committee scoring rules, either as a whole or by looking at its subclasses or specific rules. As a few examples, we mention the works of Procaccia et al. (2008), Lu and Boutilier (2011), Skowron et al. (2016), Aziz et al. (2015), and Faliszewski et al. (2017; 2018), who have shown that many committee scoring rules are NP-hard to compute, but for some there are good approximation algorithms (see also the works of Byrka et al. (2018), and Dudycz et al. (2020)), the work of Aziz et al. (2017), who introduced the axioms of justified representation and studied them for some approval-based variants of committee scoring rules (and some other rules), and the work of Debord (1992), who provided an axiomatic characterization of the k-Borda rule.

[3] For a detailed discussion of different types of proportionality we point the reader to the recent work of Peters and Skowron (2020) and, more broadly, to Lackner and Skowron's survey of approval-based voting methods (Lackner and Skowron 2020).

The classification of multiwinner voting goals into seeking individual excellence, diversity, and proportionality is due to Faliszewski et al. (2017). We often seek rules that achieve some sort of a compromise between individual excellence and diversity, but such ideas were also pursued using other approaches. For example, Faliszewski et al. (2017) considered three families of committee scoring rules that form spectra between k-Borda and Chamberlin–Courant and provided their experimental analysis, as well as some theoretical results. The main difference between their work and ours is that they specify the scoring functions and observe the behavior of the rules, whereas we fix the expected behavior of the rule and seek what scoring functions implement it.

Lackner and Skowron (2019) used approximation analysis to establish where various rules are located between individual excellence and diversity. Specifically, for a number of rules they provided guarantees regarding how well their winning committees approximate those of the multiwinner approval voting rule (which is seen as capturing individual excellence in the world of approval elections) and the approval-based Chamberlin–Courant rules (which is seen as capturing diversity). Then, Kocot et al. (2019) considered the problem of computing committees that achieve good compromises according to the measures proposed by Lackner and Skowron (2019) and evaluated such committees experimentally (the difference, however, is that Kocot et al. (2019) focused on the ordinal setting and Lackner and Skowron (2019) studied the approval one).

Our approach heavily relies on considering Euclidean elections. In Euclidean elections each candidate and voter is a point in some Euclidean space and a voter ranks the candidates with respect to their distance from him or her (the closer a candidate is to a voter, the higher this voter ranks the candidate). The idea of looking at Euclidean elections is quite old and was studied in depth, e.g., by Enelow and Hinich (1984; 1990) and many other authors, but our inspiration comes from the work of Elkind et al. (2017). They used 2D Euclidean elections to visualize the results of multiwinner rules by generating a large number of elections, where candidate and voter points were selected from given distributions, and drawing histograms showing how many candidates come from particular areas of the space. We used their results as inspiration and as a guideline in choosing our utopic distributions. More broadly, looking at Euclidean elections allows us to use geometric intuitions regarding elections. There is also a number of other geometric interpretations of voting, of which we mention the classic one, due to Saari (1994), and a more recent one, based on swaps of adjacent candidates, due to Obraztsova et al. (2020; 2013). These approaches are quite different from ours and we do not use them in our study.

The main focus of our work is on automatic design of voting rules. Similar ideas were already proposed by Xia (2013), who considered using machine learning for this task. Recently, the task of automatically designing scoring rules was also taken up by Baumeister and Hogrebe (2019). Specifically, they considered the following problem: Given a family of elections over a common candidate set, is it possible to design a single-winner scoring rule, so that a specified candidate wins in each of these elections? They showed that this problem can be solved in polynomial time, but becomes NP-hard if we put some constraints on the scoring rule to be designed.

The main differences between their work and ours is that our study is experimental and theirs is mostly theoretical (although they do provide some experiments), and we consider multiwinner voting rules while they study single-winner ones. Further, in our approach it is possible to design rules that are close to satisfying a given goal, but, nonetheless, do not meet it perfectly, whereas in their approach doing so is possible but less natural.

On a related, but somewhat different front, Boixel and Endriss (2020) proposed a framework for automatically explaining why a particular candidate should win an election. Briefly put, in their approach one first provides a set of properties that one expects from a voting rule and, then, asks who should win in given profiles based on these properties. One can either view this approach as automatically designing voting rules from axioms, or as a way for generating utopias: Given a collection of example elections, their system could explain which candidates should win and, then, we could find a voting rule which, indeed, selects them. Nonetheless, the current version of their work is focused on single-winner voting and, thus, cannot be directly applied.

Finally, our work is closely related to the line of work on the distortion of voting rules, initiated by Procaccia et al. (2006). The idea here is that each voter has some utility value for each candidate and ranks the candidates in the order of decreasing utilities. The quality of a candidate is defined as the sum of the utilities that the voters associate with him or her, and the distortion of a voting rule is the worst-case ratio between the quality of the candidate it selects and the optimal one (note that the voting rule only has access to the preference orders and not the actual utility values). Distortion was studied in quite some detail, e.g., by Caragiannis and Procaccia (2011), Boutilier et al. (2015), Anshelevich et al. (2018), Pierczynski and Skowron (2019), and a number of other authors. In particular, Anshelevich et al. (2018) considered utilities derived from metric spaces, which is very close to our approach. While they study distortion values of well-known rules, and also establish general bounds on distortion, we experimentally seek rules that achieve low distortion on a given set of elections (and, formally, we minimize the average distance from a desirable solution, instead of minimizing the worst-case one) . Unfortunately, so far the works on distortion focused on the single-winner setting, whereas we consider the multiwinner setting.

3 Preliminaries

An election $E = (C, V)$ consists of a set of candidates $C = \{c_1, \dots, c_m\}$ and a collection of voters $V = (v_1, \dots, v_n)$, where each voter v_i has a strict, linear order $\succ_{v_i}$ ranking the candidates from the one that v_i appreciates most to the one that v_i appreciates least. We refer to $\succ_{v_i}$ as the preference order of voter v_i (and, sometimes, as the vote of v_i). For a voter v and a candidate c, we write $\mathrm{pos}_v(c)$ to denote the position of c in v's preference order (the top-ranked candidate has position 1, the next one has position 2, and so on). A multiwinner voting rule is a function $\mathcal{R}$ that, given an election $E = (C, V)$ and an integer k, $1 \le k \le |C|$, outputs a family of size-k subsets of C (i.e., a family of committees) that tie as winners of this election.

Example 1 Consider an election $E = (C, V)$ with candidate set $C = \{a, b, c, d, e, f\}$ and 8 voters, $v_1, \dots, v_8$, whose preference orders are as follows:

$$v_1 \colon a \succ e \succ b \succ f \succ c \succ d, \qquad v_2 \colon b \succ d \succ e \succ c \succ f \succ a,$$
$$v_3 \colon a \succ d \succ e \succ f \succ c \succ b, \qquad v_4 \colon b \succ d \succ f \succ c \succ e \succ a,$$
$$v_5 \colon a \succ d \succ f \succ c \succ b \succ e, \qquad v_6 \colon c \succ f \succ d \succ b \succ e \succ a,$$
$$v_7 \colon d \succ c \succ e \succ a \succ f \succ b, \qquad v_8 \colon f \succ c \succ e \succ a \succ d \succ b.$$

We have that $\mathrm{pos}_{v_1}(a) = 1$, $\mathrm{pos}_{v_6}(a) = 6$, and $\mathrm{pos}_{v_8}(a) = 4$. In further examples, we will refer back to this election and, in particular, we will see that many prominent multiwinner rules provide different sets of winning committees for it.

For each integer t, we write $[t]$ to denote the set $\{1, \dots, t\}$. In particular, if m is the number of candidates, then we often interpret the set $[m]$ as the set of positions that candidates may take in a preference order. A single-winner scoring function (for an election with m candidates) is a non-increasing function $\gamma_m \colon [m] \to \mathbb{R}$ that associates each position in a vote with a score value. We define the γ_m-score of a candidate c in an election $E = (C, V)$ to be $\gamma\text{-score}_E(c) = \sum_{v \in V} \gamma_m\big(\mathrm{pos}_v(c)\big)$. As a matter of convention, we require that $\gamma_m(1) = 1$ and $\gamma_m(m) = 0$. For example, the Borda scoring function (denoted β_m), is defined as $\beta_m(i) = {}^{m-i}/_{m-1}$, and the t-Approval scoring function (denoted α_t, where $t \in [m]$ is a parameter[4]) is a function that associates score 1 with the first t positions and score 0 with the remaining ones.

Committee scoring functions are defined analogously to the single-winner ones, but for a generalized notion of a position. Let us fix a committee size k. Then, given a committee S and a vote v, we define the position of S in v, denoted $\mathrm{pos}_v(S)$, to be the sequence of positions of the members of S in v, sorted in the increasing order (i.e., we obtain $\mathrm{pos}_v(S)$ by sorting the set $\{\mathrm{pos}_v(s) \mid s \in S\}$ in the increasing order). We write $[m]_k$ to denote the set of all length-k increasing sequences of elements from $[m]$, and we interpret elements of $[m]_k$ as committee positions. We say that committee position $I = (i_1, \dots, i_k)$ weakly dominates committee position $J = (j_1, \dots, j_k)$, denoted $I \succeq J$, if for each $t \in [k]$ it holds that $i_t \le j_t$.

Example 2 Recall the election from Example 1. We have that $\mathrm{pos}_{v_5}(\{a, c, f\}) = (1, 3, 4) \in [6]_3$, because candidate a is on the first position, c is on the fourth one, f is on the third, and the positions are sorted. We also have $\mathrm{pos}_{v_5}(b, c, d) = (2, 4, 5)$ and $(1, 3, 4) \succeq (2, 4, 5)$.

A committee scoring function (for m candidates and committee size k) is a function $f_{m,k} \colon [m]_k \to \mathbb{R}$, such that for each two committee positions $I, J \in [m]_k$, if $I \succeq J$ then $f(I) \ge f(J)$. The $f_{m,k}$-score of committee S in election $E = (C, V)$ is defined

[4] As opposed to our general notation, t in α_t does not refer to the number of candidates in the election, but to the number of top-ranked approved candidates.

as $\sum_{v \in V} f_{m,k}(\mathrm{pos}_v(S))$. For a family $f = (f_{m,k})_{k \le m}$ of committee scoring functions (one for each number of candidates and committee size), we define the committee scoring rule $\mathcal{R}_f$ as follows: Given an election $E = (C, V)$ with m candidates and committee size k, it outputs all size-k committees S with the highest $f_{m,k}$-score.

Example 3 Let us fix an election E with m candidates and committee size k. The SNTV rule is defined by committee scoring functions of the form:

$$f^{\mathrm{sntv}}_{m,k}(i_1, \ldots, i_k) = \alpha_1(i_1).$$

This means that the rule selects a committee of k candidates that are ranked first most frequently (or several such committees, in case of ties). In particular, in the election from Example 1, for committee size two, SNTV provides committee $\{a, b\}$ (a is ranked first three times, b is ranked first two times, and every other candidate is ranked first at most once).

The Bloc rule uses functions of the form:

$$f^{\mathrm{bloc}}_{m,k}(i_1, \ldots, i_k) = \alpha_k(i_1) + \cdots + \alpha_k(i_k),$$

which can be interpreted as saying that each voter gives one point to each of his or her k most favorite candidates, and the k candidates with the highest score form the winning committee. For $= 2$ and the election from Example 1, the winning committees are $\{a, d\}$ and $\{c, d\}$. Indeed, d is ranked five times among the top two positions in the votes, whereas a and c take a position among the top two candidates three times each.

The k-Borda rule chooses k candidates with the highest Borda scores and is defined through the functions:

$$f^{\mathrm{kb}}_{m,k}(i_1, \ldots, i_k) = \beta_m(i_1) + \cdots + \beta_m(i_k).$$

Somewhat tedious calculations show that k-Borda winning committees for $k = 2$ and the election from Example 1 are $\{c, d\}$ and $\{d, f\}$, both with 46 points in total.

The Chamberlin–Courant rule (the CC rule) uses scoring functions of the form:

$$f^{\mathrm{cc}}_{m,k}(i_1, \ldots, i_k) = \beta_m(i_1).$$

This means that given a committee S, each voter associates it with the Borda score of this member of S that he or she ranks highest (this candidate is called the representative of the voter). In our example election, the winning committee under CC is $\{a, d\}$. In this case, a is a representative for voters v_1, v_3, v_5, and v_8 (and collects 17 points from them), whereas d represents voters v_2, v_4, v_6, and v_7 (and collects 16 points). Note that the fact that each candidate represents the same number of voters and that their scores are similar are a coincidence and this does not need to happen in general.

Finally, the Harmonic-Borda rule (Faliszewski et al. 2017) (the HB rule) and the Proportional Approval Voting rule (PAV) are defined using the following scoring functions:

$$f^{\mathrm{hb}}_{m,k}(i_1, \ldots, i_k) = \beta_m(i_1) + {}^1\!/_2\beta_m(i_2) + \cdots + {}^1\!/_k\beta_m(i_k),$$
$$f^{\mathrm{pav}}_{m,k}(i_1, \ldots, i_k) = \alpha_k(i_1) + {}^1\!/_2\alpha_k(i_2) + \cdots + {}^1\!/_k\alpha_k(i_k).$$

Intuitively, the function defining Harmonic-Borda rule is somewhat between those used by k-Borda and CC. Like k-Borda, it uses the Borda score of all the committee members, but like CC it puts most weight on the top-ranked committee member. As a consequence, it tends to choose committees that achieve a compromise between selecting individually excellent candidates (like k-Borda) and selecting a diverse committee (like CC). In our example election, Harmonic-Borda chooses committee $\{d, f\}$ with score 39. Interestingly, committees $\{c, d\}$ and $\{a, d\}$, which win under k-Borda and CC, respectively,[5] have score 38.5.

Analogously, PAV can be seen as achieving a compromise between the Bloc rule and a variant of the Chamberlin–Courant rule based on the k-Approval function. In fact, it is well-known that PAV achieves proportional representation of the voters (Kilgour 2010; Aziz et al. 2017; Brill et al. 2018). In our example, committee $\{c, d\}$ is winning under PAV with score 7.5. Note that $\{c, d\}$ also wins under Bloc, but the other Bloc winning committee, $\{a, d\}$, does not win under PAV and gets only 7 points (indeed, there are two voters who rank both a and d among the top two positions and each such voter contributes only 1.5 points toward the PAV score of the committee; for $\{c, d\}$ there is only one such voter).

Given the plethora of rules already provided in the example above (and many more discussed by Faliszewski et al. (2019)), it is often useful to focus on certain subclasses of committee scoring rules. Below we present those that we analyze in this paper; we consider a setting with m candidates, where the desired committee size is k, and where $\mathcal{R}$ is a committee scoring rule:

1. We say that $\mathcal{R}$ is *weakly separable* if its committee scoring function is of the form $f(i_1, \ldots i_k) = \gamma(i_1) + \cdots + \gamma(i_k)$, where γ is a single-winner scoring function.
2. We say that $\mathcal{R}$ is *OWA-based* if its committee scoring function is of the form $f(i_1, \ldots, i_k) = \lambda_1\gamma(i_1) + \cdots + \lambda_k\gamma(i_k)$, where $\Lambda = (\lambda_1, \ldots, \lambda_k)$ is a sequence of non-negative real numbers and γ is a single-winner scoring function (we refer to the vector Λ as the OWA vector; indeed, it acts as an ordered weighted average operator of Yager (1988)). Note that every weakly separable rule is OWA-based (with OWA vectors of the form $(1, \ldots, 1)$). Among OWA-based rules we distinguish two more classes:

   2a. We say that $\mathcal{R}$ is OWA/Borda-based if it is OWA-based and uses Borda scoring function.

[5] Note that under CC also $\{d, f\}$ wins.

2b. We say that $\mathcal{R}$ is OWA/Approval-based if it is OWA-based and uses some t-Approval scoring function (for some t).

For the purpose of this paper, we assume that the OWA vectors are non-increasing and λ_1 is always 1 (even though in some settings other OWA vectors would be natural).

Example 4 Let us consider the rules from Example 3. First, we note that all the rules are OWA-based. Further, SNTV, Bloc, and k-Borda are also weakly separable (for SNTV this follows from the fact that its scoring function can be equivalently expressed as $f_{m,k}^{\text{sntv}}(i_1, \ldots, i_k) = \alpha_1(i_1) + \cdots + \alpha_1(i_k)$). Then, SNTV, Bloc, and PAV are OWA/Approval-based, whereas k-Borda, CC, and Harmonic-Borda are OWA/Borda-based.

4 Methodology

In this section we describe our technique of designing voting rules that match given utopic distributions. The general idea of our methodology is as follows: We consider Euclidean elections, in which voters and candidates are embedded in a Euclidean space and devise several utopic distributions for such elections, as well as ways to measure the distance between a solution distribution of a certain rule and a specific utopic distribution. Then, given a parameterized family of multiwinner voting rules, we use local search heuristics to find parameter values resulting in solution distributions that are as close as possible to the utopic distribution at hand.

4.1 Euclidean Elections

In the t-dimensional Euclidean model of elections each individual u, i.e., each candidate and each voter, is represented by a point $p(u) \in \mathbb{R}^t$ in the t-dimensional space. Intuitively, the coordinates of this point may correspond to u's position regarding some t issues. For example, in the 1-dimensional model this single issue may be the acceptable level of taxation, while in the 2-dimensional model the two issues may indicate the levels of belief in personal and economic freedom (Enelow and Hinich 1984; 1990). Each voter forms his or her preference order by sorting the candidates in increasing order of the distances of the candidates' ideal points from the voter's ideal point (i.e., the closer a candidate is to a voter, the higher the voter ranks the candidate).

In our computations, we use either 1-dimensional Euclidean elections, where we generate the ideal points of candidates and voters by drawing them uniformly at random from the [0, 1] interval, or 2-dimensional elections, where we draw the ideal points uniformly at random from a disc centered at point (0.5, 0.5) with radius 0.5. We refer to the former as the *interval model* and to the latter as the *disc model*. We always generate elections with 100 candidates and 100 voters, and we seek committees of size 10. We chose these parameters to ensure that our results are comparable to those already present in the literature (Elkind et al. 2017; Faliszewski et al. 2017; Celis et al. 2018). Let us note that while we use the same distributions of points both for the candidates and for the voters, we could as well use different ones. Indeed, sometimes such settings are used, e.g., to highlight the differences between rules focused on proportionality and diversity; for examples, see the technical report version of the work of Elkind et al. (2017) or the works of Aziz et al. (2018) and Faliszewski and Talmon (2018).

We use the interval model throughout the whole process of designing voting rules, and we use the disc model to check whether the rules that we produce maintain their features after changing (and, in a sense, generalizing) the setting. Arguably, the disc model is very similar to the interval one, and the reason for choosing it was that, on the one hand, we wanted to test our rules in a (somewhat) different setting, but, on the other hand, we wanted this setting to be sufficiently similar, so we could maintain the same intuitions while observing how the behavior of our rules changes.

Following Elkind et al. (2017), we present the results of our elections visually. For a given voting rule $\mathcal{R}$ and a given election model (interval or disc), we generate a number of elections according to the model, compute the $\mathcal{R}$ winning committee for each election (if there are ties, we break them arbitrarily), and—depending on the model—present them as follows:

1. For the interval model, we partition the $[0, 1]$ interval into subintervals, count how many times a candidate from a given subinterval was in a winning committee, and present these numbers as a histogram. Note that we do not normalize the histograms; different ones have different scales.[6]
2. For the disc model, we show a scatter plot, where each member of a winning committee is indicated as a blue dot (thus, as opposed to the work of Elkind et al. (2017), our plots for the disc model are *not* histograms[7]). In addition to the blue dots, we also show the gray disc from which the candidates' and voters' points are drawn.

In Figs. 1 and 2 we show visualizations of the results of k-Borda, SNTV, Bloc, CC, and HB. The former figure shows weakly separable rules and the latter one shows OWA/Borda-based rules (k-Borda appears in both figures as it belongs to both classes). Figures 1 and 2 have the following structure:

1. In Fig. 1, the large plot on the left—marked (a)—shows the values of the scoring functions used by the respective rules (so we have 100 possible positions in a

[6] The point of these histograms is to show the "shape" of the election results, i.e., the areas where we have increased or reduced numbers of winning candidates. Normalizing all the histograms to the same scale would make many of them difficult to present.

[7] Our plotting tool draws the blue dots as "partially transparent," so areas with fewer winners appear in lighter shade of blue, whereas areas with high concentration of winners appear as dark blue. The reason to use scatter plot instead of 2D histograms is that, similarly to the histograms, it provides a good intuition on how the given rule behaves, but it requires far fewer election results.

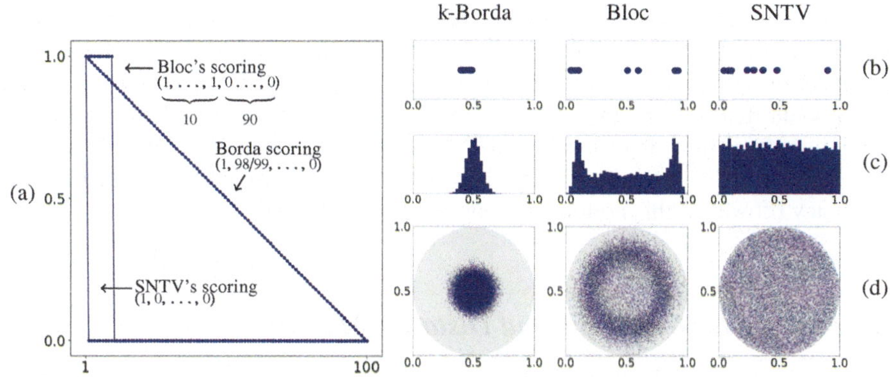

Fig. 1 Visualization of election results under weakly separable rules. Plot **a** shows the values of the scoring functions used, plots **b** show results of a sample interval election, plots **c** show histograms for interval elections, and plots **d** show scatter plots for disc elections

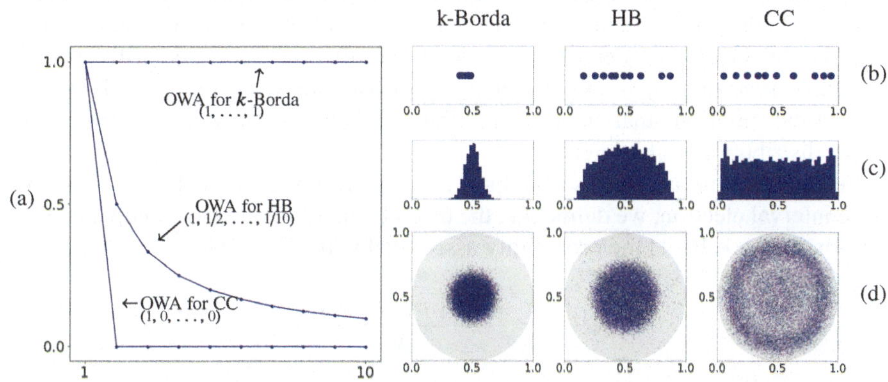

Fig. 2 Visualization of election results under OWA/Borda-based rules. Plot **a** shows the values of the OWA vectors used, plots **b** show results of a sample interval election, plots **c** show histograms for interval elections, and plots **d** show scatter plots for disc elections

vote on the x-axis). The corresponding plot in Fig. 2 shows the values of the OWA vectors used by the presented rules (so we have 10 entries).

2. The plots on the right (in the rows marked (b), (c), and (d)) have the same interpretation in both figures and form a matrix, where each column corresponds to a particular rule and each row corresponds to a type of plot. In row (b) we show results of a single example interval election (only the ideal points of the winners are marked), in row (c) we show the histograms for the interval elections, and in row (d) we show scatter plots for the disc elections. The histograms are based on 1000 interval elections and the scatter plots are based on 2000 disc elections; for 1D histograms we use 40 subintervals. We use these parameters for all the plots throughout the paper.

To compute results of weakly separable rules, we use their direct polynomial-time algorithms. For OWA-based rules, we compute winning committees by solving integer linear programs (ILPs) provided for this task by Peters (2018) (we use CPLEX, a very popular ILP solver). In particular, Peters showed that his formulations give rise to a polynomial-time algorithm for the case of single-peaked elections. Since elections generated for the interval model are single-peaked, we enjoy this guaranteed efficiency (however, this no longer holds for the disc model).

4.2 Utopic Distributions and Distance Measures

We model the requirements regarding our rules as *utopic distributions*, that is, as probability distributions over the [0, 1] interval that represent how, ideally, we would like the winners of our interval elections to be distributed (or, roughly speaking, how we would like their visualizations to appear). For example, the utopic distribution that models the goal of individual excellence associates the whole probability mass with the center of the interval, whereas the distribution associated with covering the whole spectrum is, simply, the uniform distribution over the interval. We will explain the intuitions that stand behind these definitions, as well as a few further utopic distributions, a bit later.

Let $\mathcal{U}$ be some utopic distribution. Given a committee $W = \{w_1, \ldots, w_k\}$ for some interval election, we define d_W, the discrete distribution associated with W, so that for each $x \in [0, 1]$ the probability associated with x is:

$$d_W(x) = \frac{\|\{w_i \mid p(w_i) = x\}\|}{k}.$$

To measure how closely W fits utopia $\mathcal{U}$, we use the Earth mover's distance (Peleg et al. 1989): We view the probability mass associated with each point (each interval) as the number of "grains of sand" that lie on this point (this interval). Moving a grain of sand from point x to point y costs $|x - y|$. The distance between two distributions is the lowest possible cost of moving the "grains of sand" needed to transform one of them into the other. While this intuition is discrete in its nature, our utopic distributions are sometimes continuous (in other words, sometimes we consider probability density functions). Instead of providing a general definition of our distance, below we describe the utopic distributions that we consider and for each we derive the appropriate distance measure based on the above intuition.[8]

[8] The reader can rightfully complain that we wanted to deal with arbitrary goal specifications and not only specific ones. We provide two answers. First, our reasoning, in essence, applies to all utopic distributions and what we do can be seen as deriving closed form formulas, to speed up computations. Second, instead of following our method, given a utopic distribution one may generate discrete histograms for it and for d_W and directly use the Earth mover's distance to measure their similarity.

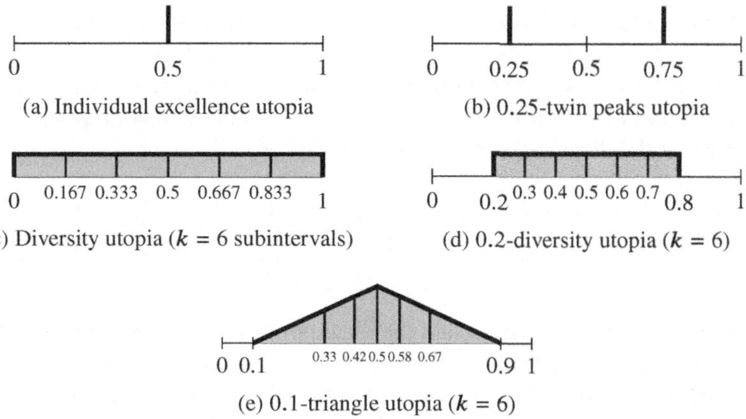

(a) Individual excellence utopia

(b) 0.25-twin peaks utopia

(c) Diversity utopia (k = 6 subintervals)

(d) 0.2-diversity utopia (k = 6)

(e) 0.1-triangle utopia (k = 6)

Fig. 3 Several utopic distributions

In the descriptions below, we set k be the committee size and $W = \{w_1, \dots, w_k\}$ be a committee, whose members have ideal points $p(w_1) \dots, p(w_k) \in [0, 1]$. We assume that these points are sorted, i.e., $p(w_1) \le p(w_2) \le \dots \le p(w_k)$.

Individual Excellence ($\mathcal{U}_{IE}$). The individual excellence utopic distribution, $\mathcal{U}_{IE}$, is defined as concentrating all the probability mass in the center of the interval, at point 0.5; see Fig. 3a. The motivation for defining the utopic distribution in this way is as follows. One of the ideas behind the individual excellence goal is that if two candidates are "similar" then either they should both be included in the committee or neither of them should be included (up to boundary conditions regarding the size of the committee). In the interval elections, a natural measure of similarity between candidates is the distance between their ideal points. Thus, if the ideal points of two candidates are close to each other, then we would like that either both of them are selected or neither of them is. Further, as the interval elections are quite symmetric, if we select a candidate somewhere to the left of the center, then we typically also have to select some candidate to the right of the center. As a consequence, if we select the candidates in the center, then the number of situations where we include in the committee some candidate but we do not include some other candidate, whose ideal point is very close, is minimized. Indeed, the k-Borda rule, which is regarded as very good for the individual excellence goal, selects the candidates in the center.

We define the distance between $\mathcal{U}_{IE}$ and d_W to be:

$$\mathrm{EMD}(\mathcal{U}, d_W) = \sum_{i=1}^{k} {}^{1}\!/_{k} |p(w_i) - 0.5|.$$

That is, for each member of the committee we pay the cost of moving him or her to the center of the interval (we multiply each $|p(w_i) - 0.5|$ by ${}^{1}\!/_{k}$ as each member of the committee is associated with probability mass ${}^{1}\!/_{k}$).

Twin Peaks ($\mathcal{U}^{\epsilon}_{TP}$). The Bloc rule motivates the study of the twin peaks utopic distributions (see Fig. 1 for the visualization of the Bloc rule and Fig. 3b for the twin peaks utopia). An ϵ-twin peaks utopic distribution for parameter ϵ, denoted $\mathcal{U}^{\epsilon}_{TP}$, places some part of the probability mass on point ϵ and the rest on point $1-\epsilon$; see Fig. 3b. We let the distance between $\mathcal{U}^{\epsilon}_{TP}$ and d_W be:

$$\mathrm{EMD}(\mathcal{U}^{\epsilon}_{TP}, d_W) = \sum_{i=1}^{k/2} {}^{1}\!/\!{}_{k} |p(w_i) - \epsilon| + \sum_{i=k/2+1}^{k} {}^{1}\!/\!{}_{k} |p(w_i) - (1-\epsilon)|$$

(we assume that k is even; recall that the ideal points of the committee members are sorted); we assign the left half of committee members to the peak on the left-hand side, and the right half to the committee members to the peak on the right-hand side.

The twin peaks utopic distributions do not have as clear an interpretation as the remaining ones. One way to look at them is that they focus on candidates that are not in the very center, but who are also not on the extremes.

Diversity ($\mathcal{U}_D$). The diversity (or, coverage) utopic distribution, denoted as $\mathcal{U}_D$ and defined to be the uniform distribution over $[0, 1]$, models the idea that a diverse committee should cover the whole interval of ideal point positions as uniformly as possible (indeed, as shown in Fig. 2, the CC rule—which is often believed to provide diverse committees—chooses the committees that cover the interval quite well). Our reasoning for the distance $\mathrm{EMD}(\mathcal{U}_D, d_W)$ is that the committee members are supposed to be distributed evenly along the interval $[0, 1]$ and, so, each of them is responsible for covering a ${}^{1}\!/\!{}_{k}$-length subinterval. We assign the subintervals to committee members so that w_1 is assigned to $[0, \frac{1}{k}]$, w_2 is assigned to $[\frac{1}{k}, \frac{2}{k}]$ and so on; see Fig. 3c.

Let $\ell = {}^{1}\!/\!{}_{k}$ be the length of the subintervals. For each w_i, $i \in [k]$, we define the cost of "spreading" his or her probability mass from d_W over the assigned subinterval $[\ell \cdot (i-1), \ell \cdot i]$ as follows:

1. If the committee member is to the left of his or her interval (i.e., $p(w_i) < \ell \cdot (i-1)$), then we need to pay the cost $(\ell \cdot (i-1) - p(w_i)) \cdot \ell$ for moving his or her probability mass (which also is equal to ℓ) to the point $\ell \cdot (i-1)$, and then the cost ${}^{1}\!/\!{}_{2} \cdot \ell^2 = \int_{\ell\cdot(i-1)}^{\ell} \left(1 - \frac{x-\ell\cdot(i-1)}{\ell}\right)dx$ for "spreading" his or her weight over the interval (note that this latter cost equals to the area of a triangle; see Fig. 4).
2. If the committee member is to the right of his or her interval (i.e., $p(w_i) > \ell \cdot i$), then we proceed analogously: Moving his or her probability weight to the point ℓi costs $\ell \cdot (p(w_i) - \ell \cdot i)$ and "spreading" this weight over the interval costs ${}^{1}\!/\!{}_{2}\ell^2$.
3. If the committee member is in his or her interval (i.e., $\ell \cdot (i-1) \le p(w_i) \le \ell \cdot i$), then it suffices to "spread" the $\frac{p(w_i)-\ell\cdot(i-1)}{\ell}$ fraction of his or her probability mass to the part of the interval left of him, at cost $\frac{1}{2}(p(w_i) - \ell \cdot (i-1))^2$ (analogously to the previous cases, this can be expressed as the area of a right triangle, with two sides of length $p(w_i) - \ell \cdot (i-1)$; see Fig. 4), and the remaining mass, to the part of the interval to the right of him or her, at cost $\frac{1}{2}(\ell \cdot i - p(w_i))^2$.

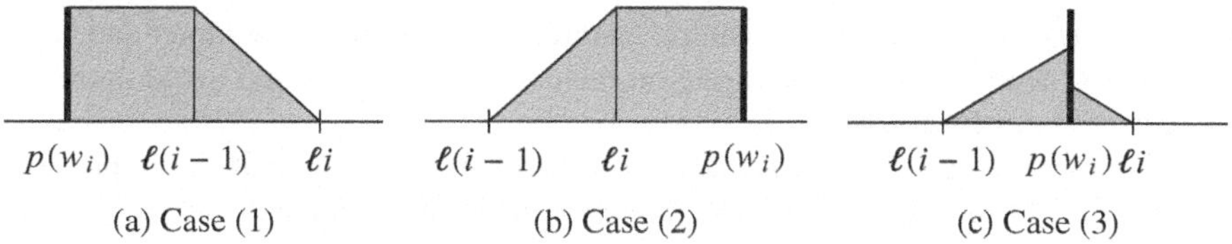

Fig. 4 The cases for the cost of EMD($\mathcal{U}_D, d_W$). The cost associated with committee member w_i is the gray area

Overall, the cost associated with w_i is:

$$\text{cost}(w_i) = \begin{cases} (\ell \cdot (i-1) - p(w_i)) \cdot \ell + \frac{\ell^2}{2} & \text{for } p(w_i) \le \ell \cdot (i-1) \\ \frac{1}{2}\left((p(w_i) - \ell \cdot (i-1))^2 + (\ell \cdot i - p(w_i))^2\right) & \text{for } \ell \cdot (i-1) < p(w_i) < \ell \cdot i \\ (p(w_i) - \ell \cdot i) \cdot \ell + \frac{\ell^2}{2} & \text{for } \ell \cdot i \le p(w_i) \end{cases}$$

and we define EMD($\mathcal{U}_D, d_W$) to be $\sum_{i=1}^{k} \text{cost}(w_i)$.

Diversity/Excellence Compromises ($\mathcal{U}_D^\epsilon$ and $\mathcal{U}_T^\epsilon$). We also consider two families of utopic distributions that achieve a certain level of compromise between the ideals of individual excellence and diversity. Let ϵ be a number in $[0, 0.5]$. We define the ϵ-diversity utopic distribution, denoted by $\mathcal{U}_D^\epsilon$, to be the uniform distribution over the interval $[\epsilon, 1-\epsilon]$.; see Fig. 3d. The idea of the compromise here is that we still seek a uniform distribution—as in the diversity utopia—but on the interval restricted to include fewer extreme candidates. In particular, for values of ϵ close to 0.5 we focus on the center candidates.

Each committee member w_i is responsible for covering interval $I_i = [\epsilon + \ell(i-1), \epsilon + \ell i]$ of length $\ell = {}^1/_k(1-2\epsilon)$. Using the same reasoning as for $\mathcal{U}_D$, we derive the cost associated with w_i to be:

$$\text{cost}(w_i) = \begin{cases} (\epsilon + \ell(i-1) - p(w_i))\frac{1}{k} + \frac{\ell}{2k} & \text{for } p(w_i) \le \epsilon + \ell(i-1) \\ \frac{1}{2k\ell}\left((p(w_i) - \epsilon - \ell(i-1))^2 + (\epsilon + \ell i - p(w_i))^2\right) & \text{for } p(w_i) \in I_i \\ (p(w_i) - \epsilon - \ell i)\frac{1}{k} + \frac{\ell}{2k} & \text{for } \epsilon + \ell i \le p(w_i) \end{cases}$$

We define EMD($\mathcal{U}_D^\epsilon, d_W$) to be $\sum_{i=1}^{k} \text{cost}(w_i)$ (note that the formulas are slightly different than in the previous case because ℓ no longer serves the double purpose of being both the length of the interval and the weight of each committee member).

Our second way of capturing a compromise between individual excellence and diversity is via a distribution whose probability density function, for a given $\epsilon \in [0, 0.5]$, is a triangle with a peak at 0.5, set over the interval $[\epsilon, 1-\epsilon]$ (the area of the triangle is always one). We call it the ϵ-triangle utopic distribution and denote it by $\mathcal{U}_T^\epsilon$. Again, each committee member is associated with an interval for which the probability mass equals exactly to ${}^1/_k$; see Fig. 3e. In the triangle utopia we focus on the center candidates—as in the individual excellence case—but we still keep quite large probability of electing non-central candidates. As we consider larger and larger values of ϵ, we focus more and more on the center candidates and avoid more and more candidates from the edges.

We derive the value $\mathrm{EMD}(\mathcal{U}_T^{\epsilon}, d_W)$ following the same logic as in the previous cases (the calculations and the formula are available in the appendix).

4.3 Search Algorithms

The final component of our method is an algorithm that given a utopic distribution $\mathcal{U}$ and one of our three families of committee scoring rules (weakly separable, OWA/Borda-based, or OWA/Approval-based rules) finds a rule $\mathcal{R}$ that is as close to $\mathcal{U}$ as possible.

General Setup. For m candidates, a weakly separable rule is defined via a non-increasing vector $X = (x_1, \dots, x_m)$, such that $x_1 = 1$ and $x_m = 0$. This vector specifies the values of the underlying single-winner scoring function γ for the possible positions in a preference order. Given a committee size k, an OWA/Borda-based rule is defined by its non-increasing OWA vector $(\lambda_1, \dots, \lambda_k)$, where $\lambda_1 = 1$ (there is no need to specify the scoring function because, by definition, it is Borda). For OWA/Approval-based rules we need to specify the OWA vector and the value $t \in [m]$ indicating that we use the t-Approval scoring function. Correspondingly, given a vector V of appropriate size, we write $\mathcal{R}_V$ to denote the rule defined by this vector (when we consider weakly separable rules, V gives the score values; when we consider OWA/Borda-based rules or OWA/Approval-based rules, it is the OWA vector; for OWA/Approval-based rules we additionally optimize the value of t, but we disregard it in the description here and we will describe how we handle it later). Given a vector V', by *normalizing* it we mean sorting it, setting its first coordinate to one, replacing all values greater than one with one and all values below zero with zero. Additionally, for weakly separable rules, we set the last coordinate to zero (so that the vector describes a legal rule from the relevant class). While this is a somewhat nonstandard definition of normalization, it will be very useful in our algorithms.

Let us fix the class of rules and the utopic distribution $\mathcal{U}$. Our goal is to find a vector V so that the winning committees under $\mathcal{R}_V$ follow $\mathcal{U}$ as closely as possible. To make this notion precise, our algorithms first compute a given number N of interval elections $E_1, \dots, E_N$ (these are fixed throughout the whole optimization process). To evaluate the rule $\mathcal{R}_V$, for each election E_i we compute the winning committee W_i (if there are ties, then we break them arbitrarily). Then we compute the average distance of these committees from the utopia:

$$\mathrm{EMD}(\mathcal{U}, \mathcal{R}_V) = {}^1/_N \sum_{i=1}^{N} \mathrm{EMD}(\mathcal{U}, d_{W_i}).$$

We refer to this value as the *score* of the rule (the lower, the better). We used two algorithms: a brute-force algorithm for basic verification and a local search procedure to obtain our main results.

(Approximate) Brute-Force Search. This algorithm tries all rules defined by vectors with values from the set $\{0, 1/r, \ldots, r-1/r, 1\}$, where r is a parameter defining the resolution of the search, and outputs the rule with the lowest score. We used it only to verify the quality of the local search algorithm. Specifically, we used this algorithm only for the case of OWA/Borda-based rules, for committee size $k = 5$ (as then the search space was sufficiently small; we used $r = 20$). All the rules that the algorithm found were, in essence, indistinguishable from those that we found using the local search algorithm for the same settings. Thus we do not report these results, but treat them as indication that the results provided by the local search algorithm are trustworthy.

Local Search. Our local search algorithm is similar to simulated annealing, but never accepts worse solutions (thus it is a form of a hill-climbing algorithm; we found local minima to not pose problems for our search space, and our approach turned out to be more effective than simulated annealing). We use the following parameters: (a) the number of iterations T, (b) the probability $\omega(i) \in [0, 1]$ of changing a given vector's coordinate, depending on the iteration number i, (c) the range parameter $r(i) \in [0, 1]$, specifying how much vector coordinates can change, depending on the iteration number i. The algorithm works as follows:

1. We select a vector V uniformly at random, with coordinates from $[0, 1]$, and normalize it.
2. We repeat the following steps T times:

   2.1. Create a vector V' using the following procedure. Set $V' = V$. Then, for each of its coordinates v'_i, compute v''_i by adding to v'_i a value drawn uniformly at random from $[-r(i), r(i)]$. With probability $\omega(i)$, replace the value of v'_i with v''_i. Normalize V'.

   2.2. If $\mathrm{EMD}(\mathcal{U}, R_{V'}) < \mathrm{EMD}(\mathcal{U}, \mathcal{R}_V)$ then replace V with V'. Otherwise, keep V as is.

3. We output the rule $\mathcal{R}_V$.

For weakly separable rules, we use $T = 3000$ iterations, $\omega(i) = \max(\frac{T-i}{2T}, 0.05)$, $r(i) = 0.5\omega(i)$, and $N = 400$ test elections. For OWA-based rules, we use $T = 300$ iterations, $\omega(i) = \max(\frac{T-i}{2T}, 0.1)$, $r(i) = 0.3 \cdot \max(\frac{T-i}{2T}, 0.05)$, $N = 40$ test elections. To speed up the algorithm for the case of OWA-based rules, we first run it for elections with 50 candidates, 50 voters, and committee size 10, and only then we re-run it for full-sized elections (with 100 candidates, 100 voters, and committee size 10), using the result of the first run as the starting vector for the second one. Our choice of parameters is quite arbitrary, and follows mostly from performing initial experiments and attempts at manually tuning the algorithm. However, all in all, the algorithm turned out to be quite robust and exact choices of the parameters were not very important (although, of course, using more iterations is better than using fewer, especially if one allows the initial iterations for changing the vector more aggressively).

For OWA/Approval-based rules we also have to search for the value of t (specifying that we use the t-Approval scoring function). We found that typically for a given utopia U and a fixed vector V, the score of R_V is unimodal with respect to t. Thus for the case of OWA/Approval-based rules we modified our algorithm as follows: Every 10 iterations we check if either by increasing or decreasing the current value of t we can decrease the score. If so, we keep in increasing/decreasing t until the score starts growing again.

Naturally, we could have chosen any other algorithm for multidimensional optimization. Indeed, we have tried several other ones, including fully-fledged simulated annealing, but we found our approach to provide high-quality results and to be faster. As our main goal is to show that the idea for multiwinner voting rule design is feasible, we chose the simplest effective solution. However, we believe that if the approach were to be used more widely, then a careful evaluation of search algorithms would be needed.

Nature of the Search Space. We conclude this section with two remarks regarding the nature of our score function. Let U be one of our utopias and, for simplicity, let us focus on weakly separable rules. Recall that for a vector V (specifying the scoring function), the score of $\mathcal{R}_V$ is defined as:

$$\mathrm{EMD}(\mathcal{U}, \mathcal{R}_V) = {}^1/_N \sum_{i=1}^{N} \mathrm{EMD}(\mathcal{U}, d_{W_i}),$$

where N is the number of elections used for the evaluation and $W_1, \ldots, W_N$ are the N committees computed according to $\mathcal{R}_V$ for these elections. As a simple consequence of this definition, we find that $\mathrm{EMD}(\mathcal{U}, \mathcal{R}_V)$ is not continuous with respect to V. Indeed, if we change vector V only very slightly, then it is quite likely that neither of the committees $W_1, \ldots, W_N$ changes and, so, the value of $\mathrm{EMD}(\mathcal{U}, \mathcal{R}_v)$ stays constant. However, if we change V so that at least one of the committees does change, then $\mathrm{EMD}(\mathcal{U}, \mathcal{R}_V)$ changes non-continuously (in essence, this follows from the fact that there is only a finite number of choices for committees $W_1, \ldots, W_N$ and, in consequence, $\mathrm{EMD}(\mathcal{U}, \mathcal{R}_V)$ assumes only finitely many different values). As our score function is non-continuous, it is also non-differentiable and we cannot directly use global optimization algorithms that require differentiability.

The above comments notwithstanding, we also note that our score function is, in some sense, quite well behaved. Indeed, if we only slightly modify V, then we expect that only a few committee members are replaced in only a few committees (in fact, with a small enough change, perhaps only one of the committees is affected). Since $\mathrm{EMD}(\mathcal{U}, \mathcal{R}_V)$ is defined as the average of the EMD values for all the N committees, if only a few committees change (and not greatly) then the value of $\mathrm{EMD}(\mathcal{U}, \mathcal{R}_V)$ changes by a small amount. While this feature still allows for many local optima to appear, we found that full-fledged simulated annealing (which can accept worse solutions with some probability, to avoid local minima) does not perform better than our algorithm (i.e., it finds solutions that are indistinguishable from those found by our algorithm, but requires more time). Thus either the local optima are very easy to avoid (and even our hill-climbing makes large enough jumps to avoid them) or are very hard to avoid (e.g., are surrounded by large volumes of search space with higher EMD scores) and full-fledged simulated annealing finds them difficult to avoid too.

5 Results

We used our search algorithm to find the best weakly separable, OWA/Borda-based, and OWA/Approval-based rules for the individual excellence, ϵ-diversity (with $\epsilon \in \{0, 0.1, 0.2\}$), ϵ-twin peaks (with $\epsilon \in \{0.167, 0.25, 0.333\}$), and ϵ-triangle (with $\epsilon \in \{0, 0.1, 0.2\}$) utopic distributions.

The results are given in Table 1 (the EMD values for the computed rules and for several well-known rules for comparison) and in Figs. 5, 6, 7, 8, 9, 10, 11, 12 and 13 (visual presentation of the computed vectors and results for the rules). Each figure has similar structure as Figs. 1 and 2, but instead of presenting results for three well-known multiwinner rules, it shows results for the best rules computed for four utopic distributions: the individual excellence utopia, which can be seen as a border case for each of the other distributions, and either the ϵ-diversity, the ϵ-twin peaks, or the ϵ-triangle distributions, for appropriate values of ϵ. The largest plot on the left of each figure, marked (a), shows vectors computed for the respective four utopias. Next to it, as Plot (b), we show graphical representation of the respective utopia (drawn as a gray area over the [0,1] interval) and a sample result of a single interval election (the blue dots). As Plot (c), we show the histograms achieved by the computed rules on interval elections (we remind the reader that different histograms have different y-

Table 1 EMD scores for the best weakly separable, OWA/Borda-based and OWA/Approval-based rules computed for our utopias, and for the rules from Figs. 1 and 2 for the same utopias

Utopia	Weakly Sep.	OWA/Borda	OWA/Approval	SNTV	Bloc	Borda	CC	HB	
$\mathcal{U}_D$	0.104	**0.042**	**0.042**	0.092	0.215	0.227	**0.044**	0.094	diversity
$\mathcal{U}_D^{0.1}$	0.098	**0.042**	0.046	0.100	0.205	0.178	0.059	**0.053**	diversity
$\mathcal{U}_D^{0.2}$	0.073	**0.044**	0.054	0.126	0.210	0.130	0.100	**0.044**	diversity
$\mathcal{U}_T$	0.118	**0.043**	0.065	0.113	0.213	0.150	0.082	**0.043**	triangle
$\mathcal{U}_T^{0.1}$	0.109	**0.045**	0.064	0.137	0.220	0.127	0.106	**0.051**	triangle
$\mathcal{U}_T^{0.2}$	0.088	**0.054**	0.060	0.174	0.238	0.105	0.160	**0.083**	triangle
$\mathcal{U}_{TP}^{0.167}$	0.158	0.135	**0.107**	0.159	0.237	0.307	**0.140**	0.172	tw. peaks
$\mathcal{U}_{TP}^{0.25}$	0.144	0.060	**0.046**	0.144	0.222	0.224	0.124	**0.116**	tw. peaks
$\mathcal{U}_{TP}^{0.333}$	0.096	0.081	**0.039**	0.156	0.221	0.141	0.136	**0.093**	tw. peaks
$\mathcal{U}_{IE}$	**0.041**	0.049	**0.041**	0.248	0.264	**0.049**	0.244	0.168	ind. ex.

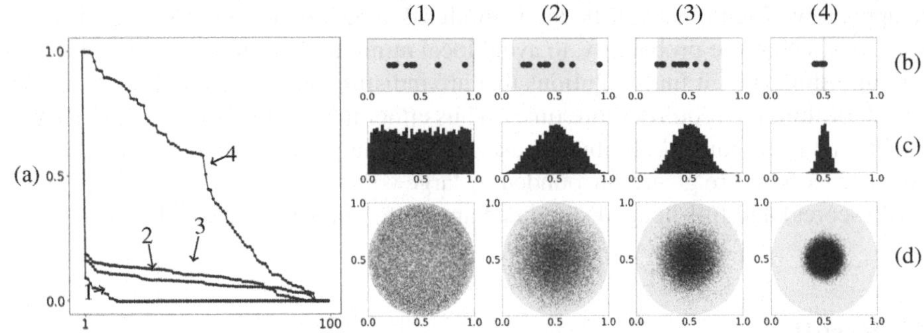

Fig. 5 Results for weakly separable rules and utopic distributions **1** $\mathcal{U}_D$, **2** $\mathcal{U}_D^{0.1}$, **3** $\mathcal{U}_D^{0.2}$, and **4** $\mathcal{U}_{IE}$

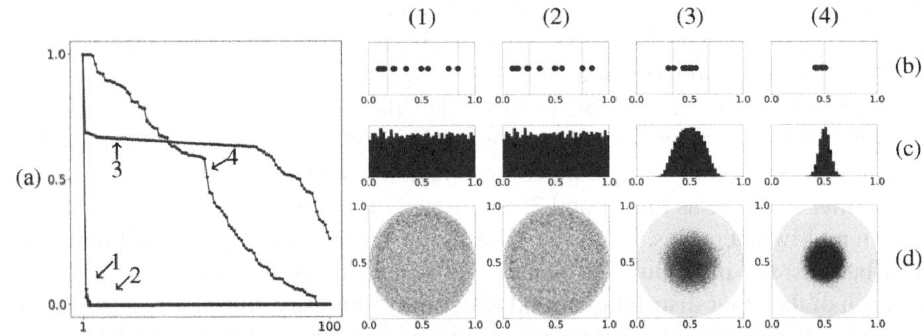

Fig. 6 Results for weakly separable rules and utopic distributions **1** $\mathcal{U}_{TP}^{0.167}$, **2** $\mathcal{U}_{TP}^{0.25}$, **3** $\mathcal{U}_{TP}^{0.333}$, and **4** $\mathcal{U}_{IE}$

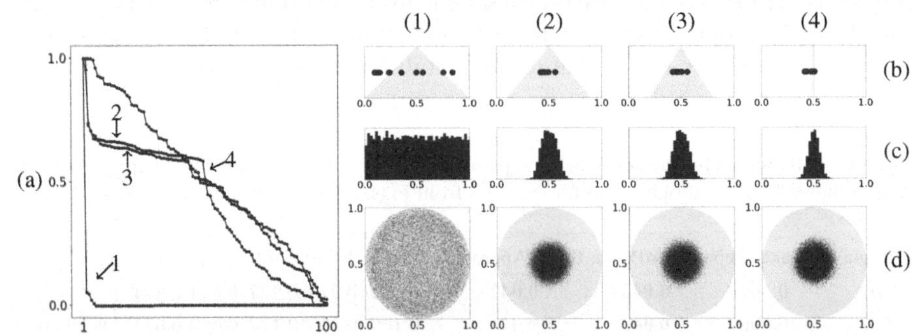

Fig. 7 Results for weakly separable rules and utopic distributions **1** $\mathcal{U}_T^0$, **2** $\mathcal{U}_T^{0.1}$, **3** $\mathcal{U}_T^{0.2}$, and **4** $\mathcal{U}_{IE}$

axis scales, as they only show the "shape" of the result). Finally, on the bottom of each figure, as Plot (d), we show the scatter plots computed for disc elections according to our four rules. The vectors computed for the utopias are marked with a number (1–

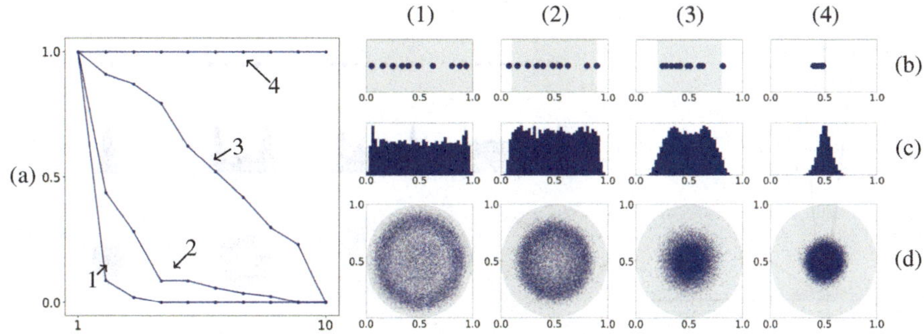

Fig. 8 Results for OWA/Borda-based rules and utopic distributions **1** $\mathcal{U}_D$, **2** $\mathcal{U}_D^{0.1}$, **3** $\mathcal{U}_D^{0.2}$, and **4** $\mathcal{U}_{IE}$

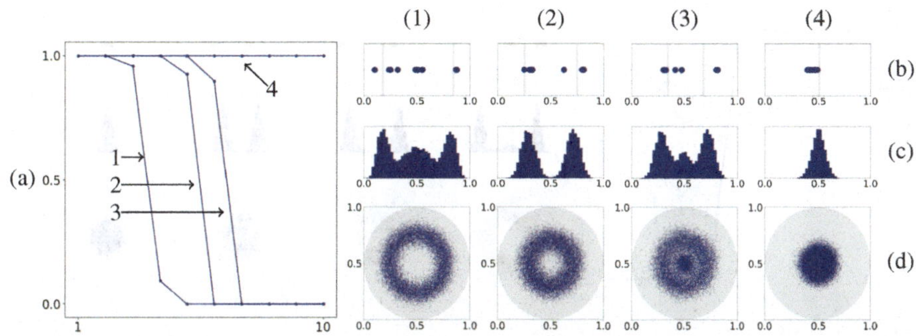

Fig. 9 Results for OWA/Borda-based rules and utopic distributions **1** $\mathcal{U}_{TP}^{0.167}$, **2** $\mathcal{U}_{TP}^{0.25}$, **3** $\mathcal{U}_{TP}^{0.333}$, and **4** $\mathcal{U}_{IE}$

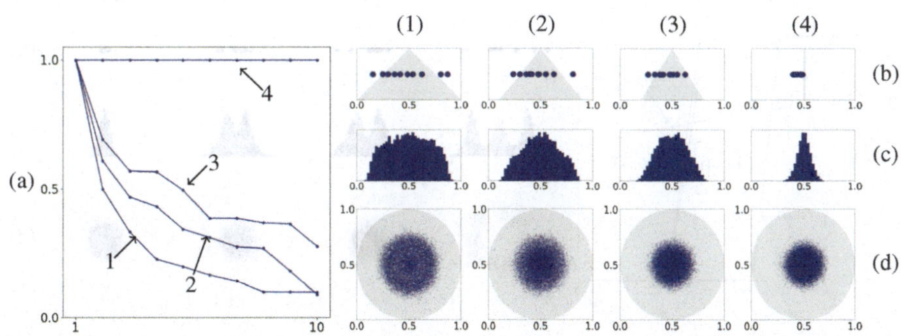

Fig. 10 Results for OWA/Borda-based rules and utopic distributions **1** $\mathcal{U}_T^0$, **2** $\mathcal{U}_T^{0.1}$, **3** $\mathcal{U}_T^{0.2}$, and **4** $\mathcal{U}_{IE}$

4) and the respective figures in (b)–(d) are marked with matching numbers. For the OWA/Approval-based rules, the vectors are also annotated with the t value specifying the t-Approval scoring function used. In the following sections we discuss the results regarding weakly separable, OWA/Borda-based, and OWA/Approval-based rules.

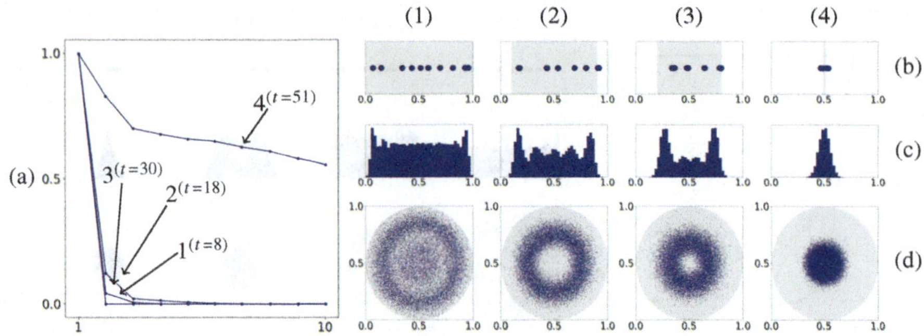

Fig. 11 Results for OWA/Approval rules and utopic distributions **1** $\mathcal{U}_D$, **2** $\mathcal{U}_D^{0.1}$, **3** $\mathcal{U}_D^{0.2}$, and **4** $\mathcal{U}_{IE}$

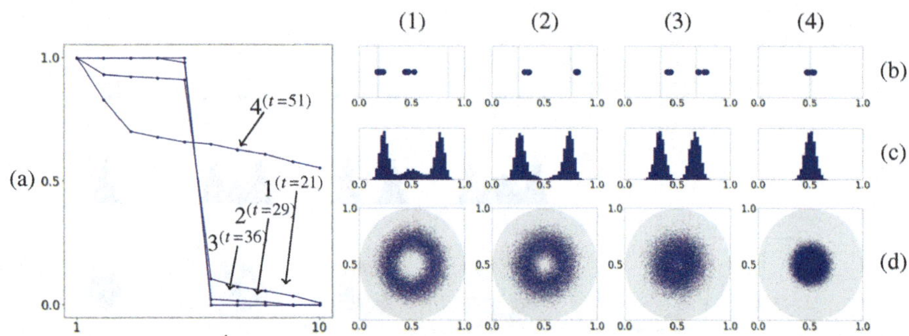

Fig. 12 Results for OWA/Approval rules and utopic distributions **1** $\mathcal{U}_{TP}^{0.167}$, **2** $\mathcal{U}_{TP}^{0.25}$, **3** $\mathcal{U}_{TP}^{0.333}$, and **4** $\mathcal{U}_{IE}$

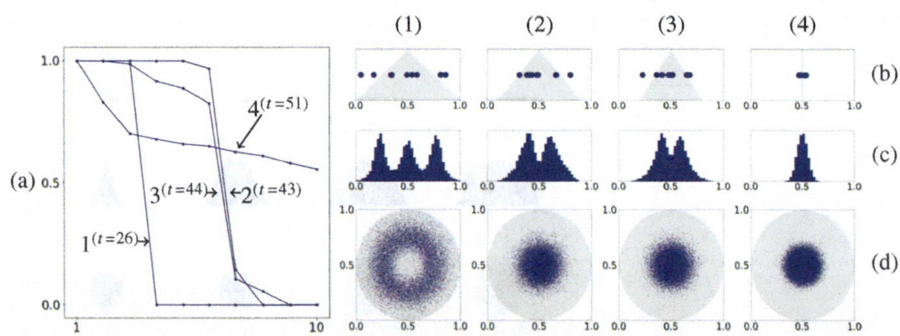

Fig. 13 Results for OWA/Approval rules and utopic distributions **1** $\mathcal{U}_T^0$, **2** $\mathcal{U}_T^{0.1}$, **3** $\mathcal{U}_T^{0.2}$, and **4** $\mathcal{U}_{IE}$

We stress that, to get a good understanding regarding how well a particular rule implements a given utopia, it is important to consider both the EMD score values from Table 1 and the histograms and scatter plots from Figs. 5, 6, 7, 8, 9, 10, 11, 12 and 13. The former gives a good insight regarding the average performance of a given rule, whereas the latter give a more direct information about its possible biases and unexpected behavior. However, as a general principle, if we compare two rules where one has a significantly lower EMD score and the other has a significantly more appealing histogram/scatter plot, then we prefer the former.

5.1 Weakly Separable Rules

The results for weakly separable rules are quite interesting, but, in some sense, also quite negative. On the one hand, they confirm that several well-known rules are among the best weakly separable rules for several utopias (e.g., this is the case for k-Borda and individual excellence, and for SNTV and the diversity utopia). On the other hand, some of these well-known rules—such as SNTV—are known to not be very appealing. Further, the EMD scores achieved by weakly separable rules are typically significantly higher than those achieved by OWA/Borda-based and OWA/Approval-based rules (individual excellence is one important exception here). Below we discuss the results in more detail.

For individual excellence (see Fig. 5, vector 4) we obtained a nearly linear scoring vector, very close to the Borda scoring function. On the other extreme, for the diversity utopia, we found a rule very close to SNTV (vector 1, which gives one point for the first position, then very quickly drops to close-to-zero scores, and for most positions gives zero points). Interestingly, the rule that our algorithm found for the case of individual excellence is slightly better than the purely linear one, used by k-Borda (its EMD score is 0.041, whereas for k-Borda it is 0.049; see Table 1), but the rule found for the diversity utopia is somewhat worse than SNTV (EMD value 0.104 versus 0.092). We believe that this is due to the fact that, for our algorithm, it is very difficult to find the vector of the form $(1, 0, \ldots, 0)$; we would need to use many more iterations to find it. Similarly, we would need much more extensive computations to test whether k-Borda is, in fact, the best rule for individual excellence and our algorithm approximates it, or, in fact, some nonlinearity in the scoring vector truly helps.

We interpret the fact that for the individual excellence utopia we found a rule very similar to k-Borda as a positive result: It confirms that k-Borda is a very good rule for this task. On the other hand, the fact that SNTV appears to be the best weakly separable for the diversity utopia is negative: Even though the histogram for SNTV suggests that it selects candidates uniformly from the whole interval, it is, in fact, only a statistical phenomenon. SNTV chooses candidates from areas with lower density of candidates and increased density of voters, which, statistically, happen equally often in each area of the interval, but in each particular election they can be located very non-uniformly (this phenomenon was already mentioned by Elkind et al. (2017); it is also visible in Fig. 5, in the results of a sample interval election for the diversity utopia). Our hope was that we could find a weakly separable rule that implements the diversity utopia more robustly, but it turned out to be unrealistic. This suggests that, if we are looking for a simple, polynomial-time computable rule that provides diverse committees, then we have to look beyond committee scoring rules (e.g., one might use the GreedyCC rule, introduced by Lu and Boutilier (2011) which is, however, sequential in nature and, thus, sometimes more difficult to analyze).

Perhaps the most interesting results are those achieved for 0.1-diversity and 0.2-diversity (vectors 2 and 3 in Fig. 5), as they show rules that, if at all, are very rarely discussed in the literature. Both vectors 2 and 3 resemble functions of the form $\gamma(i) = (1-x)\alpha_1(i) + x\beta_m(i)$, where $x \in [0, 1]$ is a parameter (and, in our case, is close to 0.2). In other words, these functions give score 1 to position 1, score $\approx$0.2 to position 2, and then decrease linearly to 0. One could say that their Borda-score component is too small to be relevant, but this is not so. In our interval elections we have 100 voters, which means that there are only 100 points to be distributed for being ranked on the first place, while there are $\approx$1000 points to be distributed for being ranked on the following places. This way, the rules described by vectors 2 and 3 achieve a compromise between SNTV and Borda. On the negative side, even though both vectors 2 and 3 resemble the same mixture of Borda and SNTV, they give quite different results. Thus the exact choice of the value x—specifying the proportion in which SNTV and k-Borda scores are mixed—is quite significant.

The results for the ϵ-twin peaks utopias (Fig. 6) are the most pessimistic ones. For $\epsilon \in \{0.167, 0.25\}$ our algorithm, in essence, found the SNTV rule, whose histogram does not resemble the respective utopias at all, but which achieves relatively low EMD score for these utopias (at least among the EMD values possible to achieve using weakly separable rules). Then, for $\epsilon = 0.333$ we found a rule whose histogram is closer to that for individual excellence than 0.333-twin peaks. Interestingly, if instead of the current cost function (which associates half of the committee members, the left-most ones, with the left peak, and half of the committee members with the right peak) we use a cost function which associates *all* the committee members from $[0, 0.5)$ with the left peak and *all* the committee members from $[0.5, 1]$ with the right peak, then we obtain rules whose histograms are much closer to the desired twin peaks shape (we do not report these results here; they were reported in the AAMAS-2018 version of the current paper and denoted, erroneously, as if they were obtained by using the original cost function). This shows that, even though in most cases our framework gives reasonable results, sometimes it may be quite non-robust and it may be difficult to specify the appropriate cost functions. Another interpretation is that our current cost function for twin peaks utopias specifies two desires: (a) the wish that the committee members are selected from the vicinity of the two peaks, and (b) the wish to obtain half of the committee members from the vicinity of one peak and half of the committee members from the vicinity of the other peak. The other cost function specifies desire (a), but not desire (b). Thus, it seems that weakly separable functions can (at least partially) fulfill desire (a), but not both desire (a) and desire (b).

The results for the triangle utopias (see Fig. 7) are disappointing as well. Either we find a rule very similar to SNTV or rules that are close to implementing the individual excellence utopias. The scoring vectors of the latter rules are, however, somewhat interesting as they seem to be of the form:

$$\gamma(i) = \begin{cases} 0.25\alpha_1 + 0.75 & \text{if } i < m/2, \\ 1.5\beta_m(i) & \text{if } i \geq m/2. \end{cases}$$

Thus, as in the case of the 0.1-diversity and 0.2-diversity, a scoring function that achieves some compromise between individual excellence and diversity can be expressed by mixing the 1-Approval and Borda scoring functions in some way.

5.2 OWA/Borda-Based Rules

The results for OWA/Borda-based rules were the most positive ones. We were able to find good rules for the diversity and triangle utopias (our rules both have low EMD scores and quite appealing histograms/scatter plots), and for the twin peaks utopias they performed acceptably, even if OWA/Approval-based rules were champions in this category.

For individual excellence (Fig. 8, vector 4), our algorithm very quickly found OWA vector of all ones, so, we obtained the k-Borda rule. For diversity (Fig. 8, vector 1), we found a vector close to that of the Chamberlin–Courant rule (CC uses vector $(1, 0, \ldots, 0)$ and our algorithm found a vector with a single one followed by values very quickly approaching zero). Thus, effectively, the algorithm obtained the CC rule. For 0.1-diversity and 0.2-diversity we found, respectively, a linearly decreasing vector 3, and vector 2 that resembles (but, admittedly, quite poorly) the harmonic sequence. These two results are intriguing. First, the linear vector is a very natural solution to finding a compromise between excellence and diversity (which, in this case, would mean finding a compromise between k-Borda and Chamberlin–Courant) that has not been considered in the literature yet (even though Faliszewski et al. (2017) looked for rules that achieve such a compromise, they did not study this rule). On the other hand, the harmonic vector has received extensive treatment, both for Harmonic Borda (Faliszewski et al. 2017) and for the PAV rule (Thiele 1895; Aziz et al. 2017; Kilgour 2010; Lackner and Skowron 2018). In fact, the OWA vectors that we have obtained for the triangle distributions (see Fig. 10) seem to be closer to the harmonic sequence. This confirms the intuition of Faliszewski et al. (2017) that this sequence achieves a good excellence/diversity compromise.

Next, let us consider the results for the twin peaks distributions (Fig. 9). In this case we found OWA vectors that consist of ones followed by zeros (the number of ones depends on the distance between the peaks). This means that the rules that we found are, in essence, the t-Borda rules of Faliszewski et al. (2017) (for a given t, the t-Borda rule uses Borda scoring function and OWA vector of t ones followed by zeros). Faliszewski et al. (2017) studied these rules in their search for excellence/diversity compromises, but concluded that they do not seem to work well for this case. The fact that they implement the twin peaks utopic distribution supports this conclusion.

It is also interesting to compare the histograms for the interval elections and the scatter plots for disc ones. In most cases they behave similarly, but sometimes strange effects appear. For example, for the OWA/Borda rule computed for the 0.167-twin-peaks utopia, in the histogram we can observe two similar peaks around positions $\frac{1}{6}$ and $\frac{5}{6}$, and a third, lower peak in the middle. Yet, in the scatter plot the center area is empty. The effect is even more interesting when we note that in the 0.333-twin-peaks case the center peak is visible both in the histogram and in the scatter plot. Possible explanation is that in the 0.333-twin peaks case the two peaks are so close to each other that they overlap. We expect that in higher dimensional spaces this "overlap effect" disappears. At the same time, in the 0.167-twin peaks case the two peaks are so far away from each other that there are some accidental winners in the center. However the problem disappears in the scatter plots.

5.3 OWA/Approval-Based Rules

OWA/Approval-based rules are quite expressive but, nonetheless, in most cases their performance (in terms of the achieved EMD scores) was between that of weakly separable rules and OWA/Borda-based ones. Yet, OWA/Approval-based rules turned out to be the best for the twin peaks utopias. Interestingly, for the case of individual excellence the OWA/Approval-based rule outperformed k-Borda, but, as we will see, the computed rule does not seem to be particularly appealing.

For each of the three twin peaks utopias our algorithm found essentially the same OWA vector, very close to $(1, 1, 1, 1, 1, 0, 0, 0, 0, 0)$, and the only difference was in the t value for the approval scoring function (see Fig. 12). Under these rules each voter can give a point to up to half of the committee members, closest to him or her. Intuitively, rules defined in this way choose the candidates from the peaks for the following reason: There are two positions, one next to each of the two edges of the interval, where the candidates are still approved by all the voters closer to the edge and a large number of the voters farther from the edge (the values of these positions depend on t). Thus the candidates in these areas have an advantage over the other ones and form the peaks. The shape of the OWA vector guarantees that the candidates supporting one peak do not interfere with those supporting the other.

For the case of the individual excellence utopia, our algorithm found a rule with an OWA vector where all values are quite high, and where $t = 51$. This value of t is particularly important: It means that the candidates in the center of the interval receive points from all the 100 voters. Thus the rule that we found achieves the specified goal, but it does so in a very unsatisfying way. Even though we still get a convincing scatter plot for disc elections, the fact that each voter should approve over half of the candidates to select the best one is intuitively unappealing; for example, if all the voters had the same preference order, there would be a tie among top 51 candidates, even though all the voters would agree which one of them were the best.

For the case of the diversity utopias, all the OWA vectors also are quite similar to each other and resemble the OWA vector for the Chamberlin–Courant rule. The difference between 0.1-diversity and 0.2-diversity is mostly implemented by choosing larger t-values (this allows us to select candidates farther from the edges, who, nonetheless, receive points from the most extreme voters). Unfortunately, OWA/Approval-based rules obtained for the 0.1-diversity and 0.2-diversity utopia promote the candidates "on the edges of the center interval" (i.e., those close to positions 0.1 and 0.9 for 0.1-diversity and those close to positions 0.2 and 0.8 for 0.2-diversity). We believe that this effect is connected to the extreme voters: On the average, the voters in the center distribute their approvals among candidates around them within a given radius, whereas the extreme voters focus on the candidate in only one direction from them (in other words, the distance between the voter and his least preferred candidate to whom he or she will still assign a t-Approval point is larger for the extreme voters, so the candidates "on the edge of the center interval" receive more points). It is also important to mention that the rules that we found for 0.1-diversity and 0.2-diversity do not seem to maintain their features under disc elections. Indeed, the scatter plots are torus-shaped, with decreasing radiuses, whereas we would expect more-or-less uniform discs (also with decreasing radiuses).

The results for the triangle utopias are quite good in terms of the EMD scores (although notably worse than in the case of OWA/Borda rules), but the histograms appear less appealing. The computed OWA vectors this time are either close to $(1, 1, 1, 0, 0, 0, 0, 0, 0, 0)$ or $(1, 1, 1, 1, 1, 1, 0, 0, 0, 0)$ and the t-values are quite high (26 for the triangle utopia, which uses the former vector, and 43 and 44 for the two other utopias). As in the discussion of the excellence utopia, we view high values of t as unappealing and not useful in practical settings.

All in all, our algorithm mostly focused on OWA vectors where a number of ones is followed by zeros. This is quite interesting as theoretical studies have focused on OWA vectors that provide a more smooth transition between 1 and 0; the harmonic vector is by far the best studied example of such a vector. This suggests that some further studies of the OWA/Approval-based rules with OWA vectors of the form $(1, \dots, 1, 0, \dots, 0)$ may be valuable.

6 Conclusions

We have developed a methodology for automatically designing multiwinner voting rules whose winning committees have properties specified via distributions on a 1D interval. Testing our method on weakly separable and OWA-based committee scoring rules, we confirmed many intuitions about the applicability of certain rules for certain tasks and discovered new rules to study. Our work is a proof of concept and shows that our approach is indeed feasible.

A natural direction for future studies is to develop better optimization-based approaches (e.g., use more powerful search algorithms, more diverse distributions of candidates and voters, and more utopias). We tested other approaches, e.g., solving linear programs with the values of the scoring function as variables, so that given candidates in given elections would be winners (e.g., candidates close to the center, for individual excellence). We got far worse results than with our main approach. Nonetheless, we believe that it is important to experiment with the methodology.

Supplied with good optimization algorithms, one should consider wider classes of rules, further utopias, and more general settings. E.g., it might be possible to automatically design methods for participatory budgeting (Cabannes 2004; Goel et al. 2019; Benade et al. 2017; Shapiro and Talmon 2017), which naturally generalizes our setting. To do so, we would consider distance measures between budgets as well as define "budgeting utopias".

Acknowledgements Preliminary version of this paper was presented at the AAMAS-2018 conference and at the COMSOC-2018 workshop. The current version contains additional discussions and an extended set of results (including all results for the OWA/Approval-based rules). Further, the current version corrects several mistakes that appeared in the AAMAS-2018 paper. We are very grateful to all the reviewers who commented on this paper. Piotr Faliszewski was supported by the National Science Centre, Poland, under project 2016/21/B/ST6/01509. Nimrod Talmon was supported by the Israel Science Foundation (ISF; Grant No. 630/19).

Appendix

In this section we describe how to compute the EMD score for the ϵ-triangle utopia. Let us assume that we have some committee W. In order to compute $\mathrm{EMD}(\mathcal{U}_T^{\epsilon}, d_W)$, we have to divide the whole triangle into k pieces, each with the same area (to simplify the algorithm description we assume that k is even). Then, each of the winning committee members will be responsible for covering one piece of the triangle. As we assume the committee members to be sorted (with respect to their ideal positions on the interval) in the ascending order, we assign them to the consecutive pieces of the triangle. By l_i and r_i we mean the left and the right border of the ith interval, respectively. For the left slope of the triangle, i.e., for $i \in \{1, \dots, \frac{k}{2}\}$, we have:

$$l_i = \epsilon + (0.5 - \epsilon)\sqrt{\frac{i-1}{\frac{1}{2}n}}, \qquad \text{and} \qquad r_i = \epsilon + (0.5 - \epsilon)\sqrt{\frac{i}{\frac{1}{2}n}},$$

whereas for the right slope, i.e., for $i \in \{\frac{k}{2} + 1, \dots, k\}$, we have:

$$l_i = 1 - \epsilon + (0.5 - \epsilon)\sqrt{\frac{k-i-1}{\frac{1}{2}n}}, \qquad \text{and} \qquad r_i = 1 - \epsilon + (0.5 - \epsilon)\sqrt{\frac{k-i}{\frac{1}{2}n}}.$$

For each w_i, where $i \in \{1, \dots, k\}$, there are two symmetric cases. Either $p(w_i)$ is located on the left slope of the triangle or on the right slope. Both of these cases have three subcases (similarly to the three cases for the diversity utopia). The main difference in computing the cost for the triangle utopia and the diversity utopia is that for the triangle utopia the intervals have different lengths and heights. In general, for points on the left slope it will cost more to spread them to the right than to the left.

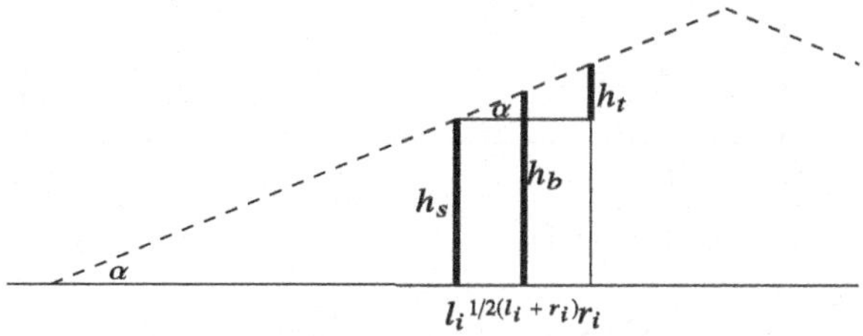

Fig. 14 Illustration of the notation used in calculations. The picture shows a fragment of the ϵ-triangle utopia, focusing on the interval between points l_i and p_i associated with committee member w_i

Analogously, for points on the right slope it will cost more to spread them to left than to the right.

For all the considered cases considered, we use the following notation (see Fig. 14 for illustration):

1. $\ell = {}^1\!/_k$ is the probability mass associated with each committee member,
2. $d_i = r_i - l_i$ is the length of the interval associated with committee member w_i,
3. $dr_i = r_i - p(w_i)$ and $dl_i = p(w_i) - l_i$ are the distances of the committee member w_i from the right and the left and of his or her interval (under the assumption that $p(w_i)$ is between l_i and r_i; we only use these values when this assumption holds),
4. h_s is the lowest value of the probability density function in the interval $[l_i, r_i]$ (so it is the value at the point l_i on the left side of the slope and it is the value at point r_i on the right slope), h_b is the value of the probability density function at point ${}^1\!/_2(l_i + r_i)$, and h_t is the difference between h_s and the highest value of the probability density function on the interval; we calculate these values as follows (Fig. 15):

$$h_t = d_i \tan\alpha, \qquad h_b = {}^{\ell}\!/_{d_i}, \qquad h_s = h_b - {}^1\!/_2 h_t.$$

(To understand how h_b is calculated, note that ℓ is the area of the trapezoid depicted on Fig. 14 between points l_i and r_i; α is the angle of the slope of our "triangle.")

With the above notation in hand, we compute the cost for each of the committee members as follows:

1. The committee member is on the left slope of the triangle (i.e., $p(w_i) \le 0.5$):

   1.1. If the committee member is to the left of his or her interval (i.e., $p(w_i) < l_i$), then we need to pay the cost $(l_i - p(w_i))$ for moving his or her probability mass to the point l_i, and then the cost $(\frac{1}{2}h_s d_i + \frac{\pi}{4}h_t d_i)$ for "spreading" his or her weight over the interval. Note that in triangle utopia "spreading" is not always symmetric.

$$cost(w_i) = (l_i - p(w_i))\ell + \frac{1}{2}h_s d_i + \frac{\pi}{4}h_t d_i$$

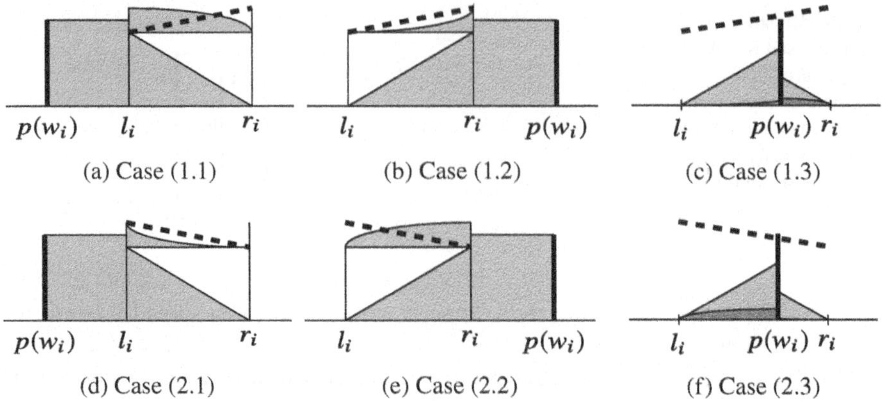

Fig. 15 The cases for the cost of EMD($\mathcal{U}_T$, d_W). The cost associated with committee member w_i is the gray area

1.2. If the committee member is to the right of his or her interval (i.e., $p(w_i) > r_i$), then we need to pay the cost $(p(w_i) - r_i)$ for moving his or her probability mass to the point r_i, and then the cost $(\frac{1}{2}h_s d_i + (1 - \frac{\pi}{4})h_t d_i)$ for "spreading" his or her weight over the interval.

$$cost(w_i) = (p(w_i) - r_i)\ell + \frac{1}{2}h_s d_i + (1 - \frac{\pi}{4})h_t d_i$$

1.3. If the committee member is inside his or her interval (i.e., $l_i \leq p(w_i) \leq r_i$), then it suffices to "spread" the $\frac{p(w_i)-l_i}{d_i}$ fraction of his or her probability mass to the part of the interval left of him, at cost $(\frac{dl_i^2}{2h_s} + (1 - \frac{\pi}{4})\frac{dl_i^3 h_t}{2d_i^2})$; and the remaining mass, to the part of the interval to the right of him or her, at cost $(\frac{dr_i^2}{2h_s} + \frac{\pi}{4}\frac{dr_i h_t (d_i^2 - dl_i^2)}{2d^2})$.

$$cost(w_i) = \frac{dl_i^2}{2h_s} + (1 - \frac{\pi}{4})\frac{dl_i^3 h_t}{2d_i^2} + \frac{dr_i^2}{2h_s} + \frac{\pi}{4}\frac{dr_i h_t (d_i^2 - dl_i^2)}{2d^2}$$

2. The committee member is on the right slope of the triangle (i.e., $p(w_i) \geq 0.5$):

2.1. If the committee member is to the left of his or her interval (i.e., $p(w_i) < l_i$), then we need to pay the cost $(l_i - p(w_i))$ for moving his or her probability mass to the point l_i, and then the cost $(\frac{1}{2}h_s d_i + (1 - \frac{\pi}{4})h_t d_i)$ for "spreading" his or her weight over the interval.

$$cost(w_i) = (l_i - p(w_i))\ell + \frac{1}{2}h_s d_i + (1 - \frac{\pi}{4})h_t d_i$$

2.2. If the committee member is to the right of his or her interval (i.e., $p(w_i) > r_i$), then we need to pay the cost $(p(w_i) - r_i)$ for moving his or her probability mass to the point r_i, and then the cost $(\frac{1}{2}h_s d_i + \frac{\pi}{4}h_t d_i)$ for "spreading" his or her weight over the interval.

$$cost(w_i) = (p(w_i) - r_i)\ell + \frac{1}{2}h_s d_i + \frac{\pi}{4}h_t d_i$$

2.3. If the committee member is inside his or her interval (i.e., $l_i \leq p(w_i) \leq r_i$), then it suffices to "spread" the $\frac{p(w_i)-\ell(i-1)}{\ell}$ fraction of his or her probability mass to the part of the interval left of him, at cost $(\frac{dl_i^2}{2h_s} + (1 - \frac{\pi}{4})\frac{dr_i^3 h_t}{2d_i^2})$; and the remaining mass, to the part of the interval to the right of him or her, at cost $(\frac{\pi}{4}\frac{dr_i^2}{2h_s} + \frac{dl_i h_t(d_i^2 - dr_i^2)}{2d_i^2})$.

$$cost(w_i) = \frac{dl_i^2}{2h_s} + (1 - \frac{\pi}{4})\frac{dr_i^3 h_t}{2d_i^2} + \frac{\pi}{4}\frac{dr_i^2}{2h_s} + \frac{dl_i h_t(d_i^2 - dr_i^2)}{2d_i^2}$$

The cost of the whole committee is the sum of the cost of its members.

References

1. Anshelevich, E., Bhardwaj, O., Elkind, E., Postl, J., Skowron, P.: Approximating optimal social choice under metric preferences. Artif. Intell. **264**, 27–51 (2018)
2. Ayadi, M., Ben Amor, N., Lang, J., Peters, D.: Single transferable vote: incomplete knowledge and communication issues (2019)
3. Aziz, H., Brill, M., Conitzer, V., Elkind, E., Freeman, R., Walsh, T.: Justified representation in approval-based committee voting. Soc. Choice Welfare **48**(2), 461–485 (2017)
4. Aziz, H., Elkind, E., Faliszewski, P., Lackner, M., Skowron, P.: The condorcet principle for multiwinner elections: from shortlisting to proportionality (2017)
5. Aziz, H., Elkind, E., Huang, S., Lackner, M., Sánchez Fernández, L., Skowron, P.: On the complexity of extended and proportional justified representation (2018)
6. Aziz, H., Faliszewski, P., Grofman, B., Slinko, A., Talmon, N.: Egalitarian committee scoring rules (2018)
7. Aziz, H., Gaspers, S., Gudmundsson, J., Mackenzie, S., Mattei, N., Walsh, T.: Computational aspects of multi-winner approval voting (2015)
8. Baumeister, D., Dennisen, S., Rey, L.: Winner determination and manipulation in minisum and minimax committee elections (2015)
9. Baumeister, D., Hogrebe, T.: How hard is the manipulative design of scoring systems? (2019)
10. Benade, G., Nath, S., Procaccia, A., Shah, N.: Preference elicitation for participatory budgeting (2017)
11. Betzler, N., Slinko, A., Uhlmann, J.: On the computation of fully proportional representation. J. Artif. Intell. Res. **47**, 475–519 (2013)
12. Boixel, A., Endriss, U.: Automated justification of collective decisions via constraint solving (2020)
13. Boutilier, C., Caragiannis, I., Haber, S., Lu, T., Procaccia, A., Sheffet, O.: Optimal social choice functions: a utilitarian view. Artif. Intell. **227**, 190–213 (2015)

14. Brandl, F., Peters, D.: An axiomatic characterization of the Borda mean rule. Soc. Choice Welfare **52**(4), 685–707 (2019)
15. Brill, M., Laslier, J., Skowron, P.: Multiwinner approval rules as apportionment methods. J. Theor. Polit. **30**(3), 358–382 (2018)
16. Byrka, J., Skowron, P., Sornat, K.: Proportional approval voting, harmonic k-median, and negative association (2018)
17. Cabannes, Y.: Participatory budgeting: a significant contribution to participatory democracy. Environ. Urban. **16**(1), 27–46 (2004)
18. Caragiannis, I., Nath, S., Procaccia, A.D., Shah, N.: Subset selection via implicit utilitarian voting (2016)
19. Caragiannis, I., Procaccia, A.: Voting almost maximizes social welfare despite limited communication. Artif. Intell. **175**(9–10), 1655–1671 (2011)
20. Celis, L., Huang, L., Vishnoi, N.: Multiwinner voting with fairness constraints (2018)
21. Chamberlin, B., Courant, P.: Representative deliberations and representative decisions: proportional representation and the Borda rule. Am. Polit. Sci. Rev. **77**(3), 718–733 (1983)
22. Cygan, M., Kowalik, L., Socala, A., Sornat, K.: Approximation and parameterized complexity of minimax approval voting. J. Artif. Intell. Res. **63**, 495–513 (2018)
23. Debord, B.: An axiomatic characterization of Borda's k-choice function. Soc. Choice Welfare **9**(4), 337–343 (1992)
24. Diss, M., Doghmi, A.: Multi-winner scoring election methods: condorcet consistency and paradoxes. Public Choice **169**(1–2), 97–116 (2016)
25. Dudycz, S., Manurangsi, P., Marcinkowski, J., Sornat, K.: Tight approximation for proportional approval voting (2020)
26. Elkind, E., Faliszewski, P., Laslier, J., Skowron, P., Slinko, A., Talmon, N.: What do multiwinner voting rules do? An experiment over the two-dimensional euclidean domain (2017)
27. Elkind, E., Faliszewski, P., Skowron, P., Slinko, A.: Properties of multiwinner voting rules. Soc. Choice Welfare **48**(3), 599–632 (2017)
28. Elkind, E., Lang, J., Saffidine, A.: Condorcet winning sets. Soc. Choice Welfare **44**(3), 493–517 (2015)
29. Enelow, J.M., Hinich, M.J.: The spatial theory of voting: an introduction. CUP Archive (1984)
30. Enelow, J.M., Hinich, M.J.: Advances in the Spatial Theory of Voting. Cambridge University Press, Cambridge (1990)
31. Faliszewski, P., Skowron, P., Slinko, A., Talmon, N.: Multiwinner rules on paths from k-Borda to Chamberlin-Courant (2017)
32. Faliszewski, P., Skowron, P., Slinko, A., Talmon, N.: Multiwinner voting: a new challenge for social choice theory. In: Endriss, U. (ed.) Trends in Computational Social Choice. AI Access Foundation (2017)
33. Faliszewski, P., Skowron, P., Slinko, A., Talmon, N.: Multiwinner analogues of the plurality rule: axiomatic and algorithmic views. Soc. Choice Welfare **51**(3), 513–550 (2018)
34. Faliszewski, P., Skowron, P., Slinko, A., Talmon, N.: Committee scoring rules: axiomatic characterization and hierarchy. ACM Trans. Econ. Comput. **7**(1), 3:1–3:39 (2019)
35. Faliszewski, P., Skowron, P., Szufa, S., Talmon, N.: Proportional representation in elections: STV vs PAV (2019)
36. Faliszewski, P., Talmon, N.: Between proportionality and diversity: balancing district sizes under the Chamberlin-Courant rule (2018)
37. Goel, A., Krishnaswamy, A., Sakshuwong, S., Aitamurto, T.: Knapsack voting for participatory budgeting. ACM Trans. Econ. Comput. **7**(2), 8:1–8:27 (2019)
38. Kilgour, M.: Approval balloting for multi-winner elections. In: Handbook on Approval Voting. Springer (2010). Chapter 6
39. Kocot, M., Kolonko, A., Elkind, E., Faliszewski, P., Talmon, N.: Multigoal committee selection (2019)
40. Lackner, M., Skowron, P.: Consistent approval-based multi-winner rules (2018)
41. Lackner, M., Skowron, P.: A quantitative analysis of multi-winner rules (2019)

42. Lackner, M., Skowron, P.: Approval-based committee voting: axioms, algorithms, and applications. Technical report arXiv:2007.01795 [cs.GT], July 2020
43. Lu, T., Boutilier, C.: Budgeted social choice: from consensus to personalized decision making (2011)
44. Monroe, B.: Fully proportional representation. Am. Polit. Sci. Rev. **89**(4), 925–940 (1995)
45. Obraztsova, S., Elkind, E., Faliszewski, P.: On swap convexity of voting rules (2020)
46. Obraztsova, S., Elkind, E., Faliszewski, P., Slinko, A.: On swap-distance geometry of voting rules (2013)
47. Peleg, S., Werman, M., Rom, H.: A unified approach to the change of resolution: space and gray-level. IEEE Trans. Pattern Anal. Mach. Intell. **11**(7), 739–742 (1989)
48. Peters, D.: Proportionality and strategyproofness in multiwinner elections (2018)
49. Peters, D.: Single-peakedness and total unimodularity: new polynomial-time algorithms for multi-winner elections (2018)
50. Peters, D., Skowron, P.: Proportionality and the limits of welfarism (2020)
51. Pierczynski, G., Skowron, P.: Approval-based elections and distortion of voting rules (2019)
52. Procaccia, A., Rosenschein, J.: The distortion of cardinal preferences in voting (2006)
53. Procaccia, A., Rosenschein, J., Zohar, A.: On the complexity of achieving proportional representation. Soc. Choice Welfare **30**(3), 353–362 (2008)
54. Saari, D.: Geometry of Voting. Springer (1994)
55. Sánchez-Fernández, L., Elkind, E., Lackner, M., Fernández, N., Fisteus, J.A., Basanta Val, P., Skowron, P.: Proportional justified representation (2017)
56. Sekar, S., Sikdar, S., Xia, L.: Condorcet consistent bundling with social choice (2017)
57. Shapiro, E., Talmon, N.: A participatory democratic budgeting algorithm. Technical report arXiv:1709.05839 (2017)
58. Skowron, P., Faliszewski, P., Lang, J.: Finding a collective set of items: from proportional multirepresentation to group recommendation. Artif. Intell. **241**, 191–216 (2016)
59. Skowron, P., Faliszewski, P., Slinko, A.: Axiomatic characterization of committee scoring rules. J. Econ. Theory **180**, 244–273 (2019)
60. Skowron, P., Lackner, M., Brill, M., Peters, D., Elkind, E.: Proportional rankings (2017)
61. Thiele, T.: Om flerfoldsvalg. In: Oversigt over det Kongelige Danske Videnskabernes Selskabs Forhandlinger, pp. 415–441 (1895)
62. Xia, L.: Designing social choice mechanisms using machine learning. In: Proceedings of AAMAS 2013, pp. 471–474 (2013)
63. Yager, R.: On ordered weighted averaging aggregation operators in multicriteria decisionmaking. IEEE Trans. Syst. Man Cybern. **18**(1), 183–190 (1988)

A Procedure for Multiattribute Reverse Auctions with Two Strategic Parameters

Gregory E. Kersten, Tomasz Szapiro, and Shikui Wu

Abstract Reverse auctions have been used in the procurement of goods and services. Single-attribute, typically price-only, auctions are well known and have been widely used. Multiattribute auction mechanisms have also been proposed and implemented. The dilemma that designers of multi-attribute auction mechanisms face is the trade-off between the buyer's preference revelation and the ability of the sellers (bidders) to construct effective offers. Effective offers are those which are preferred by the buyer over offers that were made earlier. Full transparency, which requires that the buyer gives complete preference information to the bidders, is often rejected by the buyers because of the future implications when competitors obtain this information. Zero transparency renders auctions ineffective because bidders lack the information necessary for bid construction. This paper develops a procedure in which preference information is transformed into information on reservation levels. The reservation level information is perturbed so that the bidders cannot compute the buyer's preferences. The perturbation is controlled by two parameters which have strategic character for the auction mechanism and jointly represent the owner's tradeoff between maintaining the preference secrecy and achieving an efficient winning bid.

1 Introduction

E-procurement is a key area of e-business and supply chain management in which catalogs and reverse auctions have been widely used (Anderson and Frohlich 2001;

G. E. Kersten (✉)
John Molson School of Business, Concordia University, Station H, P.O. Box 2002, Montreal, QC H3G 2V4, Canada
e-mail: gregory@jmsb.concordia.ca

T. Szapiro
Institute of Econometrics, Warsaw School of Economics (SGH), Al. Niepodległosci 162, 02-544 Warsaw, Poland

S. Wu
Faculty of Business Administration, Lakehead University, Thunder Bay, ON P7B 5E1, Canada
e-mail: shikui.wu@lakeheadu.ca

T. Szapiro and J. Kacprzyk (eds.), *Collective Decisions: Theory, Algorithms And Decision Support Systems*, Studies in Systems, Decision and Control 392,
https://doi.org/10.1007/978-3-030-84997-9_3

Jap 2003). On average, about 70% of corporate revenue is spent on purchasing; savings of 5% translate into hundreds of millions of dollars (Peleg 2003; Wagner and Schwab 2004). There are two kinds of auctions: single-shot and iterative. Iterative auctions, which allow bidders to revise their bids, are becoming prevalent in procurement (Parkes and Kalagnanam 2005). This paper discusses the design of iterative auctions. One of the limitations of auctions is that they use a single attribute (i.e., price), which leads to inefficient agreements (Strecker and Seifert 2004) and is not practical in many business transactions (Teich et al. 2004).

Two types of iterative auctions are possible: synchronous and asynchronous. An auction is synchronous if every seller makes at most one bid in each round. The buyer selects the best bid as the reference bid and presents it to the sellers. The sellers use the reference bid to construct their bids in the next round bid. In addition, the buyer may provide other information to help sellers construct their bids. Asynchronous auctions allow sellers to bid at any time until the auction's deadline. The best bid is shown to the bidders. If a better bid is submitted then it replaces the previous best bid. In this paper we focus on synchronous auctions; however, the procedure proposed can be adapted to asynchronous auctions.

Auction design has traditionally focused on the construction of rules which govern the behavior of auction participants so that their auctions lead to a desired market outcome. The outcome is the final allocation of the goods and money. The desired aspect of the outcome is the auction initiator (the buyer in our case), profit or revenue maximization, or it is the creation of an efficient market (Kittsteiner and Ockenfels 2006). The rules specify the winner determination formula, auction duration and the type of deadline (extendible or fixed), types of bids (sealed or open), and so on.

In synchronous iterative single-attribute auctions the rules determine whether: (1) all bids are open and posted to visible to all bidders; (2) only some bids are open and visible; or (3) only the best bid made in a given round is open and visible. Either of these options is sufficient for the bidders to decide on bidding in the next round. Therefore, the rule defining an acceptable bid is simple—every submitted bid must exceed the last posted bid. This rule assures that the time-order of bids is the same as the profit-order for the buyer, that is, later bids are better than earlier. The concept "better than" is easily operationalized by the explicit auction criterion, which is the single attribute.

Procurement of more complex goods and services often requires consideration of multiple attributes (e.g., total costs of ownership components, quality, risk and schedules). Multiattribute auctions cannot have a "better than" rule because there is no auction criterion that is explicit and known to all participants. Ways to overcome the lack of an explicit criterion include: (1) pre-selection of bidders so that only bidders who are known to meet the additional criteria are included; (2) giving incumbents an advantage because their qualifications are known; and (3) the use of disclaimers such as "the lowest bid may not be awarded the contract" (Bichler and Kalagnanam 2005; Engelbrecht-Wiggans et al. 2007; Schoenherr and Mabert 2007). In these types of auctions either the selection or bidding process are modified so that a single-attribute auction can be used.

The results of such auction modification are mixed because of collusion and selection of inferior offers (Elmaghraby 2004; Katok and Wambach 2011). In some situations the process becomes an auction in name only, as is the case with an auction in which neither the winner nor any other participant is awarded the contract.

Another, seemingly simple approach is to give bidders all information which the buyer uses in order to analyze and compare bids (Chen-Ritzo et al. 2005). This somewhat complicates the computation because the bidders need to optimize using both their own and the buyer's information (e.g., utility or a scoring function). It may also encourage the buyer to engage in strategic misrepresentation and announce a utility function with the aim of pushing the sellers to make favorable bids (Burmeister et al. 2002).

This approach is unacceptable when buyers do not want to disclose their preferences for strategic, competitive, or other reasons (Burmeister et al. 2002; Parkes and Kalagnanam 2005). In the context of multiattribute bidding, this means that the bidders do not know how to bid; they cannot make tradeoffs that take the buyer's preferences into account and they may misinterpret the buyer's preferential directions. The bidders may make strong assumptions about the buyer's utility and bid accordingly. This may be acceptable if their knowledge of the buyer's preferences is accurate and the buyer accepts an inefficient winning bid.

Another option has been proposed by economists. This option rests on the assumption that all attributes can be expressed in monetary terms so that only two items need to be considered: (1) price, and (2) monetized attributes, which typically represent costs—for the sellers and value (income)—for the buyer. When an assumption is added that these two terms are monotonic and the buyer compares bids using the difference between value and price, then the sellers can determine the buyer's preferential order of the alternatives.

The attribute monetization methods have been widely implemented and tested (e.g., Che 1993; Strecker and Seifert 2004; Bichler and Kalagnanam 2005), and they are considered a standard in the auction literature (Parkes and Kalagnanam 2005). These methods, based on two-attribute monetary value functions, are appealing because they allow buyers and sellers to integrate and trade off all attributes included in the cost function (Strecker and Seifert 2004). On one hand, the bidder may choose a bid among his/her indifferent alternatives (i.e., different bids which yield the same utility for this bidder) that yields the highest utility to the buyer; on the other hand, the owner evaluates bids based on the total utility of bids and chooses the highest one. The limitation of this method is the underlying assumption that all attributes can be measured with money. The assumption is questionable, if one considers such attributes as trust, brand, or color.

The design of auction mechanisms that rely on attribute monetization involves the construction of rules that help the sellers to make *progressive bids*; i.e., bids which are better for the buyer than the bids made earlier. The information conveyed to the sellers is about the buyer's preferences and it is either complete or incomplete but sufficient to assure the auction convergence. A different approach has been proposed by Teich et al. (1999) in which the sellers are informed about a path in the space of alternatives.

In this paper we build on the approach proposed by Teich et al. (1999) in the sense that any information conveyed to the sellers refers to the space of alternatives. Furthermore, rather than inform the sellers about preferential direction(s) and/or alternatives preferred over any given alternative, we restrict information to the acceptable alternatives so that no preferential information need be conveyed.

Given our focus, we are not concerned here with the complete set of auction design rules, which assures that given outcomes are met. Instead, we are concerned with rules which assure that no alternative is removed which could yield the desired outcome. Whether such an outcome is achieved depends on the sellers' behavior, which is not discussed here.

The proposed procedure is controlled by two strategic parameters. These parameters determine the acceptable difference of utility between an accepted inefficient bid and an efficient bid which a bidder could make. Following Bellosta et al. (2008, p. 402) we say that the winning bid is *efficient* when it is a feasible alternative and no seller, except for the winning seller, can provide a bid that is better than the winning bid. Strictly speaking, an efficient alternative takes into consideration the buyer's as well as the bidders' utility functions: "no better bid" means that there is no bid that would yield higher utility for at least one from a pair (buyer, seller) and not worse for any member in the pair. Because we do not consider here bidders' preferences and utilities, we use the term "efficient alternative" to describe a bid which maximizes the buyer's utility over all possible bids.

By controlling the value of the strategic parameters the buyer controls: (1) the ability of the bidders to determine her preferences and the possibility of the winning bid to be efficient; and (2) the effectiveness of bidding process determined by the bidders ability to make consecutive bids which are better for the buyer than the earlier bids. A relationship is such that the less likely is for the bidders to discover the buyer's preferences yet be able to bid effectively, the more likely it is that the winning bid is inefficient.

The motivation for the proposed procedure derives from behavioral experiments in which we compared multiattribute auctions and multibilateral negotiations (Yu et al. 2008) and the requirement to adapt the mechanisms used in these experiment to more realistic settings. The auction mechanism based on the price/costs function is inadequate for such problems as procurement of logistic services (Pontrandolfo et al. 2010) and energy trading (Block et al. 2010).

2 Multiattribute Reverse Auctions

The two key tasks in multiattribute auctions are: (1) representation of the buyer's preferences that allow for the comparison of bids; and (2) specification of the rules for feedback information which the sellers need to receive in order to construct bids. These two issues are discussed in the following sections.

2.1 Preference Representation

There two main types of preference representation methods (Fishburn 1976; Dieckmann et al. 2009):

1. Compensatory methods, which include additive value functions or more complex utility functions based on multiattribute utility theory (MAUT); and
2. Non-compensatory methods, which include attribute lexicographic ordering and the Tchebychev measure.

Compensatory methods are based on the assumption that decision makers' preferences are defined on both attributes and attribute values, and that they can formulate trade-offs between attributes and between attribute values. This assumption allows for the aggregation of preferences into some kind of function which measures the worth of an alternative. The measure is a utility.[1] Multi-attribute auction mechanisms have been designed using scoring functions or utility (e.g., Bichler 2001; Beil and Wein 2003; Engel and Wellman 2010). Additive linear functions (i.e., weighted sum) have been implemented in e-sourcing systems offered by B2emarkets.com (now bravosolutions.com) and Perfect Commerce (perfect.com), (both current December 2010).

A specific class of compensatory methods is the use of costing: all attributes and their values are transformed into monetary values. These methods can be used only when the attributes can be priced; examples include A.T. Kearney Procurement & Analytic Solutions (ebreviate.com) and CapGemini IBX (ibxeurope.com), (both current December 2010).

Non-compensatory models have been proposed to evaluate bids at the attribute level but not between the attributes. The two well-known methods are: lexicographic ordering and the Tchebychev function. Lexicographic approaches are simple heuristics in which the attributes are ordered from the most important to the least important. The alternatives are compared first using the most important attribute. If they do not differ on this attribute, then the second most important is used, and so on. These heuristics were found to perform well and sometimes better than a compensatory method (i.e., conjoint analysis) in *ranking* of alternatives but not in *rating* them (Dieckmann et al. 2009). In their study as well as earlier studies (e.g., Dhar 1996) the participants were given little time and need to choose from among several alternatives. When participants were given more time and could explore information, as well as when the problem was more complex (i.e., measured by the number of attributes and alternatives) compensatory models outperformed lexicographic strategies (Yee et al. 2007). If the number of alternatives is large, then lexicographic models may fail because they do not allow for large difference in values of several attributes of lower importance to outweigh a small difference in the value of a more important attribute.

Bellosta et al. (2004) proposed the use of Tchebychev distance to represent the buyer's preferences. The non-compensatory character of this distance allowed the

[1] Term utility covers in this text as well utility functions or value functions as simple ratings.

authors to suggest feedback based on attribute values. The sellers need not consider tradeoffs, instead their bids have to contain a value greater than the previous best bid on at least one attribute and not worse on any attribute.

Bellosta et al. (2008) proposed MERA, a procedure for mechanism design for both synchronous and asynchronous multiattribute reverse auctions in which, in addition to Tchebychev distance, lexicographic ordering and weighted sum function can be incorporated. The framework relies on the notion of reservation levels for which constructing the preference aggregation method is used. In this paper we also use reservation levels for auction design. The key difference between our proposal and MERA is the space in which these levels are constructed. While in both procedures the levels originate in the utility space, we transform the reservation from the utility space to the space of alternatives. This has an important and desirable impact on the information feedback discussed in the next section.

2.2 Feedback Rules

A common concern in multi-attribute auctions pertains to the exchange of additional information relevant to the buyer's preferences.

Several rules have been designed to provide feedback to bidders during auctions, including: complete value function, winning bids (with/without value), and all bids (with/without ranking).

In the framework proposed by Bellosta et al. (2008) the information passed by the owner depends on the way she constructs her representation. When the representation includes a linear additive utility function, then the owner passes this utility and its lower bound. When the preferences are represented as a lexicographic aggregation model or a Tchebychev function, then the owner passes bounds imposed on the attribute values. This dependency is difficult to reconcile with the requirement that the owner does not make her preference model public (Burmeiste et al. 2002; Parkes and Kalagnanam 2005). The owner's inability to keep her preference private may force her to use a different model, which she does not know, agree with, or is inappropriate to the particularities of the problem.

Teich et al. (1999) suggest a feedback rule in which the buyer prescribes a preference path, an ordered set of combinations of prices and non-priced attributes. The preference path begins with an anchor point and the rule specifies that a point further from the anchor is preferred by the owner over the point that is closer to it. This allows the sellers to decrease the worth of their bids (as seen by the buyer) by proposing a combination that is more preferred by the buyer than that combination previously proposed. Burmeister et al. (2002) note one drawback of this method which is bidders' restriction in their choices, i.e., they are only allowed to bid on the preference path. Another limitation is the possibility for sellers to use the preference path to construct the buyer's utility function.

The feedback rule described by Bellosta et al. (2008) depends on the preference representation method. For a compensatory method the feedback is the buyer's preference aggregation function. The feedback also includes the minimum scoring value which is the value of the best bid made in the earlier round plus an arbitrary increment. This feedback allows the sellers to determine if they are willing to make bids with scoring value exceeding the minimum. Informing the sellers about the buyer's preferences is the primary limitation this approach. As we have mentioned above, in many instances buyers' are not willing to provide this information. Therefore, we propose a procedure in which the buyer's preferences be not revealed yet the sellers obtain information which allows them to make progressive bids.

3 Problem Representation

3.1 Preliminaries

The proposed procedure for multiattribute auction has two types of components: (1) the auction owner component, and (2) the bidder component. There is one or more bidders and they may behave according to the same or a different set of rules. We consider here reverse auctions, therefore the owner is the buyer and bidders are sellers. The procedure may, however, be modified to standard auctions in which the owner is the seller and bidders are buyers.

Reverse auction *A* is a set of collections which describe the owner's (buyer) and all *J* bidders' (sellers') problem representations:

$$A = \{P_t, I_t, B_{jt}, O_{jt}\}, (\mathrm{t} = 1, \ldots, T; j = 1, \ldots, J) \tag{1}$$

where, for round t ($t = 1, \ldots, T$):

1. P_t is the auction owner's (buyer's) problem representation;
2. I_t is the information (feedback) which the auction owner presents to the bidders;
3. B_{jt}, ($j = 1, \ldots, J$) is the representation of individual bidder's j decision problem; and
4. O_{jt}, ($j = 1, \ldots, J$) is a bid made by bidder j constructed on the basis of solutions of B_{jt}.

Let us observe that, for example, if in a round t (in the following index t is omitted unless it is important to indicate rounds) for $J = 3$ the tuple $A = \{P, I, B_1, O_1, B_2, O_2, B_3, O_3\}$ represents ability (I) of owner to communicate the problem to three bidders and their reactions to the information to this message. Problem representations B_1, B_2 and B_3 of every of the three bidders are unknown to the owner and other bidders, and they are not considered in this paper. The bidders' reactions O_1, O_2 and O_3 are known to the bidder. In the sequel we assume an elicited additive utility for the owner. The bidders' representation remains uncovered in the analysis.

It is important to allow for the separation of the owner's and the bidders' representations. In particular, two types of separation are required:

1. Owner framing: while I is derived from P, the same information I may be obtained for different representations P, and vice versa; and
2. Bidder separation: irrespectively of the representation B_j bidder j is using, she formulates her bid O_j ($j = 1, \ldots, J$), in a manner required by the auction protocol.

The two separations are conceptually similar in that they both state that the information passed by one person (entity) to another does not depend on the way this person formulates and solves her decision problem. We are concerned here with the perspective of the buyer and the information that the buyer presents to, and obtains from, the sellers. The sellers are completely independent—they may construct any type of their own problem representation and they may make any offer they wish, providing they do it according to the auction protocol.

3.2 Representation of the Buyer's Decision Problem

The buyer's problem representation comprises a set of feasible and acceptable alternatives X and the buyer's utility function u, that is, $P = \{X, u\}$. We consider alternatives which are described using N attributes[2] ($N \geq 1$).

Let us introduce the following notation:

x_j^n—j-th value of attribute n, ($j = 1, \ldots, J_n$; $n = 1, \ldots, N$);

$$X_n = \{x_j^n, j = 1, \ldots, J_n\}\text{—set of feasible values of attribute } j;\ (j = 1, \ldots, J_n;\ n = 1, \ldots, N) \quad (2)$$

$\boldsymbol{x}_l = \left[x_{jl}^n\right]$—$l$-th alternative, $l = 1, \ldots, L$;

X—set of feasible alternatives, $X = \{\boldsymbol{x}_l = [x_{jl}^n], l = 1, \ldots, L\}$.

From (2) it follows that set X is assumed to be discrete; it has $L = |X| = \prod_{n=1}^{N} J_n$ alternatives and is bounded.[3]

The attributes may be discrete or continuous variables. If they are discrete (e.g., categorical or nominal), then their values are known to auction participants. If they represent a continuous variable (e.g., distance or weight), then we assume that only a

---

[2] Attributes serve not only to describe alternatives but they also play an important role in their evaluation.

[3] The discretization of X is not necessary but it simplifies the process. It also has little practical implications because in most, if not all, auctions bidders cannot bid below the smallest allowable units, be it dollars or cents, meters or grams.

discrete subset is considered. The *permissible attribute values* are those which differ by no more than ε^n, which is the smallest meaningful increment of attribute x^n (e.g., a centimeter, gram or dollar).

In order to describe an alternative's evaluation, we assume, without loss of generality, that the owner wants all attributes to be achieved at the highest possible levels.[4]

Consequently, we assume that there is the owner's numerical evaluation $u : X \rightarrow R$ of alternatives from the set X. We also assume that every bid, which can be accepted, is an element of set X. This means that bids O_j introduced in (1) are feasible alternatives $\boldsymbol{x}_l$. The utility of the j-th bidder's offer $\boldsymbol{x}_j$ aggregates utility attribute and then it takes into account values of attributes.

For each attribute n, $n = 1, \ldots, N$, let us denote by w^n the weight representing the owner's preference for this attribute. Partial utility $u^n\left(x_j^n\right)$ describes the owner's preference regarding j-th value x_j^n $(j = 1, \ldots, J_n)$ of the n-th attribute. It is a product of the attribute and attribute value weights, i.e.:

$$u^n\left(x_j^n\right) = w^n v\left(x_j^n\right).$$

The utility of an alternative $\boldsymbol{x}$ is the sum of attribute and attribute value weights. We assume that the utility function is additive and monotonic (it may be nonlinear), i.e.

$$u_l = u_l(\boldsymbol{x}) = \sum_{n=1}^{N} w^n v^n\left(x_j^n\right). \tag{3}$$

Useful but not necessary information is about an alternative that yields the highest utility value for the buyer. We call this ideal alternative $\boldsymbol{x}_M \in X$. This means that there is no feasible alternative that yields higher utility. In a similar manner we distinguish the worst alternative $\boldsymbol{x}_1 \in X$ as one that yields the lowest utility value.

3.3 *Reservation Levels*

Auctions literature suggests specification of reservation levels (Milgrom and Weber 1982; Walley and Fortin 2005). These are the bounds used to distinguish the acceptable attribute values from the unacceptable ones. The acceptability of an alternative depends on its feasibility and utility. A given utility value is used as a threshold so

[4] Qualitative attributes (e.g., color or mark) have no natural order (e.g., one cannot say what is the increment from attribute value "blue" to value "green". Such attributes can be ordered with a subjective scale, e.g., according to their utility. This means that the auction participants may have very different orderings of a qualitative attribute. The buyer may provide her preferential order or exclude some values of these attributes throughout the auction. To be consistent with our perspective that buyers do not want to disclose their preferences, we choose the latter option.

that alternatives which yield lower utilities are deemed unacceptable and those that meet or exceed the threshold are *acceptable.*

In single-attribute auctions, bidders are informed about the reservation level (reserve price). Bids that fell below that value are rejected. Because the first bid has to be higher than the reservation level and a subsequent bid has to be higher than the bid it precedes, the initial reservation level need not be revised. Bids play a role of a reservation levels; there are as many changes in the de facto reservation levels as there are bids.

In multiattribute auctions we need a different approach.

In the proposed procedure, in one round, the sellers submit bids not knowing about their order. That is, they do not know which bid is the winning bid and whether the buyer prefers their bid over other bids. Before they move to the next round they may be informed about the winning bid. Our procedure allows for this option. However, the announcement is not sufficient for bidders to be able to propose bids that yield higher owner's utility value than the utility of the winning bid.

At the beginning of each round, the bidders need to obtain information about the acceptable bids. This information is contained in the announced (by auction owner), revised reservation levels. The reservation levels thus play a similar role in the proposed procedure as in a single-attribute auction, but they are revised after each round. The purpose of the revisions is to guide the sellers into the subset of X which is acceptable to the owner.

For example, in a single-attribute auction the best bid in round t is \$512. This informs the bidders that in round $t + 1$ any bid below \$512 is unacceptable and any bid above \$512 is acceptable. If there are two attributes—price and delivery time—then information that the best bid in round t was \$406 and 35 days does not provide the bidders with sufficient information. If, however, the owner announces that either the price has to be higher than \$410 and delivery time 33 days, or that the price has to be higher than \$390 and delivery 29 days, then the bidders can make acceptable bids.

Assuming that the auction may be represented as a series of bidding rounds t, ($t = 1, 2, \ldots T$), reservation levels are determined at the beginning of rounds. The process of reservation-level revision resembles a single-attribute auction with the difference that here the revised values need to be computed. In a single-attribute auction the most recent bid value becomes de facto a new reservation level and the next bid cannot be below this value. In the multiattribute case, the levels need to be re-evaluated so that they reflect the buyer's preferences over multiple attributes.

Another difference between single- and multi-attribute auctions is that, in the former case, a single reservation level is sufficient to restrict biddings to all alternatives that the buyer prefers over a particular reservation levels. This is not the case in multiattribute auctions. The preference is determined by the minimum acceptable utility value which then needs to be transformed into reservation levels set for all attributes.

We say, that, set X_r is acceptable with the reservation level r, $r \in \mathbf{R}$, for the owner when all its elements meet the *reservation level condition*, i.e.:

$$X_r = \{\boldsymbol{x} \in X : u(\boldsymbol{x}) \geq r\} = u^{-1}([r, +\infty[). \tag{4}$$

where r represents the minimum acceptable utility value.

The reservation level r means that every bid which yields lower utility than r is rejected; it is considered to be an infeasible bid. To stress this characteristic, we call X_r *r-feasible set* and its elements *r-feasible alternatives.*

We assume that the buyer does not want to inform the bidders about the minimum acceptable utility value because this would be tantamount to informing them about her utility function. Therefore, she has to transform the above condition so that it is defined on the attribute rather than on the utility values. That is, the buyer has to transform utility reservation level r to the attribute reservation levels which she can pass to the bidders. We assume (see Sect. 3.2) that the owner wants all attributes to be achieved at the highest possible levels. Therefore, the attribute reservation levels are lower bounds.

Note that some attributes may be nominal. For these attributes lower bounds are not meaningful. These attributes' values are ordered by the buyer's preferences which are unknown to the bidders. Therefore, we assume that the bound, which is defined for nominal attributes, divides their values into acceptable and unacceptable ones. For example, consider color as an attribute with four feasible values (black, grey, red and white). A bound dividing the feasible set may set black and grey as unacceptable and red and white as acceptable. The bidders are told that only bids in which the color is either red or white can be accepted.

To illustrate the construction of the lower bounds based on formula (4) consider an example shown in Fig. 1a. Set X of feasible alternatives consists of all points shown in Fig. 1. Set X_r^A is the set of all points on and above line r.

Let's assume that the reservation level r is determined by alternative $\boldsymbol{x}_r$, that is, $u(\boldsymbol{x}_r) = r$. Alterative $\boldsymbol{x}_r$ is said to be the *reference alternative* because it is used to construct *r-acceptable* set X_r^A, as follows:

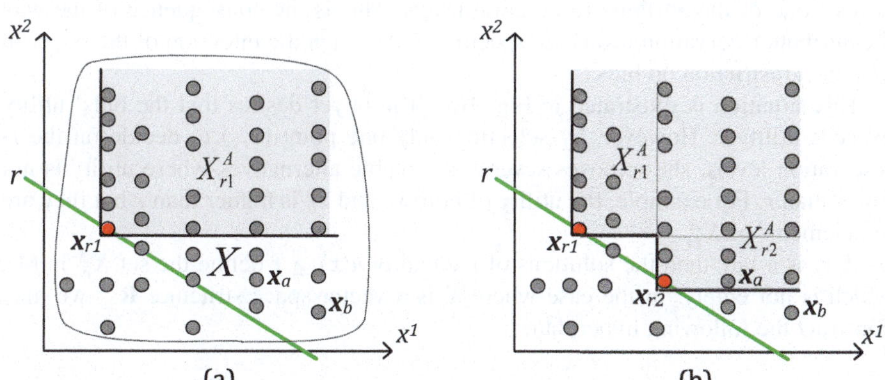

Fig. 1 **a** Single-point reservation levels; **b** Two-point reservation levels

$$X_r^A = \{x \in X : [x^n] \sqsupseteq [x_r^n]\}. \tag{5}$$

where $\sqsupseteq$denotes: (i) the relation $\geq$ if x^n is a numerical variable, or (ii) it is another preferential order that divides set X_n of attribute values into acceptable and unacceptable subsets.

This means that X_r^A comprises alternatives with utility values not lower than r. Note that information about X_r^A can be easily conveyed to the bidders. Information about X_r^A is included in I defined in (1). For numerical attributes it is sufficient to include in I the requirement that the attribute value cannot be lower than x_r^n. If the attributes are nominal, then I must include either all acceptable attribute values or all unacceptable values.

Information included in I, which describes X_r^A, is obtained by transforming the reservation level condition given in (4) from the utility space to the alternative space. To indicate that it is obtained by the process of selection of utility value r and reference alternative $\boldsymbol{x}_r$, we call this information *r-reservation* levels.

Taking into account monotonicity of the owner's utility, we obtain:

$$X_r^A \subset X_r. \tag{6}$$

In what follows, we assume, for simplicity, that alternative values are numerical. In this case X_r^A comprises points inside the rectangle ($\boldsymbol{x}_r \oplus \mathbf{R}_+{}^{\mathrm{N}}$).

3.4 Design Parameters

The construction of set X_r^A defined in (5) may result in a loss of r-feasible alternatives (i.e., in the case where $X_r - X_r^A$ is non-empty). This means that some r-feasible alternatives are not r-acceptable; their utility value is greater than r but they violate at least one of the attribute reservation levels. This is the consequence of the way the attribute reservation levels are specified rather than the intention of the buyer in placing a restriction on bids.

This situation is illustrated in Fig. 1(a). The buyer desires that the bids' utility exceeds utility r. However, by selecting only one point ($\boldsymbol{x}_{r1}$) to decide on the r-reservation levels, she removes several acceptable alternatives where utility is not lower than r. For example, the utility of both $\boldsymbol{x}_a$ and $\boldsymbol{x}_b$ is higher than r but they are not elements of X_{r1}^A.

If $\boldsymbol{x}_r$ is a bid, then the solutions of inequality $u(\boldsymbol{x}) \geq r$ define the set X_r^A in (4), which is not empty. In the case where X is a vector-space structure $\mathbf{R}^{\mathrm{N}}$, we may construct the following hyperplane:

$$H_{\mathrm{r}} = \left\{\boldsymbol{x} \in \mathbf{R}^{\mathrm{N}} : r = u(\boldsymbol{x}) = \sum_{n=1}^{N} w^n v^n(x^n)\right\}.$$

Observe that $\overline{X}^r = u^{-1}(r) = X \cap H_r$. Not all elements of $\overline{X}^r$ have to be included in the set which the owner presents to the bidders in order to receive bids with utility value not lower than r (see e.g. Fig. 1(a)) and without revealing her utility. We show that if there are r-feasible alternatives which are not r-acceptable, then they can be used to expand the set of r-acceptable bids.

Proposition. The owner can operationally split the set of r-acceptable alternatives into alternatives with attribute values exceeding attribute thresholds defined by a bid with utility r, and the remaining alternatives.

In the case of a finite set the split can be done by enumeration of bids and direct comparison of their attribute using relation $\sqsupseteq$(5). If this relation is represented by the relation $\geq$ then the condition for a split can be phrased as membership in the set of solutions of an inequality of real numbers constructed using scalar products of vectors spanned by some boundary bids.

3.4.1 Construction of r-Acceptable Sets

The fact that some r-feasible alternatives are not included in r-acceptable set X_r^A may lead to an auction which terminates with an inferior, Pareto-dominated, bid. This is the case when the bidders want to bid only on the excluded alternatives but are unable to do it. Consider again the situation illustrated in Fig. 1(a). If no bidder submits a bid that meets r-reservation levels (i.e., is an element of X_r^A), then an earlier bid with utility lower than r will be accepted. It is possible, however, that bidders would propose $\boldsymbol{x}_a, \boldsymbol{x}_b$ or some other alternative which is r-feasible but not r-acceptable.

We can increase the number of r-acceptable sets by increasing the number of reference alternatives (see (5)).

The case of two reference alternatives is illustrated in Fig. 1(b). The introduction of the second reference alternative expands the r-acceptable set, that is:

$$X_{r2}^A = X_{r1}^A \cup X_{r2}^{At} = \left\{\boldsymbol{x} \in X : \geqq \boldsymbol{x}_{r1}\right\} \cup \left\{\boldsymbol{x} \in X : \boldsymbol{x} \geqq \boldsymbol{x}_{r2}\right\}. \tag{7}$$

The addition of set X_{r2}^{At} allowed for the inclusion of alternative $\boldsymbol{x}_a$ in the acceptable set (see Fig. 1(b)). However, despite this addition alternative $\boldsymbol{x}_b$ remains unacceptable.

Let $X^{\geqq}(\boldsymbol{x}_0)$ be a set comprising of feasible alternatives such that each attribute value of its elements is no smaller than the corresponding attribute value of $\boldsymbol{x}_0$. That is:

$$X^{\geqq}(\boldsymbol{x}_0) = \left\{\boldsymbol{x} \in X : \boldsymbol{x} \geqq \boldsymbol{x}_0\right\}.$$

In (7) two reference alternatives are used to expand the r-acceptable set. In general, D such alternatives may be used. Hence we obtain:

$$X_{rD}^A = \bigcup_{d=1}^{D} X_{rd}^A = \bigcup_{d=1}^{D} \{\boldsymbol{x} \in X : \boldsymbol{x} \geq \boldsymbol{x}_{rd}\}. \tag{8}$$

where: alternative $\boldsymbol{x}_{rd}$, is d-th reference alternative ($d = 1, \ldots, D$), such that $U(\boldsymbol{x}_{rd}) \geq r$, and there is at least one reference alternative $\boldsymbol{x}_{rd*}$for which $U(\boldsymbol{x}_{rd*}) = r$. Reference alternative $\boldsymbol{x}_{rd*}$ is the alternative which was used to determine the acceptable value of utility r (see (4)).

Figure 1(b) illustrates the situation in which both reference alternatives yield the same utility value r. This is not necessary and there may be situations when there is only one such alternative, yet two or more reference alternatives need to be selected. In such case alternatives which utility value is greater but different from r as little as possible, can be selected to generate r-acceptable sets. We propose to select alternatives with utility as close to r as possible so that no alternative that is significantly better for the buyer can be removed. Only these alternatives which are marginally better can be removed.

Any choice of the number of reference alternatives influences information I_t which the bidders receive at the beginning of round t (see (1)). An increase of reference alternatives expands the set of r-acceptable alternatives for the next round by alternatives dominated from owner's point of view and it may encourage bidders to submit new offers which otherwise would be excluded.

In each round t information I_t about r-acceptable sets is presented to the bidders. The notation X_{rD}^{At} is used to describe the set of r-acceptable alternatives which is a subset of X formulated for round t and defined by D reference alternatives ($t = 1, 2, \ldots, T$). To simplify the notation, the r value, number of reference alternatives and/or the round number, are dropped when unnecessary.

Note that the assumed monotonicity of utility function allows us to use single reference alternative $\boldsymbol{x}_r$ for the construction of the r-acceptable set X_{r1}^{At} for t, ($t = 1, 2, \ldots, T$). If the utility is non-monotonic and reaches one or more optimum within the set rather than on its boundary, then the acceptable set needs to be defined by more points.

Parameter D, which defines the number of reference alternatives used to construct the r-acceptable set, is one of the *mechanism design parameters*. The buyer needs to determine its value and this requires taking into account the following two types of tradeoffs:

Tradeoff 1. The relationship between the number of alternatives which are r-feasible but not included in r-acceptable set X_D^{At} and the bidders' difficulty in selecting bids from this set. The greater the value of D, the fewer acceptable alternatives are not included but the number of sets in which the bidders need to consider increases making bidding more difficult.

Tradeoff 2. The relationship between the number of alternatives D used to specify X_D^{At} and the bidders' ability to discover the buyer's utility function. The greater the value of D, the fewer acceptable alternatives are not included but it is easier to determine the analytical form of the utility.

The second type of tradeoff should be addressed because D may be greater than the minimum number of alternatives required to determine the owner's utility function N^u. In the example shown in Fig. 1(b) the selection of two reference alternatives would allow the bidders to determine the buyer's utility function. Reduction of the number of reference alternatives from two to one leads to the situation described in Fig. 1(a), in which many acceptable alternatives are excluded. However, even in the case of two reference alternatives there may be many *r*-feasible alternatives excluded.

3.4.2 Perturbation of *r*-Acceptable Set

Design parameter D is used to control the construction of such an *r*-acceptable set, that excluded *r*-feasible alternatives are not sufficiently relevant for the buyer. The buyer knows that some alternatives are excluded but they do not differ much from some of the alternatives which are then included.

An increase of the value of D expands the set but, as Tradeoff 2 above states, it does so at the cost of increasing the possibility of preference disclosure. In general, it is not possible to avoid indirect disclosure of the owner's preferences and keep all *r*-feasible alternatives in the *r*-acceptable set (i.e., these that yield equal or higher utility than the winning bid). If, however, the owner accepts that some *r*-feasible alternatives are excluded, then a disclosure can be avoided by making a small change in some of the reference alternatives. This situation is illustrated in Fig. 2.

To address the issue of the discovery of the buyer's utility function, a deviation from the reference alternatives may be introduced. An example of this intervention is presented as the transformation of set X_2^A shown in Fig. 2(a) to set X_{2E}^A shown in Fig. 2(b). The intervention is the replacement of the reference alternatives. In our example, alternative x_{r2} is replaced with $\boldsymbol{x}_{r2E}$. The result of this perturbation may be a loss of some of the *r*-feasible alternatives but the bidders' ability of discovering the buyer's utility is diminished. In effect we have the third type of tradeoff:

> Tradeoff 3. The relationship between the deviation size which is associated with the number of removed acceptable alternatives and the difficulty in discovering the buyer's utility function. The greater the deviation, the greater the difficulty but

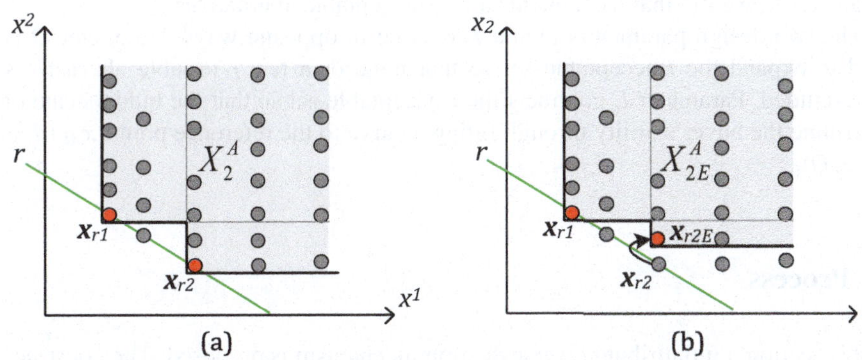

Fig. 2 **a** Acceptable set with two-point reservation levels; and **b** the same set with a single-point perturbation

this comes at the expense of an increase in the number of acceptable alternatives removed which may lead to the loss of efficient bids.

We use parameter E to control the deviation. This parameter has two parts, i.e., $E = \{\boldsymbol{e}, n^E\}$, where $\boldsymbol{e}$ is a vector of attribute increments ($\boldsymbol{e} = [e_1, \ldots, e_N]$); and n^E is the number of attributes values which are changed. (If the value of only one attribute at the time is changed, then $n^E = 1$).

E represents the second mechanism design parameter. It reflects the buyer's willingness to accept a winning bid that is not Pareto-optimal in X. Its values need to be decided externally to the procedure. In many cases the values of the elements of $\boldsymbol{e}$ may be the minimal attribute increments which are meaningful for the participants.

Design parameter E may be applied in a number of ways. Perturbation may involve every reference alternative or only some of them. We propose that first consider the number N^u of alternatives that is sufficient to uniquely determine the utility function. If $D < N^u$, then perturbation is not necessary, albeit it may be performed. If $D \geq N^u$, then there are more alternatives yield utility equal r than it is necessary to determine the utility's analytical form. In general, it is possible that D alternatives do not uniquely determine utility function. To simplify the discussion, we do not verifying if this is the case, but assume that at most N^u-1 alternatives can yield the same utility. Hence, the minimum number of alternatives that need be perturbed is $D - N^u + 1$. The perturbation involves an increase in the value of one or more attributes in each of the selected reference alternatives. The chosen attributes should be different and their selection should be such that the utility of thus constructed reference alternatives differs as little as possible from the winning bid's utility.

The application of parameter E allows replacing set of r-acceptable alternatives X_D^{At} with the subset X_{DE}^{At}, i.e.,

$$X_{DE}^{At} \subset X_D^{At}. \tag{9}$$

To indicate that X_{DE}^{At} is a subset of r-acceptable set obtained by the application of parameter E, we say that its elements are r_E-acceptable alternatives.

The two design parameters D and E operate in opposite ways. Parameter D is used to expand the r-acceptable set so that none or a few r-feasible alternatives are excluded. Parameter E contracts the r-acceptable set so that the bidders cannot determine the buyer's utility through fitting a curve to the reference points $\boldsymbol{x}_{rd}$ ($d = 1, \ldots, D$).

4 Process

In this section multiattribute reverse auction mechanism is proposed. The construction of r-feasible and r-acceptable sets is at its core.

4.1 Preliminaries

During the auction, the construction of these sets relies on reference alternative $\boldsymbol{x}_r$, which utility is r, (i.e., $u(\boldsymbol{x}_r) = r$). Before the auction begins we need a way to construct the first r-acceptable X_{DE}^{A0} set and present it to the bidders. There may be different ways to construct this set. For example, the owner may use the initial reservation levels as a reference for bidders or propose the feasible set X.

Figure 3(a) illustrates the auction at the beginning of round t. It is the continuation of the situation shown in Fig. 2 in which r-acceptable X_2^{At} set was perturbed to obtain set X_{2E}^{At}. Set X_{2E}^{At} is defined by two alternatives $\boldsymbol{x}_{r1}$ and $\boldsymbol{x}_{r2}$. In round t three bids are made. These are shown in Fig. 3(b).

Among the three bids, bid $\boldsymbol{x}_{r1}^2$, yields the highest utility value u_2. Therefore u_2 and $\boldsymbol{x}_{r1}^2$ are used to determine the r-acceptable set for round $t+1$. Because $D = 2$, an additional alternative $\boldsymbol{x}_{r2}^2$ is selected. The set X_2^{At+1} defined by these two alternatives is shown in Fig. 3(b).

The two design parameters D and E have a critical role in the mechanism and its convergence. Their values may be constant or changed during the auction according to a predefined formula. In each case, the values which are used in round t need to be verified for their feasibility.

The reason for the verification of D is that there may be no alternative which can be used as a reference alternative given by (5). That is, after $d = 1,\ldots, d_1$ alternatives were selected, there may be no alternatives $\boldsymbol{x}_{rd}$, ($d = d_1 + 1, \ldots, D$), such that the following two conditions are met:

$$\text{(i) } u(\boldsymbol{x}_{rd}) \geq r\text{; and} \qquad \text{(ii) } \boldsymbol{x}_{rd} \notin \bigcup_{d=1}^{d_1} X_{rd}^{A}. \tag{10}$$

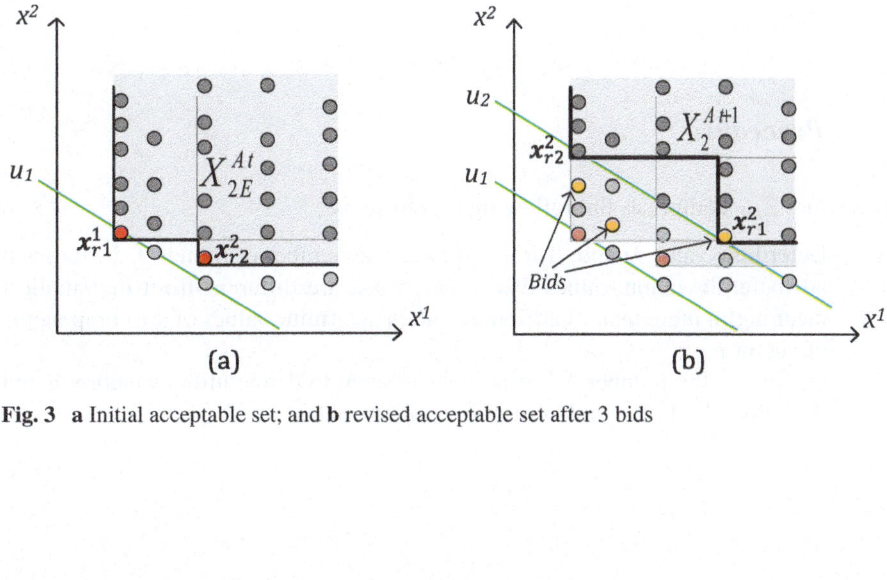

Fig. 3 **a** Initial acceptable set; and **b** revised acceptable set after 3 bids

Condition (i) above is a part of the construction of the r-feasible set. Condition (ii) needs to be met because set X^A_{ra} should be introduced only if it contains at least one alternative which is not a member of set X^A_{rb}, $(1 \leq b < a \leq D)$. Therefore, if a reference alternative $\boldsymbol{x}_{ra}$ is selected, which is already an element of r-feasible set X^A_{rb}, $(1 \leq b < a \leq D)$, then $X^A_{ra} \subset X^A_{rb}$. This means that set X^A_{ra} is redundant and should not be constructed.

The value of the deviation parameter $\boldsymbol{e} = [e_1, \ldots, e_N]$ may also need to be verified when its elements are greater than the smallest allowable increments. If the value of one or more components is greater, then its application may result in the modification of set X^A_{rd} to set X^A_{rdE} which removes all r-acceptable alternatives so that X^A_{rdE} is empty.

Note that the value of parameter e_n may depend on the characteristics of an attribute. For example, the value will be different for attribute describing price, warranty and delivery time. In such situations we denote e^n as the deviation control parameter of attribute n, $(n = 1, \ldots, N)$. If an attributes is nominal, then it cannot have a deviation parameter, that is, there is no e_j for nominal parameter j, $(1 \leq j \leq N)$. In such situations perturbation means that the attribute value which yields the smallest increase of utility is selected.

The last component of the perturbation parameter E is n^E–the number of attributes values, which values are perturbed. There may be a situation when this parameter has to be modified during the auction. If, for example, the utility function is non-monotonic, then there may be different number of points (N^u) needed to determine it in one subset of X and different another subset. This could lead to the changes of the attributes that need to be perturbed. These situations are not likely to occur in practice due to the requirements imposed on the buyer in the preference elicitation process. Therefore, we assume that n^E is constant throughout the auction.

Given the above caveats, we can simplify the description of the procedure and assume that the values of the two parameters are held constant and that they need not be changed during the auction. To stress the procedure's reliance on the two parameters and the assumption that their values are held constant, we denote the procedure $\mathbb{P}_{DE}$.

4.2 Procedure

Procedure $\mathbb{P}_{DE}$ comprises the following eleven steps:

1. Determine value D and, if it is required, the number of rounds T. If, for each attribute, deviation values $\boldsymbol{e} = [e_1, \ldots, e_N]$ are different from the smallest meaningful increment of attribute ε^n, then determine values of the components of vector $\boldsymbol{e}$.
2. Determine the number N^u of points sufficient to define utility function u and the value n^E of the number of perturbed attributes.

3. Set $t = 1$. Construct and present initial acceptable set X_{DE}^{A1} to the bidders. Request bids.
4. Terminate if: either there is at most one acceptable bid received or $t = T$. Otherwise select the best bid $\boldsymbol{x}^{t*}$ made in round t.
5. Set$t = t + 1$. Let $u\left(\boldsymbol{x}_r^{t-1*}\right) = r^t$.
6. Determine the number N^{tr} of alternatives yielding utility r^t. Select

   a .If $N^{tr} \geq D$, i.e., there are more alternatives yielding utility r^t than is required by the design parameter D, i.e., then select D alternatives $\boldsymbol{x}_d$, such that, $u(\boldsymbol{x}_d) = r^t$, $d = 1, \ldots, D$. Go to Step 8.
   b .If $N^{tr} < D$, there are fewer alternatives, then select N^{tr} reference alternatives $\boldsymbol{x}_d$, $(d = 1, \ldots)$, such that, $u(\boldsymbol{x}_d) = r^t$.

7. The value of the design parameter D being greater than N^{tr} requires the selection of D-N^{tr} alternatives with utility as close to$r$ as possible, but not smaller than r^t.
8. Use formula (7) to construct set X_D^{At}.
9. Let $\delta = \min\,(D, N^{tr})$; δ is the number of alternatives which yield utility r^t.

   a .If $\delta < N^u$, then perturbation is not necessary.[5] Set $X_{DE}^{At} = X_D^{At}$ and go to Step 11.
   b .If $\delta \geq N^u$, then go to Step 10.

10. Construct set X_{DE}^{At}. The number of alternatives which need to be perturbed is $D^E = \delta$-$N^u + 1$. Select reference alternative $\boldsymbol{x}_d$ such that $u(\boldsymbol{x}_d) = r^t$, $d = 1, \ldots, D^E$. For each selected alternative choose n^E attributes and increase their values by e_n $(n = 1, \ldots, N)$.[6]
11. Present set X_{DE}^{At} to the bidders and request bids. Go to Step 4.

4.3 *Efficiency*

In Sect. 1 we defined efficient alternatives. Correspondingly, we define auction mechanism efficiency in terms of the existence of efficient alternatives. Hence, a mechanism is efficient if its rules do not remove any efficient alternative. For procedure $\mathbb{P}_{DE}$ this means every efficient alternative which is an element in X is also an element in sets X_{DE}^{At}, $t = 1, \ldots, T$.

---

[5] Strictly speaking an additional test may be required. There may be N^u alternatives which utility is the same and higher than r^t. This may take place when D is large and N^u small. In such situations perturbation described in Step 10 must be repeated for every subset containing N^u reference alternatives with the same utility.

[6] There are different ways to implement perturbation. For example, one may begin with changing the value of the attribute which yields the smallest utility increase. Note that we assume that the buyer wants as high attribute values as possible. If she prefer smaller values over greater, then in the perturbation the attribute value needs to be decreased.

One of the roles of parameters, D and E, is to control the degree of procedure $\mathbb{P}_{DE}$ inefficiency, which is the difference between the winning bid and the utility value of an efficient alternative.

From (7) it follows that X $\supset X_D^{At}$, $t = 1, \ldots, T$. Beginning with the initial r-acceptable set X_{DE}^{A0} and using (9) repetitively we obtain decreasing family X_{DE}^{At}, $t = 1, \ldots, T$:

$$X \supset X_{DE}^{A0} \supset \ldots \supset X_{DE}^{AT-1} \supset X_{DE}^{AT}. \tag{11}$$

Proposition: Formula (11) states that procedure $\mathbb{P}_{DE}$ reduces the number of alternatives which the bidders need to consider. Using (3), (4) and (6) we can formulate a sufficient condition for X_{DE}^{AT}, (given by (11)), to include efficient solution, which is:

$$u_0 < u_1 < \ldots < u_{T1-1} < u_T \tag{12}$$

where $u_t = u\left(\boldsymbol{x}_r^{t*}\right)$, $\boldsymbol{x}_r^{t*}$ is the best bid made in round t.

The proof is straightforward – the two-dimensional version is demonstrated in the Fig. 2.

Theorem 1 If parameter D, defined above, satisfies inequality

$$\forall \boldsymbol{x} \in X_{DE}^{At} : u(\boldsymbol{x}) \geq u_t(\boldsymbol{x}), \tag{13}$$

and we have that

$$X_{DE}^{At} = X_{DE}^{At-1} \backslash \{\boldsymbol{x} : u(x) < u\left(\boldsymbol{x}^{t*}\right)\} \neq \emptyset, t = 1, \ldots, T, \tag{14}$$

then procedure $\mathbb{P}_{DE}$ does not remove any efficient alternative.

Proof: The proof results from a contradiction: An efficient alternative is one which utility is not smaller than $u\left(\boldsymbol{x}^{t*}\right)$. If such an alternative is removed so that it is not an element of X_{DE}^{At}, then formula (13) for the construction of X_{DE}^{At} is not satisfied. If (14) is obeyed, then only alternatives which are worse for the buyer than the best bid $\boldsymbol{x}^{t*}$ are excluded.

Formula (12) assures that during the construction of r-acceptable set X_{DE}^{At} no acceptable alternative (i.e., one which meets the utility condition (4)) is removed. Condition (13) assures that only alternatives which utility is lower than the best bid utility may be removed when r-acceptable sets are constructed.

Definition: Bid $\boldsymbol{x}^*$ made in round T is winning if:

1. $X_{DE}^{AT+1} = \emptyset$; and
2. $u(\boldsymbol{x}^*) = u\left(\boldsymbol{x}_{j*}\right) \geq u\left(\boldsymbol{x}_j\right), j = 1, \ldots, J_T.$

Assuming that there is at least one bid in the auction governed by the process described by Theorem 1, then this process is convergent.

5 Discussion

The design parameters D and E contribute to the process complexity and efficiency as well as the possibility of the sellers discovering the buyer's utility. If D is large, the process is complex because the buyer conveys information about many r-acceptable sets and the process efficiency decreases. If E is large, then the efficiency increases and fewer rounds are required but the possibility of removing efficient alternatives from the r-acceptable sets increases.

In general, there is no process in which: (1) no efficient alternative is removed; and (2) the sellers are unable to determine the buyer's utility. In this section we propose two procedural tactics which can be implemented for monotonic utility functions.

5.1 *Distance-Minimizing Strategy*

Utility theory posits that the buyer is not interested in the particularities of an alternative but in the alternative's utility value. The buyer may, however, be interested in the efficiency of the process. The efficiency may be increased if the procedure directs the bidders towards the shortest path from any given point to the ideal alternative $\boldsymbol{x}_M$.

Figure 4 illustrates the case when the best bid $\boldsymbol{x}^{t*}$ was made in round t. According to (12), the utility reservation level in round $t+1$ becomes $u_{t+1} = u\left(\boldsymbol{x}^{t*}\right)$.

The distance-minimizing strategy means that an alternative where utility is u_{t+1} and which is the closest to $\boldsymbol{x}_M$ becomes the reservation point $\boldsymbol{x}_{rl}$ used to construct r-acceptable set X_{r1}^A. The buyer may wish to reduce the number of acceptable alternatives that are not included in this set and add the winning bid thus constructing set X_{r1}^A as illustrated in Fig. 4.

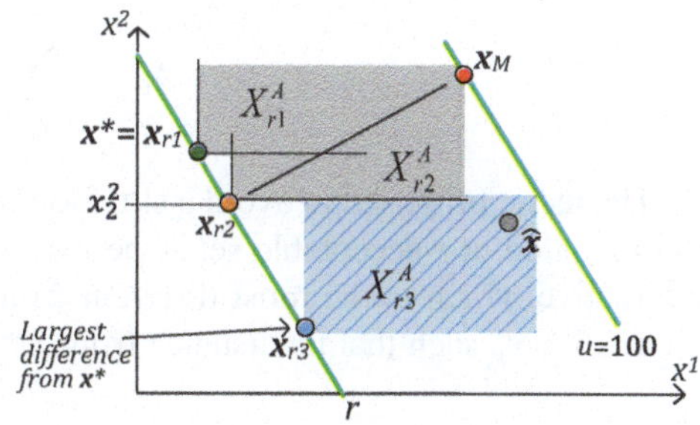

Fig. 4 Best bid $\boldsymbol{x}^*$, ideal alternative, and reservation levels for strongly non-uniform preferences

5.2 Structurally Different Alternatives

There are situations when a buyer's preferences are strongly non-uniform. Some attributes are much more important to the buyer than others. In a two-dimensional space, this situation is illustrated in Fig. 4; the angles of utility isoquants are significantly different so that x^1 is approximately twice as important as x^2.

The best bid ($\boldsymbol{x}$*) may be relatively close to the ideal point so that there are many alternatives not included in the set $X_r^A = X_{r1}^A \cup X_{r2}^A$. This is shown in Fig. 4. Limiting the bids to the union of sets X_{r1}^A and X_{r2}^A may result in removing a large part of the r-feasible alternatives (i.e., all alternatives yielding utility not lower than r). This may lead to a loss of efficient alternatives when, for example, there is a seller who bid $\hat{\boldsymbol{x}}$, and no other seller made a bid yielding more than $u(\hat{\boldsymbol{x}})$.

In order to avoid this situation, we propose to use alternatives where utility is equal or close to r but where they are significantly different from the best bid $\boldsymbol{x}$*. To determine the structural difference between elements we firsts define the difference between two values of the same attribute n ($n = 1, \ldots, N$). Let $\boldsymbol{x}_p$ and $\boldsymbol{x}_s$ be two alternatives in which n-th attribute takes values x_p^n and x_s^n, respectively. The difference between values x_p^n and x_s^n is a natural number i_{ps}^n, which is the number of minimum increments ε^n of attribute n. For example if the attribute's minimum increment is 2 and the two values of this attribute are 4 and 14, then the difference between them is 5 (14 = 4 + 2*5).

Note: For qualitative attributes and also attributes which increment value is not fixed, we need to calculate the number of intermediate attribute values. In the case of qualitative attributes the intermediate values are obtained through the ordering of attribute values according to their utility value (see formula (3)).

Definition: The difference δ between two alternatives $\boldsymbol{x}_p$ and $\boldsymbol{x}_s$ is the sum of absolute differences between the alternative's attribute values measured by the number of the intermediary values:

$$\delta_{p,s} = \sum_{n=1}^{N} |i_{p,s}^n|. \tag{15}$$

The tactic proposed to avoid a situation in which many r-feasible alternatives are not included in r-acceptable sets (see Fig. 4) is based on the selection of a feasible alternative which is the most different from the winning bid $\boldsymbol{x}$*. That is, we seek alternative $\boldsymbol{x}_p$ such that its distance from $\boldsymbol{x}$* is *, where:

$$\delta^* = \max_{\{l:u(x_l)=u(x*)\}} \delta_{*,l}. \tag{16}$$

Note: If there are no alternatives yielding utility r, then we may search for alternatives in the neighborhood providing that they are r-feasible.

In the example illustrated in Fig. 4, the largest difference computed with (17) corresponds to point $\boldsymbol{x}_{r3}$.

5.3 Auction Rounds and Closure

We consider auctions which have T rounds. An auction may end in a *failure* if there was no bid in the first round ($t = 1$). If T is known from the outset, then an auction may end *earlier*, when no more than one bid was made in round t, $t < T$.

The buyer may want to control the process and decide about the minimum utility value increment. The increment $\Delta u_{t+1} = u_{t+1} - u_t$, defined by (12), is sufficiently small so that no acceptable alternative that could be a winning bid is removed. If the buyer increases the minimum increment, then the process' efficiency may increase and fewer rounds are required. The downside of such increment change is that the winning bid may be inefficient. This means that minimum utility increment, denoted as Δ_u, constitutes a process-defining parameter.

Figure 5 illustrates the move from round t to round $t + 1$. In Fig. 5(a) the new minimum increment (u_{t+1}–u_t) does not remove any efficient alternative. Bid $\boldsymbol{x}_{r2}$ made in round t, is an element of the r-acceptable set and only alternatives which this bid dominates are removed. This process follows formulae (13) and (14).

Consider the situation when the minimum utility increment is greater than Δu_{t+1} defined by (12). The reason for setting a higher increment may be the buyer's need to get the best bid by the self- or externally-imposed deadline or within a given number of rounds. In the case of a fixed length auction the minimum increment may be defined by:

$$\Delta_u = \left(u\left(x^M\right) - u_0\right)/\mathrm{T} \tag{17}$$

where: u_0 – the minimum acceptable utility value (used to construct the initial set X_{DE}^{A0} described in Step 3 of procedure $\mathbb{P}_{DE}$).

Actual bids are not taken into account in (18), making it not useful when, for example, the best bid utility in one round significantly exceeds the increment required in the next round.

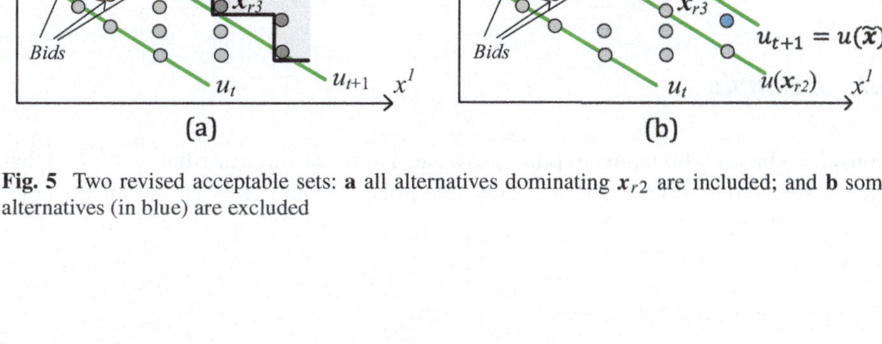

Fig. 5 Two revised acceptable sets: **a** all alternatives dominating $\boldsymbol{x}_{r2}$ are included; and **b** some alternatives (in blue) are excluded

Increment may also be a function which initially it takes a large value which with each subsequent round decreases (e.g. exponential Brigui-Chtioui and Pinson 2010). Because the bidders are likely to make bids that exceed the utility reservation value, an adaptive rule which is a generalization of (18) can be used:

$$\Delta_{ut} = \max\{\Delta_u; (u(\boldsymbol{x}^M) - u(\boldsymbol{x}^{t*}))/(T - t + 1), \quad (18)$$

where: $\boldsymbol{x}^{t*}$ – the best bid obtained in round t,$t = 1, \ldots, T - 1$.

Taking into account utility of bids' allows using the minimum utility increment only when it is necessary. Formula (18) does not require changing procedure $\mathbb{P}_{DE}$, (see Sect. 4.2) as long as the biding process progresses according to the buyer's expectations, that is the best bid utility exceeds minimum utility increment, that is:

$$u(\boldsymbol{x}^{t*}) \geq u_t + \Delta_{ut}, t = 1, \ldots, T - 1. \quad (19)$$

If, however, the best bid utility is lower, that is inequality (19) does not held, then the minimum increment Δ_{ut} is introduced. Because the utility of bid $\boldsymbol{x}^{t*}$ is lower than required $u_t + \Delta_{ut}$ another reference point has to be introduced. Let $\tilde{\boldsymbol{x}}^t \in X_{DE}^{At}$ be the reference point replacing $\boldsymbol{x}^{t*}$. Following this replacement, Step 5 in $\mathbb{P}_{DE}$, is modified as follows:

Step 5 (revised). Select the best bid $\boldsymbol{x}^{t*}$ made in this round and apply (18). If $\Delta_u < \Delta_{ut}$, then go to Step 6, otherwise replace $\boldsymbol{x}^{t*}$ with $\tilde{\boldsymbol{x}}^t$ and set: $r = \tilde{\boldsymbol{x}}^t$.

The situation in which (19) does not held is shown in Fig. 5(b). In round t, the best bid is $\boldsymbol{x}_{r2}$ but $u(\boldsymbol{x}_{r2}) < u_t + \Delta_{ut} = u_{t+1}$. Alternative $\tilde{\boldsymbol{x}}$ with utility equal to u_{t+1}. Additional reference points are selected and set X_4^{At+1} is constructed. In the situation when there are no more bids, $\boldsymbol{x}_{r2}$ becomes the winning bid. This bid, however, may be inefficient because we cannot exclude the possibility the bidders could make a bid that yields utility higher for the owner than $\boldsymbol{x}_{r2}$ and lower than $\tilde{\boldsymbol{x}}$. The bidders could choose one of the alternatives shown in blue that dominate $\boldsymbol{x}_{r2}$, but they were unable to do so by the set of constraints they had to obey.

6 Illustration

We illustrate the proposed reverse auction procedure with the example used by (Bellosta 2008, p. 403).

6.1 Example

Consider a buyer who wants to purchase a car. There are three attributes ($N = 3$) that she is interested in: trademark, warranty, and price. Trademark is a nominal attribute.

Table 1 Value function compositions

Attribute	Trademark						Warranty (month)		Price ($)	
Weight of attribute (v_j)	$v_1 = 0.3$						$v_2 = 0.2$		$v_3 = 0.5$	
Attribute values (x_l^j)	Star	Citron	Pejo	Betha	Reno	Lux	0 0	60	10,000	50,000
Weight of attribute value (w_l^j)	0	20	40	60	80	100		100	100	0

Table 2 Two alternative examples

	Alternative 1 ($\boldsymbol{x}_1$)			Alternative 2 ($\boldsymbol{x}_2$)		
	x_1^j	w_1^j	u^j	x_2^j	w_2^j	u^j
1. Trademark	Reno	80	24	Pejo	40	12
2. Warranty	36	60	12	24	40	8
3. Price	40,400	24	12	30,800	48	24
Utility	$u(\boldsymbol{x}_1) = 48$			$u(\boldsymbol{x}_2) = 44$		

Warranty and price are numerical attributes; the minimum increment for guarantee is one month and for price it is $10.

The attributes and their values (ranges) are shown in Table 1. In the table weights for each attribute and weights for the nominal values of the trademark and the extreme values of the numerical attributes are also given.

Partial utilities of the two numerical attributes are respectively defined by the following linear functions:

- Partial utility of Warranty: $w_l^2 = 100x_l^2/60$; and
- Partial utility of Price: $w_l^3 = 100(50,000 - x_l^3)/(50,000 - 10,000)$.

Using the partial utility functions for two attributes and the information given in Table 1 we can calculate partial utilities. The utility calculation for two alternatives: $\boldsymbol{x}_1 =$ [Reno; 36; 40,400] and $\boldsymbol{x}_2 =$ [Pejo; 24; 30,800] is shown in Table 2.

The buyer prefers $\boldsymbol{x}_1$over $\boldsymbol{x}_2$because $u(\boldsymbol{x}_1) = 48 > u(\boldsymbol{x}_2) = 44$.

The buyer decides to initiate a reverse auction; its process is described in the next section.

6.2 *Auction*

In this section we illustrate the reverse auction procedure.

6.2.1 Preparation

The buyer formulates the following initial reservation levels:

1. Every trademark except for "Star" is acceptable;
2. The minimum warranty period is 12 months;
3. The maximum price is $50,000.

In order to initiate the auction several design parameters must be set.

Step 1. The parameter D value is equal to the number of attributes (i.e. $D = 3$). The values of parameter $\boldsymbol{e} = [e_1, e_2, e_3]$ are determined as follows. e_1 corresponds to the nominal variable (trademark) and therefore it is not set. e_2 is 12 months, and e_3 is $800. This selection of the attribute value increment reduces the space of alternatives from 1.44 million to 1,836 distinct alternatives.

Step 2. The number N^u of points sufficient to define the buyer's utility function u is determined. Two attributes (Price and Warranty) have linear partial utility functions.

The attribute Trademark is nominal and neither the order of its values nor the preferential distance between the different trademarks is known. Therefore, knowledge of as many alternatives of equal utility value as there are attribute values (i.e., trademarks) is required to determine partial utility. This means that $N^u = 2 + 5$; because trademark Star is not acceptable, $N^u = 6$.

If perturbation is required, then one attribute value will be changed, i.e., $n^E = 1$.

Step 3. Round counter is set to 1, ($t = 1$). The initial reservation levels are used to construct the initial set of acceptable bids X_{DE}^{A1}.

$$X_{DE}^{A1} = \left\{X_1^1; x_2 \geq 12; x_3 \leq 50,000\right\},$$

where $X_1^1 = \{\text{Citron; Pejo; Betha; Reno; Lux}\}$. (Note that the minimum acceptable utility value is 10; it is the utility of [Citron; 12; 50000]).

The auction is now open and the bidders are asked to submit their bids.

6.2.2 Exchange of Bids and Reservation Levels

Three sellers ($J = 3$) submitted bids.

Step 4. The following bids were submitted:

O_{11} = [Pejo; 24; 46,000]; O_{21} = [Reno; 12; 50,000]; and O_{31} = [Betha; 24; 47,600].

The buyer's utility of each bid is respectively: $u(O_{11}) = 25$; $u(O_{21}) = 22$; $u(O_{31}) = 29$. Hence, the best bid was made by the third bidder ($j = 3$). We set $\boldsymbol{x}_r^{1*}$ = [Betha; 24; 47,600].

Step 5. The auction moves to the next round, i.e., $t = 2$. The reservation level in round 2 is 29, ($r^2 = 29$).

Step 6. In order to select reference alternatives, first the number N^{2r} of alternatives yielding utility $r^2 = 29$ needs to be determined. There are three such alternatives:

$u(\boldsymbol{x}_1) = u$[Betha; 24; 47,600] = 29;

$u(\boldsymbol{x}_2) = u$[Pejo; 24; 42,800] = 29; and
$u(\boldsymbol{x}_3) = u$[Pejo; 36; 46,000] = 29.

Because $N^{1r} = 3 \geq D = 3$, the above three alternatives are selected for the construction of set X_D^{A2}. The process moves to Step 8.

Step 8. Each of the alternatives $\boldsymbol{x}_1, \boldsymbol{x}_2$ and $\boldsymbol{x}_3$ is used to construct set of r-acceptable alternatives.

From Table 1 we see that, if Betha is the worst acceptable alternative, then there are only two trademarks preferred over Betha, that is X_1^2 = {Betha; Reno; Lux}. Similarly, for Pejo trademark we obtain $X_2^2 = X_3^2$ = {Pejo; Betha; Reno; Lux}.

The set of acceptable alternatives X_D^{A2} is the union of three sets:

$$X_D^{A2} = X_{D1}^{A2} \cup X_{D2}^{A2} \cup X_{D2}^{A2} = \left\{X_{11}^2; x_2 \geq 24; x_3 \leq 47,600\right\} \cup$$
$$\cup \left\{X_{12}^2; x_2 \geq 24; x_3 \leq 42,800\right\} \cup \left\{X_1^2; x_2 \geq 36; x_3 \leq 46,000\right\}.$$

Step 9. In this step we need to determine perturbation. Because $\delta = \min(D, N^{1r}) = 3 < N^u = 6$, perturbation is not required. For illustrative purposes we show how one of the alternatives selected in Step 6 can be perturbed.

Step 10. Select one ($n^E = 1$) attribute n for which an increase of its value by e_n increases the alternative utility value the least. Any change for Trademark increases the utility by 6 (0.3×20). An increase of Warranty by 12 months increases the utility by 4 (0.2×20) and a decrease of Price by \$800 increases the utility by 1 (0.5×2). Therefore attribute Price is selected. (Note that the preferential direction for price is decreasing.)

From among the three alternatives ($\boldsymbol{x}_1, \boldsymbol{x}_2$ and $\boldsymbol{x}_3$) determined in Step 3 select one for which attribute Price takes the highest value.[7] This is alternative $\boldsymbol{x}_1$ and its Price attribute is decreased by e_3 = \$800. This perturbation of $\boldsymbol{x}_1$ is used to replace set.

$$X_{D1}^{A2} = \left\{X_{12}^2; x_2 \geq 24; x_3 \leq 47,600\right\} \text{with}$$
$$X_{D1E}^{A2} = \left\{X_{12}^2; x_2 \geq 24; x_3 \leq 46,800\right\}.$$
$$\text{set } X_{DE}^{A2} = X_{D1E}^{A2} \cup X_{D2}^{A2} \cup X_{D3}^{A2}.$$

Step 11. Present the bidders with the new reservation levels, that is set X_{DE}^{A2}, and request bids.

Step 11 completes the second auction round. By moving to Step 4, the next, third, round is initiated. The rounds continue until there is only one bidder left.

[7] The highest value is selected because of the preferential direction. If the minimum partial utility was minimal for Warranty, then the lowest value would be selected. This rule is appropriate for linear partial utility functions and when the partial utility values between any two values of a nominal attribute are equal or very close.

References

Anderson, J., Frohlich, M.: FreeMarkets and online auctions. Bus. Strateg. Rev. **12**(2), 59–68 (2001)

Beil, D., Wein, L.: An inverse-optimization-based auction mechanism to support a multiattribute RFQ process. Manag. Sci. **49**(11), 1529–1545 (2003)

Bellosta, M., Kornman, S., et al.: A unified framework for multiple criteria auction mechanisms. Web Intell. Agent Syst. **6**(4), 401–419 (2008)

Bellosta, M.J., Brigui, I., et al.: A multicriteria model for electronic auions. In: ACM Symposium on Applied Computing (SAC 2004) (2004)

Bichler, M., Kalagnanam, J.: Configurable offers and winner determination in multi-attribute auctions. Eur. J. Oper. Res. **160**(2), 380–394 (2005)

Bichler, M.: The Future of E-markets: Multidimensional Market Mechanisms. Cambridge University Press, Cambridge (2001)

Block, C., Collins, J., et al.: Exploring retail energy markets through competitive simulation. In: ACM EC 2010 Workshop on Trading Agent Design and Analysis, Harvard University (2010)

Brigui-Chtioui, I., Pinson, S.: A variable bid increment algorithm for reverse English auction. In: Progress in Artitficial Economics, Springer,Cham (2010)

Burmeister, B., Ihde, T., et al.: A practical approach to multi-attribute auctions. In: 13th Workshop on Database and Expert Systems Applications (DEXA), IEEE (2002)

Che, Y.K.: Design competition through multidimensional auctions. Rand. J. Econ. **24**(4), 668–680 (1993)

Chen-Ritzo, C.H., Harrison, T.P., et al.: Better, faster, cheaper: an experimental analysis of a multi-attribute reverse auction mechanism with restricted information feedback. Manage. Sci. **51**(12), 1753–1762 (2005)

Dhar, R.: The effect of decision strategy on deciding to defer choice. J. Behav. Decis. Mak. **9**, 265–281 (1996)

Dieckmann, A., Dippold, K., et al.: Compensatory versus noncompensatory models for predicting consumer preferences. Judgm. Decis. Mak. **4**(3), 200–213 (2009)

Elmaghraby, W.: Auctions and pricing in E-marketplaces. In: Simchi-Levi, D., Wu, D., Shen, M. (eds.) Handbook of Quantitative Supply Chain Analysis: Modeling in the E-Business Era. Kulwer, Norwell (2004)

Engel, Y., Wellman, M.P.: Multiattribute auctions based on generalized additive independence. J Artif. Intell. Res. **37**(1), 479–526 (2010)

Engelbrecht-Wiggans, R., Haruvy, E., et al.: A Comparison of buyer-determined and price-based multi-attribute mechanisms. Mark. Sci. **26**(5), 629–641 (2007)

Fishburn, P.C.: Noncompensatory preferences. Synthese **33**, 393–403 (1976)

Jap, S.: An exploratory study of the introduction of online reverse auctions. J. Mark. **67**(3), 96–107 (2003)

Katok, E., Wambach, A.: Collusion in dynamic buyer-determined reverse auctions. Manag. Sci. (2011, forthcoming)

Kittsteiner, T., Ockenfels, A.: Market design: a selective review. Z. Betriebswirt. **5**, 121–143 (2006)

Milgrom, P., Weber, R.: A theory of auctions and competitive bidding. Econometrica **50**(5), 1089–1122 (1982)

Parkes, D.C., Kalagnanam, J.: Models for iterative multiattribute procurement auctions. Manage. Sci. **51**(3), 435–451 (2005)

Peleg, B.: The value of procurement via online bidding. Whitepaper **3** (2003)

Pontrandolfo, P., Wu, S., et al.: Auctions and Negotiations in Transportation Service Procurement. Group Decision and Negotiations (2010)

Schoenherr, T., Mabert, V.A.: Online Reverse auctions: common myths versus evolving reality. Bus. Horiz. **50**(5), 373–384 (2007)

Strecker, S., Seifert, S.: Electronic sourcing with multi-attribute auctions. HICSS 37, Hawaii (2004)

Teich, J.E., Wallenius, H., et al.: Multiple issue auction and market algorithms for the world wide web. Decis. Support Syst. **26**, 49–66 (1999)

Teich, J.E., Wallenius, H., et al.: Emerging multiple issue E-auctions. Eur. J. Oper. Res. **159**(1), 1–16 (2004)

Wagner, S.M., Schwab, A.P.: Setting the stage for successful electronic reverse auctions. J. Purch. Supply Manag. **10**(1), 11–26 (2004)

Walley, M., Fortin, D.: Behavioral outcomes from online auctions: reserve price, reserve disclosure, and initial bidding influences in the decision process. J. Bus. Res. **58**(10), 1409–1418 (2005)

Yee, M., Dahan, E., et al.: Greedoid-based Noncompensatory Inference. Mark. Sci. **26**, 532–549 (2007)

Yu, B., Kersten, G.E., et al.: Multi-attribute auctions and negotiations in service procurement: an experimental comparison. In: Group Decision and Negotiation Conference (2008)

Aggregation of Stochastic Rankings in Group Decision Making

Miłosz Kadziński, Grzegorz Miebs, Dariusz Grynia, and Roman Słowiński

Abstract We propose a novel method for group decision making. It is applicable in the context of multiple criteria ranking problems, where alternatives need to be ordered from the best to the worst by multiple Decision Makers (DMs). In the first stage, incomplete preference information of each DM is analyzed within the framework of stochastic analysis. Specifically, the Monte Carlo simulation is applied for exploiting the space of preference model parameters compatible with each DM's preferences. In this way, we estimate the values of stochastic acceptability indices that quantify the support given to the preference, indifference, and incomparability relations for each pair of alternatives. In the second stage, such stochastic rankings are aggregated into a group compromise recommendation that minimizes either an average or a maximal distance from each DM's input. Apart from accounting for the utilitarian and egalitarian perspectives, the dedicated mathematical programming models deal with the processing and constructing of complete or partial rankings. The proposed method is coupled with the robust variants of PROMETHEE I and II methods, however, it can be combined with any method from the broad family of Stochastic Multicriteria Acceptability Analysis (SMAA) techniques. Its applicability

M. Kadziński (✉) · G. Miebs · D. Grynia · R. Słowiński
Institute of Computing Science, Poznan University of Technology, Piotrowo 2, 60-965 Poznań, Poland
e-mail: milosz.kadzinski@cs.put.poznan.pl

G. Miebs
e-mail: grzegorz.miebs@cs.put.poznan.pl

D. Grynia
e-mail: dariusz.grynia@cs.put.poznan.pl

R. Słowiński
e-mail: roman.slowinski@cs.put.poznan.pl

R. Słowiński
Systems Research Institute, Polish Academy of Sciences, Newelska 6, 01-447 Warsaw, Poland

T. Szapiro and J. Kacprzyk (eds.), *Collective Decisions: Theory, Algorithms And Decision Support Systems*, Studies in Systems, Decision and Control 392,
https://doi.org/10.1007/978-3-030-84997-9_4

for supporting real-world group decision making is demonstrated in an illustrative case study concerning the ranking of project proposals by a research funding agency.

Keywords Multiple criteria ranking · Rank aggregation · Stochastic acceptability index · Robustness analysis · Binary linear programming · PROMETHEE

1 Introduction

In Multiple Criteria Decision Aiding (MCDA), we consider decision problems concerning a finite set of alternatives (e.g., potential actions, projects, candidates, or patients) evaluated on a consistent family of criteria (e.g., attributes, impacts, skills, aspects, benefits, or risks) (see, e.g., Cinelli et al. 2020; Greco et al. 2016). The criteria represent heterogeneous viewpoints on the quality of alternatives. Many such decision problems fall into the category of ranking, where the relevant options need to be ordered from the best to the worst. The constructed rankings can be complete or partial. In the former, all pairs of alternatives need to be comparable. Opposite to the former, the latter ones admit incomparability, indicating that none alternative in a specific pair can be judged at least as good as the other (Roy 1991).

MCDA offers a plethora of tools for supporting Decision Makers (DMs) in carrying forward the process of ranking of a set of alternatives. These approaches ask the DM for preference information, indicating the subjective priorities that should be considered when imposing the order on the set of alternatives. Such information is used to construct a preference model, that aggregates the performances on multiple criteria according to the DM's value system. In most ranking methods, the parameters of the preference model need to be specified directly by the DM (see, e.g., MAVT (Keeney and Raiffa 1993), TOPSIS (Hwang and Yoon 1981), ELECTRE III (Roy 1978), or PROMETHEE I and II (Brans and Vincke 1985)). The application of such methods results in a deterministic ranking, leaving no doubt on the relation holding for each pair of alternatives.

In recent years, we have been able to observe an increasing interest paid to the methods that admit incomplete or partial preference information. They constitute a growing family of so-called Robust Ordinal Regression methods (Greco et al. 2008). These methods question the assumption that the DM can provide reliable estimates of the preference model parameters due to a high cognitive effort, misunderstanding of their interpretation, or a lack of confidence in specifying precise numbers (Miebs and Kadziński 2021). Incomplete preferences take the form of implicitly or explicitly provided constraints on the parameters' values. For example, they may be defined by ordering all criteria in terms of their importance (Figueira and Roy 2002), providing admissible value intervals (Salo and Hamalainen 1995), or specifying holistic preference information in the form of, e.g., pairwise comparisons (Greco et al. 2012) or rank-related requirements (Kadziński et al. 2013). Under such a scenario, many compatible preference model instances exist. Their application on the set of alternatives usually leads to an ambiguous ranking.

Two major methodological streams exist for quantifying the robustness of recommendation given a set of compatible preference model instances. On the one hand, one may investigate the extreme outcomes confirmed by all, at least one (see, e.g., Greco et al. 2008; Salo and Hamalainen 1995), or the most and the least advantageous (Kadziński et al. 2012) feasible parameter values. Such an analysis is usually conducted with mathematical programming techniques that consider infinitely many compatible model instances defined by the set of linear constraints. On the other hand, one could incorporate the Monte Carlo simulations to sample a large representative set of possible parameter values (Tervonen et al. 2013). Then, the method ranks the alternatives for each preference model instance, and the robustness of results is quantified through stochastic acceptability indices. This way of proceeding has been implemented, e.g., in Lahdelma and Salminen (2001), Kadziński and Tervonen (2013), and Tervonen et al. (2008). The acceptability indices indicate the share of compatible parameter values supporting a given part of the recommendation, e.g., attaining some rank by an alternative or the truth of some specific relation for a pair of alternatives. When conducting such a simulation-based analysis, the ranking is not deterministic, in turn, revealing uncertainty for the recommended order of alternatives. We will focus on such stochastic rankings in this chapter.

An ever-growing complexity of real-world multiple criteria decision-making problems implies that many decisions are made by a group of DMs (see, e.g., Kilgour et al. 2010, Jelassi et al. 1990). Examples applications of this kind can be found in Norese (2006), Chen et al. (2007), Govindan et al. (2017), and Ciomek et al. (2018). The main challenge in group decision-making problems derives from the potentially substantial differences between the DMs regarding their aspirations, value systems, or visions (Leyva-Lopez and Fernandez-Gonzalez 2003). The process of transforming individual preferences into a collective outcome can be conducted in two ways. In the input-oriented approaches (Salminen et al. 1998), the DMs need to act as a unit, agreeing on the problem structure and preferences. Even though such methods are appealing as they give space to the application of MCDA methods initially designed for a single DM, it is often not realistic to assume that the DMs can reach a comprehensive agreement. In turn, in the output-oriented techniques (see, e.g., Chen 2014, Kilgour et al. 2010), the individual recommendation is first constructed for each DM, and the consensus outcome is attained by applying some pre-defined procedure.

When it comes to rank aggregation, all existing methods focus only on exploiting deterministic rankings. In particular, GDSS PROMETHEE (Mareschal et al. 1998) and ELECTRE GD (Leyva-Lopez and Fernandez-Gonzalez 2003) use individual orders for, respectively, deriving consensus flows of alternatives or constructing an outranking relation representing the entire group. Furthermore, VIP–G (Dias and Climaco 2005) and UTA^{GMS}–GROUP (Greco et al. 2012) focus on quantifying the support given to different elements of the ranking by the group members. Finally, the construction of a group compromise ranking was advocated in Govindan et al. (2017), Miebs and Kadziński (2021) in the context of, respectively, small- and large-scale problems. For this purpose, the former approach incorporates mathematical programming techniques, whereas the latter uses heuristic methods for rank aggregation.

In this chapter, we focus on the aggregation of stochastic rankings. These rankings are represented using the acceptability indices for the preference, indifference, and incomparability relations. Specifically, we adopt the Binary Linear Programming (BLP) models presented in Govindan et al. (2017) to allow for uncertain inputs. In particular, we present the models dealing with partial and complete rankings oriented toward constructing either utilitarian or egalitarian rankings. The former minimizes an average distance from the input rankings, whereas the latter minimizes the maximal distance from the stochastic ranking of any DM. We illustrate the applicability of the proposed methods in the context of PROMETHEE I and II (Brans and Vincke 1985). These approaches are parameterized with the weights compatible with the incomplete preferences on the criteria importance expressed in the revised Simos (SRF) procedure. Nevertheless, the rank aggregation techniques proposed in this chapter are also applicable in the context of such ranking approaches as SMAA-2 (Lahdelma and Salminen 2001), SMAA-III (Tervonen et al. 2008), Stochastic Ordinal Regression (Kadziński and Tervonen 2013), SMAA-Choquet (Angilella et al. 2012), SMAA-PROMETHEE (Corrente et al. 2014), or ELECTRE III with a hierarchy of interacting criteria, imprecise weights and stochastic analysis (Corrente et al. 2017).

This chapter is organized in the following way. In Sect. 2, we first remind the PROMETHEE ranking methods and their stochastic variants admitting imprecision in the specification of weights. Then, we discuss the algorithms for constructing group compromise rankings. Section 3 illustrates the use of these methods on a problem concerning ranking a set of project proposals considered by a research funding agency. We incorporate the preferences of three DMs for constructing either a utilitarian or an egalitarian ranking of projects while tolerating incomparability or imposing a strict order. The last section concludes and provides avenues for future research.

2 Construction of a Group Compromise Ranking Based on the Stochastic Rankings

Let us consider a finite set of alternatives $A = \{a_1, \ldots, a_i, \ldots, a_n\}$ evaluated in terms of a consistent family of criteria $G = \{g_1, \ldots, g_j, \ldots, g_m\}$. Without loss of generality, we assume that all criteria are of gain type, i.e., greater performances are more preferred. The performance of alternative $a_i \in A$ on criterion $g_j \in G$ is denoted by $g_j(a_i)$. We account for a set of DMs $\mathcal{D} = \{D_1, \ldots, D_h, \ldots, D_t\}$. The aim is to construct a single compromise ranking for all DMs in $\mathcal{D}$, using an output-oriented approach.

In this section, we first discuss the ranking methods used for constructing either a complete or a partial ranking. In this basic setting, we consider a precise weight vector $w = [w_1, \ldots, w_j, \ldots, w_m]$ associated with the criteria. Then, we generalize these approaches to account for a set of weight vectors compatible with the DM's incomplete preference information. The results of this step are quantified with stochastic acceptability indices, indicating the share of weights confirming a given type of relation (e.g., preference, indifference, or incomparability) for each pair of alternatives. Finally, we present the mathematical programming models for constructing a group compromise ranking given individual DMs' stochastic rankings.

2.1 Reminder on the PROMETHEE Ranking Methods

In this section, we remind the PROMETHEE I and II methods for ranking a finite set of alternatives evaluated in terms of multiple criteria (Brans and Vincke 1985). For each criterion $g_j \in G$, we shall consider a preference function $\pi_j(a_i, a_k)$, such that for all pairs $a_i, a_k \in A$:

$$\pi_j(a_i, a_k) = F_j(d_j(a_i, a_k)) \in [0, 1],$$

where $d_j(a_i, a_k) = g_j(a_i) - g_j(a_k)$. In PROMETHEE, six types of particular preference functions have been proposed. However, we will refer to a so-called usual function, such that $\pi_j(a_i, a_k) = 1$ if $d_j(a_i, a_k) > 0$, and $\pi_j(a_i, a_k) = 0$, otherwise. To express the degree in which a_i is preferred to a_k over all criteria, we will refer to an aggregated preference index:

$$\pi(a_i, a_k) = \sum_{j=1}^{m} w_j \pi_j(a_i, a_k) \text{ for all } (a_i, a_k) \in A \times A,$$

where w_j is a weight expressing relative importance of g_j in set G. Further, the positive flow $\Phi^+(a)$ expresses how much alternative a_i is outranking all the other $n-1$ alternatives:

$$\Phi^+(a) = 1/(n-1) \sum_{a_k \in A \setminus \{a_i\}} \pi(a_i, a_k),$$

whereas the negative flow $\Phi^-(a_i)$ expresses how much alternative a_i is outranked by all the others:

$$\Phi^-(a_i) = 1/(n-1) \sum_{a_k \in A \setminus \{a_i\}} \pi(a_k, a_i).$$

The balance between the positive and the negative outranking flows is reflected in the net flow score:

$$\Phi(a_i) = \Phi^+(a_i) - \Phi^-(a_i).$$

Obviously, the higher the net flow score, the better the alternative.

In PROMETHEE II, a complete ranking of alternatives is imposed by the application of net flow scores in the following way:

$$a_i \text{ is preferred to } a_k \ (a_i P^+ a_k) \text{ iff } \Phi(a_i) > \Phi(a_k),$$
$$a_i \text{ is indifferent with } a_k \ (a_i I a_k) \text{ iff } \Phi(a_i) = \Phi(a_k).$$

Note that an inverse preference P^- can be defined as follows: $a_i P^- a_k \Leftrightarrow a_k P^+ a_i$. In PROMETHEE I, an incomplete ranking is constructed based on the positive and negative flows. In particular, $a_i P^+ a_k$ if a_i is at least as good as a_k in the rankings imposed by both types of flow and strictly better in at least one of them. Furthermore, $a_i I a_k$ if a_i and a_k attain the same values of the positive and negative flows. Finally, a_i and a_k are incomparable ($a_i R a_k$) if each of them is more preferred in the ranking imposed by either Φ^+ or Φ^-.

2.2 *Stochastic Analysis of the Rankings Delivered by PROMETHEE*

For each $DM_h \in \mathcal{D}$, the use of PROMETHEE requires specification of a weight vector w^h, expressing the importance of each criterion g_j, $j = 1, \ldots, m$, for this DM. There exist various techniques that support the elicitation of meaningful weights. We will refer to the SRF procedure (Figueira and Roy 2002).

This method expects DM_h to first assign some importance rank $l^h(j)$ to each criterion g_j, while admitting some criteria to be assigned the same rank. Then, the DM is allowed to differentiate the intensity of preference between the neighboring subsets of indifferent criteria, L_s^h and L_{s+1}^h. This is attained by inserting e_s^h white (empty) cards between these groups so that greater e_s^h implies a greater difference between weights. Finally, (s)he needs to specify ratio Z^h between the importances of the most and the least significant criteria denoted, respectively, by $L_{v(h)}^h$ and L_1^h (Corrente et al. 2017).

The original SRF procedure derives precise weight values. We will apply, however, its stochastic variant exploiting the polyhedron of weights compatible with the preferences expressed by the DM in the SRF procedure (see, e.g., Corrente et al. 2017; Kadziński et al. 2018). Such a space, denoted by $w^h(SRF)$ for $DM_h \in \mathcal{D}$, is defined by the following constraint set $E^h(SRF)$:

$$\left.\begin{array}{l}
[O1]\ w_i^h > w_j^h, \text{ for all } g_i \in L_t^h, g_j \in L_s^h, \text{ and } t > s,\\
[O2]\ w_i^h = w_j^h, \text{ for all } g_i, g_j \in L_s^h,\\
[O3]\ w_i^h = Z^h w_j^h, \text{ for all } g_i \in L_{v(k)}^k,\ g_j \in L_1^k,\\
[O4]\ w_{j+1}^h - w_j^h > w_{p+1}^h - w_p^h, \text{ if } e_j^h > e_p^h,\\
[O5]\ \sum_{j=1}^m w_j^h = 1.
\end{array}\right\} E^h(SRF)$$

The above constraints respect the compatibility with the ranking of criteria (O1 and O2), the ratio between the most and the least important criteria provided by DM_h (O3), intensities of preference (O4), and normalization of weights (O5) (Kadziński et al. 2018).

There exist multiple weight vectors compatible with such incomplete preference information. Thus, we apply the Monte Carlo simulation to sample a large set of uniformly distributed compatible weight vectors $w \in w^h(SRF)$ (Tervonen et al. 2013). For each of them, we run PROMETHEE I or PROMETHEE II. The relation observed for a pair $(a_i, a_k) \in A \times A$ under such a scenario is denoted by $S_{ik}^{h,w} \in \{P^+, P^-, I, R\}$. The robustness of particular elements of the ranking obtained for thus generated weights is quantified with the stochastic acceptability indices.

Let us define a pairwise relation acceptability index $PRAI^h(a_i, a_k, S)$ for DM_h and $S \in \{P^+, P^-, I, R\}$ as the share of feasible weights $w \in w^h(SRF)$ for which the relation holds for the comparison of a_i and a_k in the final ranking derived with PROMETHEE, i.e.:

$$PRAI^h(a_i, a_k, S) = \int_{w \in w^h(SRF)} m^h(w, a_i, a_k, S)\, dw,$$

where $m^h(w, a_i, a_k, S)$ is the confirmation function of a specific relation $S \in \{P^+, P^-, I, R\}$:

$$m^h(w, a_i, a_k, S) = \begin{cases} 1, & \text{if } a_i S_{ik}^{h,w} a_k, \\ 0, & \text{otherwise.} \end{cases}$$

The definition of $PRAI^h(a_i, a_k, S)$ can be adapted to the case of P^+, P^-, I, and R. The respective indices are called pairwise winning PWI, losing PLI, indifference PII, and incomparability PRI indices. Obviously, for each pair $a_i, a_k \in A$ and each $DM_h \in \mathcal{D}$, $PWI^h(a_i, a_k) + PLI^h(a_i, a_k) + PII^h(a_i, a_k) + PRI^h(a_i, a_k) = 1$. Moreover, PRI is considered in the context of PROMETHEE I only.

Hence, for each $DM_h \in \mathcal{D}$, we derive a stochastic ranking resulting from the application of PROMETHEE for different feasible weight vectors. Such a ranking takes the form of matrices of acceptability indices indicating the shares of weights confirming a given relation for all pairs of alternatives.

2.3 Construction of a Group Compromise Ranking

The stochastic rankings obtained for each DM form the input for the construction of a group compromise ranking. We consider a pair of settings in terms of accounting for the preferences of individual DMs. On the one hand, the utilitarian ranking minimizes an average distance from the rankings of all DMs. On the other hand, the egalitarian ranking minimizes the maximal distance from any DM's ranking (Govindan et al. 2017).

The distances $\delta(S_{ik}^{h'}, S_{ik}^{h''})$ between relations $S_{ik}^{h'}, S_{ik}^{h''} \in \{P^+, P^-, R, I\}$ are presented in Table 1. These distances were justified by a set of logical and semantic conditions (Roy and Słowiński 1993). Specifically, the greatest distance is attributed to comparing inverse preference P^- and preference P^+. It is twice as big as for the

Table 1 Definition of distances $\delta(S_{ik}^{h'}, S_{ik}^{h''})$ between different pairs of relations (Roy and Słowiński 1993)

$S_{ik}^{h'}/S_{ik}^{h''}$	$a_i P^+ a_k$	$a_i P^- a_k$	$a_i I a_k$	$a_i R a_k$
$a_i P^+ a_k$	0	4	2	3
$a_i P^- a_k$	4	0	2	3
$a_i I a_k$	2	2	0	2
$a_i R a_k$	3	3	2	0

indifference I and either preference (P^+ or P^-) or incomparability R. The value selected for the comparison of preference and incomparability is intermediate. The comprehensive distance between two rankings $R^{h'}$ and $R^{h''}$ composed of well-established, deterministic relations can be computed as a sum of distances between relations observed for all pairs of alternatives (Roy and Słowiński 1993):

$$\sum_{i,k\ i<k} \delta(S_{ik}^{h'}, S_{ik}^{h''}).$$

In the considered setting, one of the rankings $R^{h'}$ is not definite, being represented by stochastic acceptability indices $PRAI^{h'}(a_i, a_k, S_{ik}^{h'})$ for $S_{ik}^{h'} \in \{P^+, P^-, I, R\}$ for all $a_i, a_k \in A$. In turn, the other ranking $R^{h''}$ is yet to be determined, and for each pair of alternatives $a_i, a_k \in A$, one of the four relations $S_{ik}^{h''} \in \{P_{ik}^+, P_{ik}^-, I_{ik}, R_{ik}\}$ can be instantiated. The distance between such rankings can be defined in the following way:

$$\Delta^{inc}(h', h'')$$
$$= \sum_{i,k\ i<k} \sum_{S_{ik}^{h'} \in \{P^+, P^-, I, R\}} \sum_{S_{ik}^{h''} \in \{P_{ik}^+, P_{ik}^-, I_{ik}, R_{ik}\}} PRAI^{h'}(a_i, a_k, S_{ik}^{h'}) \cdot S_{ik}^{h''} \cdot \delta(S_{ik}^{h'}, S_{ik}^{h''}),$$

where $S_{ik}^{h''} \in \{P_{ik}^+, P_{ik}^-, I_{ik}, R_{ik}\}$ are binary variables that are equal to one if the respective relation (from among $\{P^+, P^-, I, R\}$) was selected for a pair $a_i, a_k \in A$ in the constructed ranking, and they are equal to zero, otherwise (i.e., when some other relation was instantiated). For example, when considering a pair $a_i, a_k \in A$ with $PWI(a_i, a_k) = 0.5$, $PLI(a_i, a_k) = 0$, $PII(a_i, a_k) = 0.3$, and $PRI(a_i, a_k) = 0.2$ in the stochastic ranking, and relation $a_i P^- a_k$ in the constructed ranking (i.e., $P_{ik}^- = 1$ and $P_{ik}^+ = I_{ik}^+ = R_{ik}^+ = 0$), the distance for this pair is equal to:

$$0.5 \cdot 0 \cdot 0 + ... + 0.5 \cdot 1 \cdot 4 + 0 \cdot 1 \cdot 0 + 0.3 \cdot 1 \cdot 2 + 0.2 \cdot 1 \cdot 3 + ... + 0.2 \cdot 0 \cdot 0 = 3.2.$$

The procedures for constructing the incomplete compromise ranking incorporate the above-defined distance measure. Moreover, for each pair of alternatives $a_i, a_k \in A$, they consider a binary variable r_{ik} that should be equal to one in case a_i is weakly preferred to a_k (Govindan et al. 2017). The variables r_{ik} and r_{ki} are used to instantiate one of the four relations for the comparison between a_i over a_k in the compromise ranking while imposing at the same time the transitivity of a weak preference relation. For example, when $r_{ik} = 1$ and $r_{ki} = 1$, then $I_{ik} = 1$, whereas in case $r_{ik} = 0$ and $r_{ki} = 0$, then $R_{ik} = 0$. Moreover, for $i, k, p = 1, 2, \ldots, n$, such that $i \neq k \neq p$, when $r_{ip} = 1$ and $r_{pk} = 1$, then $r_{ik} = 1$. The compromise utilitarian ranking R^U admitting incomparability can be constructed by solving the following Binary Linear Programming model:

$$Minimize\text{: } \frac{1}{t} \sum_{h=1,\ldots,t} \Delta^{inc}(h, U)$$

s.t.

$$\left.\begin{array}{l}
[R] \text{ for all } i, k = 1, 2, \ldots, n : \ i \neq k : \\
\quad [R1] \ r_{ik} - r_{ki} \leq P^+_{ik}, \\
\quad [R2] \ r_{ki} - r_{ik} \leq P^-_{ik}, \\
\quad [R3] \ r_{ik} + r_{ki} - 1 \leq I_{ik}, \\
\quad [R4] \ 1 - r_{ik} - r_{ki} \leq R_{ik}, \\
\quad [R5] \ P^+_{ik} + P^-_{ik} + I_{ik} + R_{ik} = 1, \\
\quad [R6] \ r_{ik}, P^+_{ik}, P^-_{ik}, I_{ik}, R_{ik} \in \{0, 1\}, \\
[T] \text{ for all } i, k, p = 1, 2, \ldots, n : \ i \neq k \neq p : \\
\quad r_{ik} \geq r_{ip} + r_{pk} - 1.5.
\end{array}\right\} E^{utilitarian}_{partial}$$

The above problem aims at minimizing an average distance between the stochastic rankings obtained for DM_h, $h = 1, \ldots, t$, and the constructed compromise ranking R^U composed of relations $S^U_{ik} \in \{P^+_{ik}, P^-_{ik}, I_{ik}, R_{ik}\}$ for all pairs of alternatives $a_i, a_k \in A$. For an exact explanation of the role of each constraint, see Govindan et al. (2017). In turn, the compromise egalitarian ranking R^E can be constructed by selecting such relations S^E_{ik} for all pairs $a_i, a_k \in A$ that altogether minimize the maximal distance from the stochastic ranking of any DM. This can be attained by solving the following BLP problem:

$$Minimize\text{: } \alpha$$

s.t.

$$\left.\begin{array}{l}
E^{utilitarian}_{partial}, \\
\text{for } h = 1, \ldots, t : \\
\quad \alpha \geq \Delta^{inc}(h, E).
\end{array}\right\} E^{egalitarian}_{partial}$$

When considering complete rankings, only preference, inverse preference, and indifference relations need to be considered. For this purpose, for each pair $a_i, a_k \in A$, we will incorporate the binary variables r_{ik}, r_{ki}, and z_{ik} (Govindan et al. 2017). The role of r_{ik} and r_{ki} is to reflect the truth of a weak preference relation, whereas z_{ik} should be equal to one in case a_i and a_k are indifferent, i.e., when $r_{ik} = r_{ki} = 1$. Note that $a_i P^+ a_k$ in case $r_{ik} = 1$, $r_{ki} = 0$ and $z_{ik} = 0$, whereas $a_i P^- a_k$ holds when $r_{ik} = 0$, $r_{ki} = 1$ and $z_{ik} = 0$. Given a stochastic ranking $R^{h'}$ represented by $PRAI^{h'}(a_i, a_k, S)$

for $S \in \{P^+, P^-, I\}$ and a ranking $R^{h''}$ to be constructed, the distance between them can be computed as follows:

$$\Delta^{com}(h', h'')$$
$$= \sum_{i,k\ i<k} \sum_{S_{ik}^{h'} \in \{P^+, P^-, I\}} \sum_{S_{ik}^{h''} \in \{P_{ik}^+, P_{ik}^-, I_{ik}\}} PRAI^{h'}(a_i, a_k, S_{ik}^{h'}) \cdot S_{ik}^{h''} \cdot \delta(S_{ik}^{h'}, S_{ik}^{h''})$$
$$= \sum_{i,k\ i<k} \sum_{S_{ik}^{h'} \in \{P^+, P^-, I\}} [PRAI^{h'}(a_i, a_k, S_{ik}^{h'}) \cdot (r_{ik} - z_{ik}) \cdot \delta(S_{ik}^{h'}, P_{ik}^+)$$
$$+ (r_{ki} - z_{ik}) \cdot PRAI^{h'}(a_i, a_k, S_{ik}^{h'}) \cdot \delta(S_{ik}^{h'}, P_{ik}^-) + z_{ik} \cdot PRAI^{h'}(a_i, a_k, S_{ik}^{h'}) \cdot \delta(S_{ik}^{h'}, I_{ik})].$$

For example, when considering a pair $a_i, a_k \in A$ with $PWI(a_i, a_k) = 0.4$, $PLI(a_i, a_k) = 0.1$, and $PII(a_i, a_k) = 0.5$ in the stochastic ranking, and relation $a_i P^+ a_k$ in the constructed ranking (i.e., $r_{ik} = 1$ and $r_{ki} = z_{ik} = 0$), the distance between these relations is equal to:

$$0.4 \cdot 1 \cdot 0 + 0.4 \cdot 0 \cdot 4 + ... + 0.1 \cdot 1 \cdot 4 + ... + 0.5 \cdot 1 \cdot 2 + ... + 0.5 \cdot 0 \cdot 0 = 1.4.$$

Overall, a complete utilitarian ranking R^U can be computed by solving the following BLP problem:

$$Minimize: \frac{1}{t} \sum_{h=1}^{t} \Delta^{com}(h, U)$$

s.t.

$$\left.\begin{array}{l} [R'] \text{ for all } i, k = 1, 2, \ldots, n : i \neq k : \\ \quad [R1'] \; r_{ik} + r_{ki} \geq 1, \\ \quad [R2'] \; z_{ik} = r_{ik} + r_{ki} - 1, \\ \quad [R3'] \; r_{ik}, z_{ik} \in \{0, 1\}, \\ [T] \text{ for all } i, k, p = 1, 2, \ldots, n : i \neq k \neq p : \\ \quad r_{ik} \geq r_{ip} + r_{pk} - 1.5. \end{array}\right\} E^{utilitarian}_{complete}$$

In turn, when constructing a complete egalitarian ranking R^E, we minimize the maximal distance from any DM's stochastic ranking, i.e.:

$$Minimize: \alpha,$$

s.t.

$$\left.\begin{array}{l} E^{utilitarian}_{complete}, \\ \text{for } h = 1, \ldots, t : \\ \quad \alpha \geq \Delta^{com}(h, E). \end{array}\right\} E^{egalitarian}_{complete}$$

Table 2 Project proposals evaluated by the National Science Centre

Project	$g_1 \downarrow$	$g_2 \uparrow$	$g_3 \uparrow$	$g_4 \uparrow$	$g_5 \uparrow$
$P1$	210	700	4	2	0
$P2$	170	750	5	3	2
$P3$	220	850	8	1	1
$P4$	260	825	9	3	1
$P5$	205	775	6	4	2
$P6$	225	650	6	4	1

3 Illustrative Study

This section illustrates the use of the proposed approach by considering a decision problem concerning the ranking of six project proposals submitted to the National Science Center. The proposals denoted as $P1$–$P6$ are evaluated on the following five criteria:

- g_1 (cost)—the comprehensive cost of the project (in thousands PLN); to be minimized;
- g_2 (gain)—evaluation of the Principal Investigator (PI) (in points for the PI's scientific papers according to the list published by the Ministry of Science);
- g_3 (gain)—scientific merits of the project (on a 0–10 scale);
- g_4 (gain)—justification for forming a new team (on a 0–5 scale);
- g_5 (gain)—possibility of realizing the project within a designated time frame (on a 0–2 scale).

The performances of the six considered projects are given in Table 2. When analyzing the performances for all pairs of projects, one can note that $P2$ is more preferred to $P1$ on all criteria ($g_1(P1) > g_1(P2)$, $g_2(P1) < g_2(P2)$, $g_3(P1) < g_3(P2)$, $g_4(P1) < g_4(P2)$, $g_5(P1) < g_5(P2)$), whereas $P5$ is at least as good as $P6$ on all criteria, and strictly better on at least one criterion. As a result, $P2$, $P3$, $P4$, and $P5$ are strongly non-dominated, whereas $P6$ is additionally weakly non-dominated.

The objective of the study is to rank the projects from the best to the worst. In this perspective, the dominance relation is too poor to impose a sufficiently decisive order. Since the dominance leaves many projects incomparable, it should be enriched by eliciting preference information and building a preference model that would aggregate each alternative's performances on multiple criteria. Three DMs, denoted as DM_1, DM_2, and DM_3, are involved in the study. The comprehensive ranking should account for their individual preferences expressed in the form of a ranking of criteria, as required by the SRF procedure (Figueira and Roy 2002). These rankings are presented in Table 3.

The order of criteria provided by each DM is enriched with blank cards inserted in-between the subsets of criteria deemed equally important, as well as the ratio Z between the most and the least important criteria. The preferences provided by the three DMs differ with respect to the importance assigned to various criteria, as well as

Table 3 The order of criteria and inserted blank cards provided by DM_1 ($Z^1 = 10$), DM_2 ($Z^2 = 5$), and DM_3 ($Z^3 = 5$) in the SRF procedure ($l(i)$—rank (the higher, the better), e_s—the number of inserted blank cards)

DM_1			DM_2			DM_3		
g_i	$l(i)$	e_s	g_i	$l(i)$	e_s	g_i	$l(i)$	e_s
g_1	1		g_3	1		g_2, g_5	1	
		1			0			0
g_3	2		g_1	2		g_4	2	
		2			0			0
g_2	3		g_2	3		g_1	3	
		1			0			0
g_4, g_5	4		g_4	4		g_3	4	
					0			0
			g_5	5				

the intensity of preference. For example, g_5 was judged as one of the most important criteria by DM_1, and the least important by DM_3. Furthermore, DM_1 differentiated the intensity of preference by means of blank cards, while DM_2 and DM_3 opted for equal intensities between subsets of criteria assigned to different ranks. Finally, the ratio value between the weights of the extremely ranked criteria selected by DM_1 was twice as great as Z indicated by both DM_2 and DM_3.

To derive a final recommendation, we will implement an output-oriented aggregation procedure. That is, the preferences of each DM are analyzed individually, and the respective results are used as the basis for constructing a compromise ranking. For illustrative purposes, we will apply PROMETHEE I, admitting incomparability, and PROMETHEE II, imposing a complete order on the set of alternatives (Brans and Vincke 1985). Instead of deriving a precise weight vector, we will conduct a robustness analysis to investigate the stability of rankings observed for the weights compatible with the preferences expressed by each DM in the SRF procedure. We will also discuss the rankings obtained when implementing both utilitarian and egalitarian perspectives (Govindan et al. 2017).

3.1 *Partial Compromise Rankings*

Let us start with the analysis of results obtained for PROMETHEE I. Table 4 presents the stochastic acceptability indices for the preference and incomparability relations for all pairs of projects given 10, 000 weight vectors compatible with each DM's preferences and sampled with the Hit-And-Run algorithm (Tervonen et al. 2013). Note that the PLIs can be obtained from PWIs (e.g., $PWI(P6, P3) = 0.182$ implies $PLI(P3, P6) = 0.182$). Moreover, an indifference relation was not observed for

Table 4 Stochastic acceptability indices for all pairs of project proposals based on the partial rankings derived with PROMETHEE I for compatible weight vectors for DM_1, DM_2, and DM_3

PWIs							*PRIs*						
	$P1$	$P2$	$P3$	$P4$	$P5$	$P6$		$P1$	$P2$	$P3$	$P4$	$P5$	$P6$
DM_1													
$P1$	0.0	0.0	0.0	0.0	0.0	0.0	$P1$	0.0	0.0	0.0	0.0	0.0	0.0
$P2$	1.0	0.0	1.0	0.419	0.0	1.0	$P2$	0.0	0.0	0.0	0.489	0.0	0.0
$P3$	1.0	0.0	0.0	0.0	0.0	0.224	$P3$	0.0	0.0	0.0	0.0	0.0	0.594
$P4$	1.0	0.092	1.0	0.0	0.0	1.0	$P4$	0.0	0.489	0.0	0.0	0.0	0.00
$P5$	1.0	1.0	1.0	1.0	0.0	1.0	$P5$	0.0	0.0	0.0	0.0	0.0	0.0
$P6$	1.0	0.0	0.182	0.0	0.0	0.0	$P6$	0.0	0.0	0.594	0.0	0.0	0.0
DM_2													
$P1$	0.0	0.0	0.0	0.0	0.0	0.0	$P1$	0.0	0.0	0.0	0.0	0.0	0.0
$P2$	1.0	0.0	1.0	1.0	0.0	1.0	$P2$	0.0	0.0	0.0	0.0	0.0	0.00
$P3$	1.0	0.0	0.0	0.0	0.0	0.318	$P3$	0.0	0.0	0.0	0.099	0.0	0.413
$P4$	1.0	0.0	0.901	0.0	0.0	0.889	$P4$	0.0	0.0	0.099	0.0	0.0	0.053
$P5$	1.0	1.0	1.0	1.0	0.0	1.0	$P5$	0.0	0.0	0.0	0.0	0.0	0.0
$P6$	1.0	0.0	0.269	0.058	0.0	0.0	$P6$	0.0	0.0	0.413	0.053	0.0	0.0
DM_3													
$P1$	0.0	0.0	0.0	0.0	0.0	0.0	$P1$	0.0	0.0	0.0	0.0	0.0	0.0
$P2$	1.0	0.0	0.441	0.316	0.0	0.771	$P2$	0.0	0.0	0.069	0.06	0.0	0.229
$P3$	1.0	0.49	0.0	0.08	0.0	0.782	$P3$	0.0	0.069	0.0	0.134	0.11	0.218
$P4$	1.0	0.624	0.786	0.0	0.066	0.991	$P4$	0.0	0.06	0.134	0.0	0.179	0.009
$P5$	1.0	1.0	0.89	0.755	0.0	1.0	$P5$	0.0	0.0	0.11	0.179	0.0	0.0
$P6$	1.0	0.0	0.0	0.0	0.0	0.0	$P6$	0.0	0.229	0.218	0.009	0.0	0.0

any pair of projects for any feasible weight vectors, and hence $PII(P_i, P_k) = 0$ for $i, k = 1, \ldots, 6$, and $i \neq k$.

For some pairs of projects, the observed relations are robust. For example, since $P5$ dominates $P6$, $PWI(P5, P6) = 1$ for all DMs. However, such unanimity is also observed for pairs not related by the dominance (see, e.g., $(P5, P2)$ and $(P3, P1)$). For some other pairs, the observed relations are not definite. In particular, when analyzing the outcomes for DM_2, $P4$ is preferred to $P3$ for 90.1% of feasible weights, whereas they are incomparable for the remaining 9.9% of weights. An even greater diversity of relations is noted in the results obtained for DM_1 for $P2$ and $P4$ ($PWI(P2, P4) = 0.419$, $PWI(P4, P2) = 0.092$, and $PRI(P2, P4) = 0.489$).

When it comes to the comprehensive evaluation of projects, $P5$ proves its superiority over all remaining ones. In particular, for DM_1 and DM_2, it is preferred over the other five projects for all feasible weights, whereas for DM_3, the least PWI for $P5$ compared with other alternatives is 75.5%. On the other extreme, $P1$ is less preferred than all other projects for all feasible weights and all DMs. A fully robust ranking of projects cannot be observed for any DM. Specifically, there are 2, 3, and

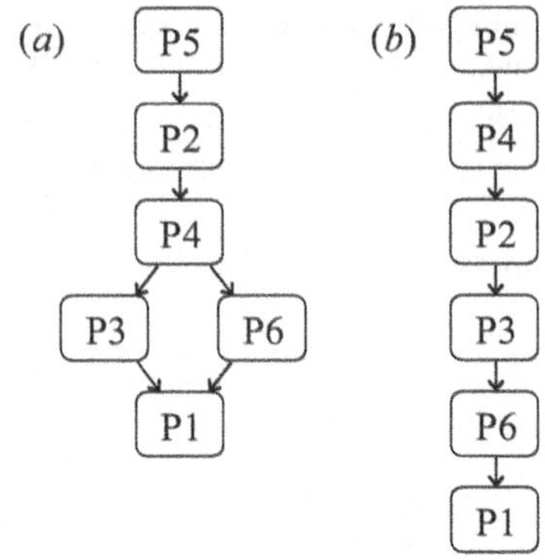

Fig. 1 Group compromise rankings admitting incomparability, built on the stochastic rankings derived with PROMETHEE I: **a**—utilitarian, **b**—egalitarian

8 pairs of alternatives for which the observed relation depends on the weight vector compatible with the preferences of, respectively, DM_1, DM_2, and DM_3. However, only for a single pair $(P3, P6)$, the observed relation is uncertain for all three DMs.

The group compromise rankings are presented in Fig. 1. When it comes to the utilitarian variant, the recommended ranking is incomplete. In particular, $P5$ and $P1$ are the top and bottom-ranked projects, respectively. Such an outcome is supported by the high superiority or inferiority of these projects against all other alternatives, reflected in the stochastic acceptability indices. However, $P3$ and $P6$ are incomparable. Again, such a relation is admitted in the outcomes observed for all DMs (the PRIs for $(P3, P6)$ are: for DM_1 – 0.594, for DM_2 – 0.413, and for DM_3 – 0.218). Overall, the compromise utilitarian ranking is close to the stochastic rankings observed for DM_1 and DM_2, while being more distant from the outcomes obtained for DM_3. The respective distances supporting this conclusion are: for DM_1 – 3.053, for DM_2 – 2.450, and for DM_3 – 9.756. Hence, to optimize an average distance, the method prioritized the preferences of two DMs, who were more consistent with each other, while slightly neglecting the preferences of DM_3.

Such negligence is prevented when constructing an egalitarian ranking (see Fig. 1). It differs with respect to the utilitarian one in terms of an inverse preference between $P4$ and $P2$ and imposing a preference of $P3$ over $P6$ instead of incomparability. Such relations can be explained in view of a high support given by DM_3 $(PWI^3(P4, P2) = 0.624$ and $PWI^3(P3, P6) = 0.782)$. To minimize the maximal distance from any DM's stochastic ranking, the method constructed the compromise ranking by combining the relations confirmed by large shares of weights compatible with preferences of different DMs. As a result, the respective distances between the stochastic rankings and the compromise one are: for DM_1 – 5.653, for DM_2 – 7.003, and for DM_3 – 6.832. Consequently, the maximal distance (7.003) is significantly lower than for the utilitarian ranking (9.756) at the cost of increasing an average distance (6.496 for the egalitarian ranking as compared to 5.086 for the utilitarian one).

Table 5 Stochastic acceptability indices for all pairs of project proposals based on the complete rankings derived with PROMETHEE II for compatible weight vectors for DM_1, DM_2, and DM_3

PWIs for DM_1							PWIs for DM_2						
	$P1$	$P2$	$P3$	$P4$	$P5$	$P6$		$P1$	$P2$	$P3$	$P4$	$P5$	$P6$
DM_1													
$P1$	0.0	0.0	0.0	0.0	0.0	0.0	$P1$	0.0	0.0	0.0	0.0	0.0	0.0
$P2$	1.0	0.0	1.0	0.712	0.0	1.0	$P2$	1.0	0.0	1.0	1.0	0.0	1.0
$P3$	1.0	0.0	0.0	0.0	0.0	0.547	$P3$	1.0	0.0	0.0	0.006	0.0	0.548
$P4$	1.0	0.288	1.0	0.0	0.0	1.0	$P4$	1.0	0.0	0.994	0.0	0.0	0.918
$P5$	1.0	1.0	1.0	1.0	0.0	1.0	$P5$	1.0	1.0	1.0	1.0	0.0	1.0
$P6$	1.0	0.0	0.453	0.0	0.0	0.0	$P6$	1.0	0.0	0.452	0.082	0.0	0.0
PWIs for DM_3							PIIs for DM_1, DM_2, and DM_3						
$P1$	0.0	0.0	0.0	0.0	0.0	0.0	$P1$	1.0	0.0	0.0	0.0	0.0	0.0
$P2$	1.0	0.0	0.475	0.342	0.0	0.988	$P2$	0.0	1.0	0.0	0.0	0.0	0.0
$P3$	1.0	0.525	0.0	0.129	0.003	0.939	$P3$	0.0	0.0	1.0	0.0	0.0	0.0
$P4$	1.0	0.658	0.871	0.0	0.152	1.0	$P4$	0.0	0.0	0.0	1.0	0.0	0.0
$P5$	1.0	1.0	0.997	0.848	0.0	1.0	$P5$	0.0	0.0	0.0	0.0	1.0	0.0
$P6$	1.0	0.012	0.061	0.0	0.0	0.0	$P6$	0.0	0.0	0.0	0.0	0.0	1.0

3.2 Complete Compromise Rankings

In this section, we focus on constructing complete rankings. Hence, PROMETHEE II is applied for working out a ranking for each feasible weight vector, and incomparability is not admitted in the compromise ranking. In Table 5, we present the stochastic acceptability indices for the preference and indifference relations for the three DMs. Since no indifference was observed for any compatible weight vector, the results for various DMs differ only in terms of the shares of weights for which some project proposals are preferred over another.

Some observations are consistent across the results of robustness analysis for all DMs. Specifically, $P5$ is preferred over the remaining projects for all or the vast majority of feasible weights. Similarly, $P1$ proved to be univocally less preferred than the remaining alternatives. The greatest disagreement in terms of indicating a more preferred project can be observed for ($P2$, $P4$) and ($P3$, $P6$). Specifically, when comparing $P2$ and $P4$, the shares of weights supporting the preference of the former are 71.2%, 100%, and 34.2% for, respectively, DM_1, DM_2, and DM_3. As far as $P3$ and $P6$ are concerned, $P3$ is ranked better for 54.7%, 54.8%, and 93.9% of compatible weight vectors. Overall, when considering the results obtained for the three DMs, the preference relation is not robust for 2, 3, and 7 pairs of projects.

The utilitarian and egalitarian compromise rankings are presented in Fig. 2. The utilitarian ranking imposes a strict preference relation on the set of alternatives. It reflects the pairwise relations which are supported for the majority of weights compatible with the preferences of DM_1 and DM_2 (see, e.g., ($P2$, $P3$) and ($P2$, $P4$)).

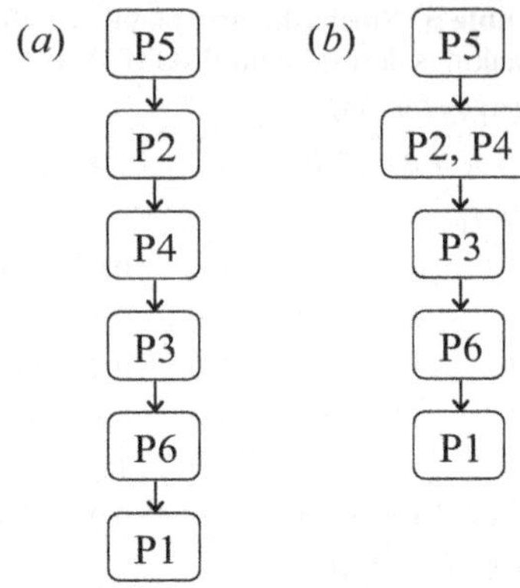

Fig. 2 Group compromise rankings built on the stochastic rankings derived with PROMETHEE II: **a**—utilitarian, **b**—egalitarian

In turn, some relations indicated as more robust by DM_3 are inverse. As a result, the distances from the stochastic rankings of the three DMs are as follows: for DM_1 – 2.964, for DM_2 – 2.16, and for DM_3 – 6.16. Thus, an average distance is equal to 3.761 and the maximal one to 6.16.

The egalitarian ranking differs from the utilitarian one only by replacing the preference of $P2$ over $P4$ with an indifference. Such a relation was found as the best compromise in view of conflicting indications of different DMs for this pair of alternatives (e.g., for $DM_1 - PWI(P2, P4) = 0.712$ and for $DM_3 - PWI(P4, P2) = 0.658$). The respective distances of the stochastic rankings from the egalitarian one are: for DM_1 – 3.812, for DM_2 – 4.16, and for DM_3 – 5.528. Hence, an average distance (4.5) is higher than for the utilitarian ranking, but the maximal distance (5.528) is lower.

4 Conclusions

In this chapter, we proposed a novel approach to group decision making. It is applicable in the context of multiple criteria ranking problems, which are frequently faced in such various fields as finance, management, engineering, economy, or education. We implement an output-oriented perspective for aiding a group of DMs. As a result, the proposed framework can be coupled with the existing ranking methods that deliver a recommendation for a single DM. We demonstrated how it could be used jointly with the robust variants of PROMETHEE I and II, but it can also be combined with numerous other methods from the family methods involving SMAA. These approaches incorporate the Monte Carlo simulations for conducting the robustness analysis given incomplete DM's preferences.

The originality of our method consists in processing stochastic rankings of the individual DMs. Such rankings are represented by matrices of the acceptability indices quantifying the support given to various pairwise relations in the space of compatible preference model instances. These matrices constitute an input for the aggregation procedure that is oriented toward constructing a group compromise ranking. In the utilitarian setting, one minimizes an average distance from the input rankings, whereas the egalitarian perspective minimizes the maximal distance. These objectives are optimized in the dedicated Binary Linear Programming models. The respective constraints define the admissible relations for each pair of alternatives, depending on whether we deal with partial or complete rankings, while ensuring at the same time the transitivity of a weak preference relation.

The applicability of the proposed approach was demonstrated in an illustrative case study concerning the ranking of project proposals by a research funding agency. We simulated the three DMs' incomplete preferences using a robust variant of the revised Simos procedure for eliciting the criteria weights. We discussed all possible combinations by considering jointly different types of compromise and rankings. When a utilitarian perspective is used, the method favors relations supported by most DMs, while possibly neglecting outlying DMs' preferences. In turn, in the egalitarian setting, the ranking is composed of relations supported by various DM so that each individual DM's viewpoint is maximally reflected in the constructed compromise ranking.

The proposed method can be incorporated into a negotiation process involving communication between the DMs (Kersten 1997) and become part of a negotiation support system (Kersten and Lai 2007). In such a setting, the method's outcome can be treated as compromise proposals discussed by the DMs, stimulating some reaction on their part (e.g., fixing relation for some pairs of alternatives in the final ranking). Then, the process of finding a compromise decision may require many iterations.

We envisage the following directions for future research. In this chapter, we analyzed the stability of results given pairwise relations. However, it is also possible to design analogous procedures that would build on the rank acceptability indices, indicating the share of feasible preference model instances for which an alternative attains a given rank. Furthermore, the proposed methods are based on Binary Linear Programming. For problems involving numerous alternatives, the number of binary variables would be prohibitively large. To deal with such problems effectively, one needs dedicated heuristic approaches for constructing a group compromise ranking.

Moreover, we accounted for two perspectives of a group compromise, utilitarian and egalitarian. It would be possible to integrate them within a single framework oriented toward finding a trade-off solution minimizing the average and maximal distances from the individual rankings at the same time. Finally, it would be useful to conduct an experimental verification of the results delivered by the proposed approaches on some real-world group decision problems. Such verification should focus on quantifying the perception of justice and satisfaction level of the involved DMs to check if they are correlated with the distances of their individual rankings from the constructed group compromise.

Acknowledgements Miłosz Kadziński and Dariusz Grynia acknowledge support from the Polish National Science Center under the SONATA BIS project (grant no. DEC-2019/34/E/HS4/00045). Grzegorz Miebs acknowledges support from the Polish Ministry of Science and Higher Education under the Diamond Grant project (grant no. DI2018 004348).

References

1. Angilella, S., Corrente, S., Greco, S.: SMAA-Choquet: stochastic multicriteria acceptability analysis for the Choquet integral. In: Greco, S., Bouchon-Meunier, B., Coletti, G., Fedrizzi, M., Matarazzo, B., Yager, R. (eds.) Advances in Computational Intelligence, pp. 248–257. Springer, Heidelberg (2012)
2. Brans, J., Vincke, P.: A preference ranking organisation method. Manage. Sci. **31**(6), 647–656 (1985)
3. Chen, T.Y.: An ELECTRE-based outranking method for multiple criteria group decision making using interval type-2 fuzzy sets. Inf. Sci. **263**, 1–21 (2014)
4. Chen, Y., Witmer, J., Hipel, K., Kilgour, D.: Strategic decision support for brownfield redevelopment. In: 2007 IEEE International Conference on Systems, Man and Cybernetics, pp. 1860–1865 (2007)
5. Cinelli, M., Kadziński, M., Gonzalez, M., Słowiński, R.: How to support the application of multiple criteria decision analysis? Let us start with a comprehensive taxonomy. Omega **96**, 102261 (2020)
6. Ciomek, K., Ferretti, V., Kadziński, M.: Predictive analytics and disused railways requalification: insights from a Post Factum Analysis perspective. Decis. Support Syst. **105**, 34–51 (2018)
7. Corrente, S., Figueira, J., Greco, S.: The SMAA-PROMETHEE method. Eur. J. Oper. Res. **239**(2), 514–522 (2014)
8. Corrente, S., Figueira, J., Greco, S., Słowiński, R.: A robust ranking method extending ELECTRE III to hierarchy of interacting criteria, imprecise weights and stochastic analysis. Omega **73**, 1–17 (2017)
9. Dias, L., Climaco, J.: Dealing with imprecise information in group multicriteria decisions: a methodology and a GDSS architecture. Eur. J. Oper. Res. **160**(2), 291–307 (2005)
10. Figueira, J., Roy, B.: Determining the weights of criteria in the ELECTRE type methods with a revised Simos' procedure. Eur. J. Oper. Res. **139**(2), 317–326 (2002)
11. Govindan, K., Kadziński, M., Sivakumar, R.: Application of a novel PROMETHEE-based method for construction of a group compromise ranking to prioritization of green suppliers in food supply chain. Omega **71**, 129–145 (2017)
12. Greco, S., Ehrgott, M., Figueira, J.: Multiple Criteria Decision Analysis - State of the Art Surveys. International Series in Operations Research & Management Science. Springer, New York (2016)
13. Greco, S., Kadziński, M., Mousseau, V., Słowiński, R.: Robust ordinal regression for multiple criteria group decision: UTA-GMS-GROUP and UTADIS-GMS-GROUP. Decis. Support Syst. **52**(3), 549–561 (2012)
14. Greco, S., Mousseau, V., Słowiński, R.: Ordinal regression revisited: multiple criteria ranking using a set of additive value functions. Eur. J. Oper. Res. **191**(2), 416–436 (2008)
15. Hwang, C., Yoon, K.: Multiple Attribute Decision Making: Methods and Applications - A State-of-the-Art Survey. Lecture Notes in Economics and Mathematical Systems, vol. 186. Springer (1981)
16. Jelassi, T., Kersten, G., Zionts, S.: An introduction to group decision and negotiation support. In: Bana e Costa, C.A. (ed.) Readings in Multiple Criteria Decision Aid, pp. 537–568. Springer, Heidelberg (1990)
17. Kadziński, M., Greco, S., Słowiński, R.: RUTA: a framework for assessing and selecting additive value functions on the basis of rank related requirements. Omega **41**(4), 735–751 (2013)
18. Kadziński, M., Greco, S., Słowiński, R.: Extreme ranking analysis in robust ordinal regression. Omega **40**(4), 488–501 (2012)
19. Kadziński, M., Rocchi, L., Miebs, G., Grohmann, D., Menconi, M., Paolotti, L.: Multiple criteria assessment of insulating materials with a group decision framework incorporating outranking preference model and characteristic class profiles. Group Decis. Negot. **27**(1), 33–59 (2018)

20. Kadziński, M., Tervonen, T.: Robust multi-criteria ranking with additive value models and holistic pair-wise preference statements. Eur. J. Oper. Res. **228**(1), 169–180 (2013)
21. Keeney, R., Raiffa, H.: Decisions with Multiple Objectives: Preferences and Value Trade-Offs. Cambridge University Press (1993)
22. Kersten, G.E.: Support for group decisions and negotiations an overview. In: Clímaco, J. (ed.) Multicriteria Analysis, pp. 332–346. Springer, Heidelberg (1997)
23. Kersten, G.E., Lai, H.: Negotiation support and e-negotiation systems: an overview. Group Decis. Negot. **16**(6), 553–586 (2007)
24. Kilgour, D., Chen, Y., Hipel, K.: Multiple Criteria Approaches to Group Decision and Negotiation, pp. 317–338. Springer, Boston (2010)
25. Lahdelma, R., Salminen, P.: SMAA-2: stochastic multicriteria acceptability analysis for group decision making. Oper. Res. **49**(3), 444–454 (2001)
26. Leyva-Lopez, J., Fernandez-Gonzalez, E.: A new method for group decision support based on ELECTRE III methodology. Eur. J. Oper. Res. **148**(1), 14–27 (2003)
27. Mareschal, B., Brans, J., Macharis, C.: The GDSS PROMETHEE procedure: a PROMETHEE-GAIA based procedure for group decision support. J. Decis. Syst. **7**, 283–307 (1998)
28. Miebs, G., Kadziński, M.: Heuristic algorithms for aggregation of incomplete rankings in multiple criteria group decision making. Inf. Sci. **560**, 107–136 (2021)
29. Norese, M.: ELECTRE III as a support for participatory decision-making on the localisation of waste-treatment plants. Land Use Policy **23**(1), 76–85 (2006)
30. Roy, B.: ELECTRE III?: Un algorithme de classements fondé sur une représentation floue des préférences en présence de critères multiples. Cahiers du CERO **20**(1), 3–24 (1978)
31. Roy, B., Słowiński, R.: Criterion of distance between technical programming and socio-economic priority. RAIRO Recherche Opérationnelle **27**(1), 45–60 (1993)
32. Roy, B.: The outranking approach and the foundations of ELECTRE methods. Theor. Decis. **31**(1), 49–73 (1991)
33. Salminen, P., Hokkanen, J., Lahdelma, R.: Comparing multicriteria methods in the context of environmental problems. Eur. J. Oper. Res. **104**(3), 485–496 (1998)
34. Salo, A., Hamalainen, R.: Preference programming through approximate ratio comparisons. Eur. J. Oper. Res. **82**(3), 458–475 (1995)
35. Tervonen, T., Figueira, J.R., Lahdelma, R., Salminen, P.: SMAA-III: a simulation-based approach for sensitivity analysis of ELECTRE III. In: Linkov, I., Ferguson, E., Magar, V.S. (eds.) Real-Time and Deliberative Decision Making, pp. 241–253. Springer, Dordrecht (2008)
36. Tervonen, T., van Valkenhoef, G., Basturk, N., Postmus, D.: Hit-And-Run enables efficient weight generation for simulation-based multiple criteria decision analysis. Eur. J. Oper. Res. **224**(3), 552–559 (2013)

Quantitative Measures for Recognition of Negotiation Style and Activity

Bilyana Martinovski

Abstract Group Decision and Negotiation Support Systems need to provide support for different types of communication activities. The purpose of this study is to identify and test cross-linguistic measures for recognition of communication activities in order to provide support for various forms of group decision and negotiation support systems, including e-negotiation support systems. The study uses quantitative cross-activity and cross-linguistic analysis in order to test if complex linguistic measures, such as caution, liveliness, and stereotypicality, can be used to recognize spoken language negotiation activity and style. The data indicate that automated linguistic analysis is a useful tool for recognition of activity type in spontaneous speech and that the activity factor influences the interaction in a more profound way than the language factor or the national culture factor.

Keywords Negotiation · Activity · Quantitative measure · Speech · Corpus linguistics

1 Introduction

Negotiation support systems, especially negotiation software agents and negotiation agent-assistance, involve communication analysis tools and provide "intelligent and independent advice, critique and support to one or more negotiation parties" (Kersten and Lai 2008) across cultures and languages (Kersten 2005). Group decision taking and negotiations embed and can be embedded in many different types of activities with various types of formality and style. Linguistic features, such as communicative acts, modality, intonation, reformulations, etc., influence the interaction (Martinovski 2000) but they can also be a product of factors such as activity, culture, and language pattern. The purpose of this study is to develop qualitative methods for quantitative recognition of activity type, which can be used to improve Negotiation Support Systems and to establish which one of these three factors has strongest impact on the interaction and that way improve negotiation support and e-negotiation systems

B. Martinovski (✉)
Department of Cogntive Semiotics, University of Lund, 221 00 Lund, Sweden

T. Szapiro and J. Kacprzyk (eds.), *Collective Decisions: Theory, Algorithms And Decision Support Systems*, Studies in Systems, Decision and Control 392,
https://doi.org/10.1007/978-3-030-84997-9_5

(Kersten and Lai 2008; Kersten 2005). Three complex stylistic measures are designed and applied to a linguistic corpus and used in:

- Cross-activity comparison
- Cross-linguistic comparison.

The study is based on Göteborg Spoken Language Corpus (GSLC), which contains 22 activity types in Swedish with a total of 300 recordings and 1,221,264 word tokens. The data sets are too small for strong generalizations but suitable to test the qualitative measures and comparisons. The analysis will proceed in the following way:

- Cross-activity comparison of all the collected activities in Swedish.
- Cross-activity comparison of court-of-law negotiations and formal over-dinner negotiations in Swedish.
- Cross-linguistic comparison of court-of-law negotiations in Swedish and in Bulgarian, according to selected criteria.

The criteria of comparison in the list below are chosen because they can be counted automatically:

- Word length
- Utterance length
- Word types/tokens
- Overlap
- Emphasis
- Pauses
- Parts of speech
- Repetitions and reformulations.

2 Cross-activity Comparison of the Swedish Data

Two types of complex measures have been applied to the whole corpus, namely measures of liveliness and stereotypicality. Let us observe how court-of-law negotiations are related to these measures (Allwood and Hagman 1994).

Interactive Liveliness

The measure of interactive liveliness is created by comparison of simple measures such as stress (StressPTok—the percent of stressed words as tokens in relation to the total amount of words as tokens), overlap (ToOVPtok—percent of overlapped words as tokens in relation to the total amount of words as tokens), mean length utterance (MLU—the mean of the number of tokens in an utterance) and pauses (PausPTok—the percent of pauses as tokens in relation to the total amount of words as tokens). The applied formula is:

$$\frac{((2.\text{ToOVPtok}) + \text{OvePutt} + (2.\text{StressPTok}) - (2.\text{PausPTok}) - \text{MLU})}{\text{Total number of tokens}}$$

Table 1 Cross-activity comparison of stereotypicality and liveliness measures applied to Swedish data

Complex measure activity	Average number of speakers	Liveliness	Stereotypicality
All	9	30.1	25.178/0.0
Sermon	5	−82.6	18.5/0.3
Auction	7	−70.3	35.7/0.02
Courts	6	9.4	25.4/0.0
Task-oriented	3	11.3	28.15/0.025
Formal meeting	11	33.9	27.7/0.0
Formal dinner	9	91.2	26.6/0.0

Stereotypicality

This measure reflects the syntactic and lexical uniformity of utterances based on types of collocations. It reflects the repetitiveness of combinations of words starting from one word and ending with ten words in a collocation. All activities tend to be more stereotypical, looking at shorter collocations, and become less stereotypical with the increase of the number of words in a collocation. A critical point is the collocation consisting of five words because at that point the most dramatic measure changes ‘happen’ in most of the 22 activities. Those, which do not have zero value for example, for ten levels of collocations, are the most stereotypical ones. In the last column, I give only the values that correspond to the repetitiveness of one word (in front of the slash) and of ten-word collocations (behind the slash) Table 1.

I have selected six activity types out of 22. The corpus includes court-of-law trial negotiation data only in Bulgarian and Swedish, I therefore chose these languages for comparison. The Bulgarian data covered only trial interactions. The Swedish data covered a wider range of activities for comparison. The sermon is one of the least lively of all because it is the most ‘monologous’. The auction, another form of negotiation, which consists mainly of offers and confirmations, is also not very lively. The task-oriented group decision and negotiation activity is not very lively but more stereotypical than the sermon. In fact, the auction appears to be the most lexically and syntactically stereotypical activity of all, with a distribution of uniform collocations with respect to five-word collocations (but even on the ten-word collocations). The court examinations, including the reading of charges and final pleading, are neither extremely stereotypical nor very lively. The formal negotiation, meeting and the dinner negotiation are more stereotypical than the court data and less so than the auction, but the dinner conversations are far livelier than both the court examinations and the formal meeting. The formal meeting is similar to the talk-over-dinner activity with respect to stereotypical use of syntactic constructions but the latter is nevertheless much livelier. Thus we may say that if we imagine the activities as a continuum on a scale of liveliness and stereotypicality, the auction represents one pole and the dinner the other pole (see Fig. 1 below). The court talk style appears to be in the middle of this scale with a slight tendency towards the auction.

Can these measures indicate something about the functions of the utterances, the predominant type of communicative acts in an activity? We may rather speculatively observe that question-answer fixed activities (here I assume that auctions are more question-answer oriented than dinner talks) are less lively but that this characteristic seems to be less relevant when it comes to their stereotypicality. However, the correspondence between the function of utterances and the overall impression of the style of the activity has yet to be studied.

Since dinner talk is an example of comparatively casual conversation, which is expected to be different from formal situations (Atkinson and Drew 1979; Ilie 1999), and since it is the most lively activity in the corpus we may compare it with examinations. In this way we may get some general characteristics of the court interaction style.

As expressed by the complex measures above, the formal dinner talk contains more than eight times more overlapped tokens than the court data but has much fewer pauses and stressed words. In the Table 2 below I have given information about both types of recordings, including the percent values of some of the simple measures.

There is a difference between the average amount of tokens in an utterance (assuming that there is no great difference in talk duration):

Court: 14.8 tokens in utterance.
Dinner: 9.4 tokens in utterance.

The table indicates that:

- Court examinations have more words in an utterance, more pauses, more stressed words, and less overlap. The combination of all these features may be interpreted as an indication of a more regulated interactive behavior.

Table 2 Number of tokens, utterances, overlaps, pauses, and stressed tokens in the Swedish dinner and court data

Activity	Token	Type	Utterance	Turn	OvePUtt a.	TovPTok b.	StrPTok c.	PauPTok d.
Dinner	30,738	3971	3251	2365	1998 (84.5%)	7503 (24.4%)	142 (0.5%)	1637 (5.3%)
Court	33,409	3667	2220	2045	637 (28.7)	1032 (3.1%)	1127 (3.8%)	2827 (8.5%)

Note
a. OvePUtt—percent of overlapped words as tokens in relation to the total number of utterances in a group
b. TokPTok—percent of overlapped words as tokens in relation to the total amount of words as tokens in a group
c. StrPTok—percent of stressed words as tokens in relation to the total amount of words as tokens in a group
d. PauPTok—percent of pauses in relation to the total amount of words as tokens in a group

Table 3 The length of words as tokens in the Swedish court and dinner data and percent of words with N number of letters in relation to the total number of tokens in a group

Letters in word	1 and 2 letters	3 letters	4 letters	5 letters	6 letters	7 letters	8 letters	9 letters	10 letters
Dinner	13 118 (42.7%)	7024 (22.9%)	3248 (10.6%)	2812 (9.2%)	1975 (6.4)	873 (2.8%)	569 (1.9%)	398 (2.3%)	721 (2.3%)
Court	11,499 (34.4%)	8050 (24.1%)	4081 (12.2%)	3093 (9.3%)	2228 (6.7%)	1318 (4.0%)	1054 (3.2%)	669 (2.0%)	1417 (4.2%)

- Dinner talk has fewer words in an utterance, fewer pauses, fewer stressed words, and more overlap, and thus seems to be interactively more dynamic.

The lower rate of overlap is due to the prefixed question–answer turn order of the examinations and the general very formal nature of turn-taking in courtrooms, but this is not a sufficient explanation for the rest of the characteristics. Let us observe the length of the words. In the next Table 3 we can compare the real numbers of letters in a word in both activities starting from three letters in a word.

In both activities, the most numerous are the short one to two letter words, which include pronouns, feedback expressions, conjunctions, own communication management expressions (OCM), and prepositions. The result above is also a consequence of the great number of pronouns (see Table 8 below). It indicates a language-specific feature, namely the fact that, in Swedish, person is expressed syntactically and that it is a pro-form language. The dinner talk seems to contain fewer instances of the above parts of speech, which are also more characteristic of informal speech. There is a smaller difference between the activities in terms of the amount of three-letter long words but ten-letter long words are more frequent in court examinations, i.e. the difference between the activities grows with the increase in the length of the words.

- Courtroom negotiations contain longer words, which indicates a more complicated lexicon than in dinner talk negotiations.

The professional speakers' use of legalese may explain this observation. The same progression occurs in terms of tokens in collocations. The following Table 4 starts with collocations consisting of three words and ends with ten-word collocations.

This difference contributes to the stereotypicality of the court examinations. However the court data contain an average of 63 word types in 100 words whereas

Table 4 Number of collocations in the Swedish dinner and court data according to the number of words (as tokens) in a collocation and percent of N number of words in collocation in relation to total number of word-tokens

Collocation	3	4	5	6	7	8	9	10
Dinner	4147 (13.5%)	985 (3.2%)	276 (0.9%)	127 (0.4%)	72 (0.2%)	50 (0.16%)	36 (0.1%)	26 (0.08%)
Court	5579 (17.0%)	1516 (4.5%)	467 (1.4%)	199 (0.6%)	118 (0.6%)	80 (0.2%)	63 (0.18%)	50 (0.15%)

the dinner talk contains an average of 22 word types in 100 words. Thus we may conclude that.

- Courtroom negotiations tend to exhibit more lexical variety than formal negotiations over dinner.

Turning to the use of emphasis (see Table 2 above), its use may be interpreted as indication of greater involvement and/or pathos. Court data include 50,2 stressed words in 100 words whereas the dinner talk consists of 4,3 stressed words per 100 words. Even if this may be due to the transcribers' attention the difference is very big. It indicates that despite the fact that court examinations were rated as less lively than dinner talk they nevertheless may have high level of involvement or pathos. The extreme use of emphasis may be interpreted also as an indication of a stronger intention to influence the beliefs of the co-interactants (Martinovski 2000) and/or as persuasion device and/or as clearer instruction device.

If we assume that overlap (see Table 2 above) is a potential measure of dynamics of communication and of distribution of power one may observe that it is not typical for Swedish court examinations, as mentioned earlier. However, the lower value of the overlaps indicates the opposite, i.e. lower level of interactive involvement.

The position of the overlap is significant. Final overlaps may indicate a tendency to be interrupted as well as feedback-giving initiated by the following utterance. Initial overlap (i.e. overlap of two utterance-initial units) stands for simultaneous starts and indicates that the person speaking may be interrupting. Overlap in the middle of the utterances may be associated more with backchannels and assessments, which are normally non-competitive. Completely overlapped utterances (for instance, cases in which the second utterance comes in the middle of the first one) are generally more common and typical of backchannels. Completely overlapping utterance (i.e. when all speakers speak simultaneously and for equally long) are generally more rare. In such situations, the speakers may have fewer chances to hear each other and the utterances can hardly be associated with feedback. In the next Table 5 we can see the distribution of overlap according to position in the utterance.

In both activities, the complete overlaps are more than all other types of overlap, which means that in both activities there is intensive feedback-giving, which is more probable than in the competitive equal simultaneous speech with regard as well to the great amount of feedback-giving in Swedish courts (Martinovski 2000). This result is less expected of the court examinations with their pre-fixed turn order than of the casual talk-over-dinner. There is, however, a significant difference between the percentage of middle and final overlaps. If we associate middle overlaps with

Table 5 Distribution of overlap according to position in an utterance in Swedish court and dinner data

Overlap	Initial	Middle	Final	Complete
Dinner	446/22.3	424/21.2	548/27.4	580/29.0
Court	152/23.9	163/25.6	139/21.8	183/28.7

Table 6 Verbal equality applied to Swedish data

Activity	Verbal Equality
	Tokens (words)
Court	70.3
Dinner	59.1
All activities	57.5

backchannels and assessments, we can see that there are more of those in the courts than at dinner. On the contrary, in courts, there are less final overlaps, which may indicate a weaker tendency for interruptions in this activity than in dinner talk. This combination of characteristics may be explained by the asymmetrical social status of the speakers in courtrooms and with face-work. It also suggests that Swedish courtroom negotiations as kinds of very formal, turn-regulated situations tend to result in more co-operative interaction and less combative or competitive styles (i.e. styles in which the parties strive to increase their interactional power). In other words, there seems to be a tendency that the more formal the situation is the more co-operative and less competitive for interactional power the Swedish speech communication is.

According to the general complex measure of verbal equality (Allwood and Hagman 1994), which uses only the amount of words and no other features, the courtroom speech is more equally distributed among the speakers than is the case in the dinner talk data, as indicated in the next Table 6:

In comparison with the mean of the measure of all the 22 activities in GSLC, the court data exhibit a higher degree of verbal equality. This result indicates once again that the socially asymmetric situation of examinations in the Swedish juridical institution manages to distribute the interactive power in a rather equal way. In fact, the court examinations are called "courtroom negotiations" in Swedish, and the appropriateness of this term is confirmed by the above measures since they describe an activity that is more cooperative than combative. The actual interaction is symmetric while the social and juridical roles of the participants are not. Thus the concrete interaction may influence the social institution in such a way that it becomes less asymmetric than it is defined to be. In other words, one may notice a manifestation of the dynamic character of social interaction where the concrete practice of an activity gradually changes the activity or where the participants actively negotiate the definition of the activity as such.

One may suppose that the higher number of hesitations point to a greater care in the formulation of utterances (Martinovski 2000). The same is indicated by the higher number of pauses in courts. Both features point to a more careful and cautious style of interaction. Since feedback-giving elicitations are some of the main spoken language devices for joint realization of communicative projects in which there is need of constant checking of what is perceived and/or understood as well as of demand for attention (especially in matters of great importance, such as testimonies), the greater amount of feedback expressions may be an indication of stronger need of such co-ordination and thus be an indication of interactive caution. Use of numerals such as exact references to time or amount and similar as well as personal names instead of pronouns can also be an indication of caution. Indeed, we can create a complex measure of how cautious a conversation (in a broad sense) is. The more OCM expressions, pauses, stressed words, expressions of precision, and feedback expressions there are, the more cautious, careful and precise the style of talking is. The more overlap and pronouns (instead of personal names) there are, the least cautious is the interaction style. It would have been good also to check expressions that are not digitally exact, such as adjectives and adverbs, but the distinction between exact and not exact adjectives and adverbs is dependant to a large extent on the context and requires a special study of its own. So, the formula is:

$$\frac{(2.\text{Pauses}) + \text{Stress} + (2.\text{OCM}) + (2.\text{FB}) + \text{Numerals} - (2.\text{Pronouns}) - (2.\text{Overlap})}{\text{total number of tokens}}$$

Thus we get the following result:

Caution
Dinner: −0.7
Court: −0.1

Court examinations appear to be more cautious than dinner talk and by implication may be interpreted as a more precision-oriented type of verbal activity. The caution is related to the commissiveness of the activity. Thus the caution and the precision in references (the large number of numerals and personal names) is a reflection of this activity's purpose, namely objective decision making based on justice and fairness.

3 Cross-linguistic Comparison

The Bulgarian corpus is smaller than the Swedish one (see Table 7 below), which precludes a direct comparison of simple values. We have to keep in mind that the mean values give only an approximate picture of the relations studied. The measures used

Table 7 Vocabulary richness of Bulgarian and Swedish court data

Group	Tokens	Utterances	TokRC a.	UttRC b.	MLU c.	Vocab d.	Types
Bulgarian	10,533	1115	23.9%	33.4%	9.4	2276.6	2345
Swedish	33,409	2220	76.1%	66.6%	15.1	1920.9	3733

Note
a. TokRC—percent of tokens in respective group in relation to total amount of tokens in both groups
b. UttRC—percent of utterances in respective group in relation to total amount of tokens in both groups
c. MLU—average number of words per utterance
d. Vocab—"Theoretical vocabulary" measure.

Table 8 Cross-linguistic and cross-activity comparison of average amount of parts of speech in court and dinner data (FB-Feedback, OCM-Own Communication Management (Allwood and Hagman 1994))

Part of speech	tokens	adj	adv	pron	Verb	FB	int	conj	prep	noun	num	ocm
Swedish dinner	30,738	3,8	17,0	23,1	20,8	5,2	0,3	7,3	6,1	11,8	0,5	1,3
Swedish court	33,409	3,1	15,3	21,5	19,3	4,8	0,0	8,6	8,6	14,5	1,5	2,07
Bulgarian court	10,533	2,2	4,7	11,9	30,1	2,8	0,0	10,0	10,1	23,0	3,3	1,7
Bulgarian casual	100,000	3,8	8,8	18,3	22,8	7,5	1,3	10,4	8,0	15,8	3,2	1,3

here are produced by a program called TRASA (Grönqvist and Sofkova-Hashemi 1998).

To evaluate the richness of vocabulary one can measure (with Vocab) the number of word types where the set of words is chosen randomly. One may also use the average of the number of tokens and the percent of utterances over the whole corpus as well as the average number of words per utterance, which are related to the richness of the vocabulary.

- The average Swedish utterance has more words than the average Bulgarian one.

The value of the MLU above shows that the Swedish speakers' utterances are longer than the Bulgarian speakers'. However, also despite the fact that the Bulgarian corpus is only 23.9% of the total Swedish-Bulgarian corpus measure of word tokens and only 33.4% of the total measure of utterances, its vocabulary is a bit richer. Thus the wordiness of the utterances does not contribute to the richness of the vocabulary.

The frequency of parts of speech is expected to be influenced by the specifics of the respective language and therefore not all comparisons between the language groups are informative as concerns those aspects of the activity culture that are not created by language differences. As indicated in the table below, Bulgarian court data contain fewer pronouns but this is mainly due to grammatical conditions, namely, noun and verb suffixes include person deixis and lower the amount of pronouns in subject position. We can check the influence of the activity factor by using Nikolova's casual speech frequency dictionary (Nikolova 1987) (the last row in Table 8 below). The corpus of conversational speech she used consists of 100 000 words, which is about ten times more than the Bulgarian court corpus. The Bulgarian court corpus is coded according to parts of speech manually by a native speaker and checked by another native speaker following only partially the definition criteria used in Nikolova's dictionary. Feedback-giving and elicitation expressions were counted as particles together with some other grammatical categories, which are not typical feedback expressions. I have counted only the feedback expressions in this group as a special part of speech. To my knowledge, such a part of speech is not an accepted category in Bulgarian linguistics.

By observing the activity and the language-dependent percentage we may notice that the activity is sometimes a stronger influencing factor and sometimes the stronger influence is the language. Thus adjectives, adverbs, pronouns, feedback, and interjections decrease in the court data in both language environments and the rest of the parts of speech increase, except for the verbs and the conjunctions in Bulgarian. Feedback-giving and elicitation is more frequent in the Bulgarian casual talk than in the Swedish informal talk but it decreases much more in the Bulgarian court environment than in the Swedish court talk. This is related to the dictation activity and to the form of the procedure where the judge controls the floor. It seems, however, that this result is also influenced by the purist attitude that feedback words, such as 'be', 'de', 'a' (used only as feedback elicitor, similar to the Swedish 'va', "what did you say" and also "don't you think so"), which are the most frequent words in the everyday speech [Nikolova 1987: 24], are parasitic expressions and are part of so called 'street language' ("ulichen ezik"), i.e. indicate a lower status speaker.

The percent of pauses and stress per token in Table 9 below indicates that the Swedish court transcriptions tend to have more pauses than the Bulgarian, but the latter tend to have more stressed words. This greater number of pauses may be explained partly by the longer utterances (the MLU value, see Tables 4 and 7 above) in the Swedish court data.

Furthermore, the percent of overlapping tokens in the Bulgarian court examination data is four times greater than in the Swedish data. Although the Bulgarian utterances are shorter, an average of 50.6% of them consist of overlap. It is normally expected that the overlaps in middle position are more frequent since it could be anything between the first and the last token. This is true for the Swedish type of overlap, which in comparison to the Bulgarian is either medial or complete. This is illustrated in the following Table 10.

However, the final and the initial overlap is more typical of the Bulgarian examinations, which in two-party conversations means that there are more starts before the previous speaker has finished an utterance and thus a greater number of opportunities for interruption/fight for the floor and/or eagerness in feedback-giving. This result may also be due to the fact that Bulgarian utterances are on average shorter.

Table 9 Pause and stress in relation to total amount of word-tokens in the Bulgarian and the Swedish court data

Group	PauPTok a.	StrPTok b.
Bulgarian	3.9%	6.2%
Swedish	8.4%	4.0%

Note

a. PauPTok—percent of pauses in relation to the total amount of words as tokens in a group

b. StrPTok—percent of stressed words as tokens in relation to the total amount of words as tokens in a group

Table 10 Overlap according to psoition in the Bulgarian and the Swedish court data

Group	OmedPO a.	OiniPO b.	OfinPO c.	OcomPO d.	TovPTok e.	OvePUtt f.
Bulgarian	18.7	30.7	30.2	20.1	13.4	50.6
Swedish	26.0	23.6	21.6	28.6	3.1	28.3

Note
a. OmedPO—percent of overlap in middle position in relation to total number of overlap per language group
b. OiniPO—percent of overlap in initial position in relation to total number of overlap per language group
c. OfinPO—percent of overlap in final position in relation to total number of overlap per language group
d. OcomPO—percent of complete overlap in relation to total number of overlap per language group
e. TovPTok—percent of overlapped words as tokens in relation to the total amount of words as tokens in a group
f. OvePUtt—percent of overlapped words as tokens in relation to the total number of utterances in a group

- Since Bulgarian court negotiation talk has more stress, overlap, shorter utterances and fewer pauses, it is more dynamic than the Swedish court talk.

The material from a Bulgarian over-tea discussion suggests that the high frequency of overlap may be a specific feature of the court setting and not of the interactive culture as a whole since the mean of overlap over tokens in the discussion is even lower than the overlap in the Swedish courts (Table 11).

The above measures also suggest that the interactive repetition/reformulation is a more preferred form of feedback in courts in comparison to informal discussions.

Table 11 Feedback, repetition/reformulation, and overlap in Bulgarian discussion and court examinations and in Swedish dinners and court examinations

Feature activity	Utterances	Tokens	Feedback %of words a.	Repetition-reformulation % of utterances b.	TovPTok in % c.
Bulgarian court	1115	10,533	4.8	16.4	13.4
Bulgarian discussion	470	5966	6.3	9.3	2.5
Swedish court	2220	33,409	4.8	6.5	3.1
Swedish dinner	3251	30,738	5.2	-	6.5

Note
a. Percent of feedback expression-token in relation to total amount of tokens
b. Percent of repetition/reformulation instances in relation to total amount of utterances
c. TovPTok—percent of overlapped words as tokens in relation to the total amount of words as tokens in a group

Whereas most of the non-repetitive feedback expressions during the discussion are positive, more than half of the repetitions/reformulations were part of disagreements or negative feedback. Thus we may conclude that the Bulgarian discussion exhibits a stronger need of positive feedback turn-taking co-ordination than the court examination. We have no data about the frequency of repetitions/reformulations in the Swedish dinner and thus cannot conclude more strongly that interactive repetition/reformulation is more common in Bulgarian than in Swedish discourse, as is suggested in the fifth column above. Similar to the Bulgarian discussion data, the Swedish dinner talk contains more feedback (excluding repetitions/reformulations) than the Swedish or Bulgarian examinations. The opposite seems to be true for overlap: speakers overlap more in Bulgarian courts than in friendly discussions, which are informal interactive occasions, whereas they overlap less in Swedish courts than at formal dinner. However, since our data are not completely comparable, it must suffice here simply to note these observations as tendencies that need more elaborate study.

Liveliness, Stereotypicality, Caution

According to the measure of liveliness, the Bulgarian court talk is described as more similar to a Swedish dinner than to a Swedish court examination mainly because it contains more words, overlap and stressed words and fewer pauses. In the last column I give only the values that correspond to the repetitiveness of one word (in front of the slash) and of ten-word collocations (behind the slash) (see also Table 1 above and its comments) (Table 12).

However, the Bulgarian courts are far more stereotypical (see also Table 1) than the Swedish dinner or court talk and, in this respect, they are similar to auctions, as illustrated in Fig. 1 below. Indeed, it is the routine and dictation repetitions/reformulations whose function may also be viewed as a precaution against misunderstanding that contribute to this.

It is thus not surprising that, despite the fact that the caution measure introduced earlier does not consider the amount of repetitions/reformulations nor the types of speech acts, the Bulgarian court talk is described as more cautious than the Swedish court and dinner talk (Table 13):

However, I bring the topic up for methodological reasons, namely that one and same measure may not be valid or meaningful if applied cross-linguistically. The caution formula uses feedback and pronouns, which are comparatively underrepresented in the Bulgarian court data (see also Tables 3 and 8 above) and influence

Table 12 Cross-linguistic and cross-activity comparison according to liveliness and stereotypicality

Complex measure	Liveliness	Stereotypicality
Bulgarian courts	71.2	37.7/2.1
Swedish auction	−70.3	35.7/0.02
Swedish courts	9.4	25.4/0.0
Swedish dinner	91.2	26.6/0.0

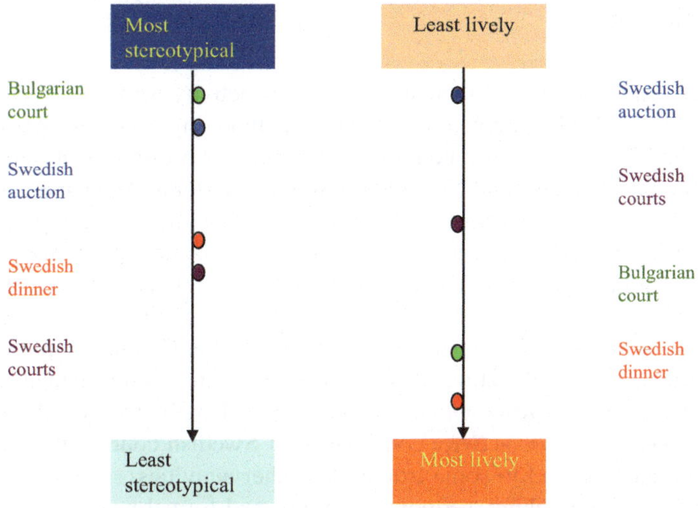

Fig. 1 Cross-activity and cross-linguistic scale based on liveliness and stereotypicality measures

Table 13 Cross-linguistic and cross-activity comparison according to caution

Activity/language	Caution
Bulgarian court	−0.02
Swedish court	−0.1
Swedish dinner	−0.7

the final measure. Use of feedback contributes to caution and use of pronouns indicates less caution, according to the formula. Pronouns are used less since Bulgarian expresses person morphologically and because it is a pro-drop language. Feedback is used comparatively less in Bulgarian, independent of activity. These contribute to the lower result of the measure for Bulgarian courts in Table 13, so it is not only the activity but also the language factor that affects this measure. Thus, the only way to test to what extent the language factor outscores the influence of the activity with regard to this measure is to add data from other types of Bulgarian activities. In the case that non-formal Bulgarian activities have a lower caution score than in courts, one may conclude that the activity is indeed an influencing factor independent of language.

However, bearing in mind this reservation, it may be concluded that:

- The Bulgarian court talk tends to be more stereotypical and more lively than the Swedish court talk. With respect to liveliness, Bulgarian court talk is similar to Swedish dinner talk. With respect to stereotypicality, the Bulgarian courts are similar to auctions.
- Complex measures are not always applicable cross-linguistically.

4 Conclusions

This paper examined cross-linguistic and cross-activity stylistic tendencies by development of hybrid qualitative-quantitative methodology and complex linguistic measures, which aim to contribute to the efficiency and usefulness of multi-lingual Decision Support Systems and Negotiation Support Systems. The paper shows how the three complex linguistic measures - interactive liveliness, stereotypicality and caution - can be utilized for the recognition and characterization of activity type, negotiation type, culture type, relation type, and interactive intensity degree. These and other such measures can be used to study not only how activity affects social interacion but also how interaction changes activity over time. Certain simple measures, such as word length or collocation tokens, can also indicate activity change: the difference between activities grows with the increase in the length of the words and tokens in collocations. The cross-activity comparison of Swedish courtroom examination data and dinner talk, and to some extent also other activities, such as face-to-face task-oriented decision taking, sermons, auctions, and formal negotiation meetings, found the following general style tendencies:

- Courtroom negotiations tend to contain more words in an utterance, more complicated lexicon, more pauses, more stressed words, and less overlap and thus tend to be less lively than dinner talk (but more lively than sermons and auctions).
- Court examinations' speech tends to be more stereotypical (but less stereotypical than auctions) and precise than dinner talk.
- Courtroom negotiations tend to contain fewer interruptions but more acknowledgements, feedback-givers, which may indicate a more regulated, more cautious, and less competitive style of verbal interaction.

The basic purpose of the court examination is implied by the constellation of linguistic features pointing out the underlined caution- and objectivity-striving character of the activity.

The main but tentative conclusions about relations between the cross-linguistic corpora are:

- The Bulgarian court talk tends to be more stereotypical and more lively and to have a richer vocabulary than the Swedish court talk.
- Swedish admissions are produced more slowly, with a greater number of pauses, than most other sequences, with less overlap and average amount of stress.
- Bulgarian admissions are much less carefully received, since there is a greater amount of overlap and speech is less carefully produced since there are fewer pauses and stressed words in comparison to the Swedish data, as well as to the rest of the Bulgarian data.
- In both language corpora, reformulations are more frequent than repetitions and have a diversity of functions.
- The total amount of repetitions and reformulations in percent is cross-linguistically almost identical, which indicates that the language factor is much less important than the activity factor.

In general, according to the above observations, the data indicate that the activity factor influences the interaction in a more profound way than the language factor or the national culture factor. The relation between the activity defined power and status and the interactively displayed power is indexical, i.e. those judicially more powerful and competent also exhibit their power interactively but use different linguistic and discursive means.

Acknowledgements Shortly before his untimely passing in year 2020, Professor Gregory Kersten and I planned a publication on the topic of this paper, which is based on data and ideas presented in my PhD thesis relevant to the improvement of Group Decision and Negotiation Support Systems. It is thus dedicated to the memory of Professor Gregory Kersten with gratitude and appreciation for his work, inspiration and caring manner.

References

Allwood, J., Hagman, J.: Some simple automatic measures of spoken interaction. In: Proceedings of the XIV Conference of Scandinavian Linguistics and Gothenburg Papers in Theoretical Linguistics, vol. 72. University of Gothenburg, Gothenburg (1994)

Atkinson, J.M., Drew, P.: Order in Court. The Organisation of Verbal Interaction in Judicial Settings. MSSRC, London (1979)

Grönqvist, L., Sofkova-Hashemi, S.: Towards multimodal spoken language Corpora. In: Proceedings of the Workshop on Partially Automated Techniques for Transcribing Naturally Occurring Continuous Speech - Transcribe 1998, Quebec, Canada (1998)

Ilie, C.: Question - Response argumentation in talk shows. J. Pragmat. **3**(8), 975–999 (1999)

Kersten, G.E.: E-negotiation systems: interaction of people and technologies to resolve conflicts. Magnus J. Manag. **1**(3), 71–96 (2005)

Kersten, G., Lai, H.: Negotiation support and E-negotiation systems. In: Handbook on Decision Support Systems 1, pp. 469–508. International Handbooks Information System. Springer, Heidelberg (2008)

Martinovski, B.: The Role of Repetitions and Reformulations in Court Proceedings – A Comparison of Sweden and Bulgaria. Gothenburg Monographs in Linguistics. University of Gothenburg, Gothenburg (2000)

Nikolova, Z.: Chestoten Rechnik Na Bulgarskata Razgovorna Rech. Nauka i Izkustv, Sofia (1987)

How Well Agents Represent Their Principals' Preferences: The Effect of Information Processing, Value Orientation, and Goals

Gregory E. Kersten, Ewa Roszkowska, and Tomasz Wachowicz

Abstract There is an implicit assumption that the agents who negotiate on behalf of principals prepare for the negotiations. This study focuses on the negotiation preparation phase during which the agents are given information in both text and graphical forms, which unambiguously expresses their principal's preferences. The analysis of data obtained from the online bilateral negotiation experiments shows that the majority of the agents make both ordinal and cardinal errors in the reconstruction of the principal's preferences. It shows that the agents' intrinsic motivation contributes to the errors and has a negative effect on their preparation. To determine how different types of motivation contribute to the agents' inaccuracy, a hierarchical framework of the epistemic, social, and identity motivation is proposed. The framework is analyzed with a structural equation model (SEM) and the resulting model shows relations among the motivation types and between motivation and accuracy. The results show that the rational motivation system has a significant direct effect on social motivation (accommodating, competing, and collaborating intentions) and identity motivation (learning and substantive goals). The experiential motivational system has no significant effect on social motivation but it affects two goals of identity motivation (substantive and relational). The model also shows that only identity motivation has a direct effect on the accuracy of the preferences formulated by the agents in comparison with their principal's preferences.

G. E. Kersten
John Molson School of Business, Concordia University, 1450 Guy Street, Montreal QC H3H 0A1, Canada
e-mail: Gregory.kersten@concordia.ca

E. Roszkowska (✉)
Faculty of Computer Science, Bialystok University of Technology, Wiejska 45A, 15-351 Bialystok, Poland
e-mail: erosz@o2.pl

T. Wachowicz
Department of Operations Research, University of Economics in Katowice, 1 Maja 50, 40-287 Katowice, Poland
e-mail: tomasz.wachowicz@ue.katowice.pl

T. Szapiro and J. Kacprzyk (eds.), *Collective Decisions: Theory, Algorithms And Decision Support Systems*, Studies in Systems, Decision and Control 392,
https://doi.org/10.1007/978-3-030-84997-9_6

Keywords Behavioral agency theory · Intrinsic motivation · Epistemic motivation · Social motivation, Identity motivation · Negotiation preparation · Representative negotiation · Preferences · Accuracy measurement

1 Introduction

Business, governments and other organizations negotiate through agents whom they employ. The agents are either their own either employees or hired outside experts. The design of incentives and rewards so that the agents represent their principals well is a widely studied problem in agency theory (Eisenhardt 1989). This paper focusses on a less studied problem, namely the agents' intrinsic motivation to learn about and correctly represent the principals' interests prior to the negotiation. It proposes an intrinsic motivational framework within which the agents think and make decisions.

Agents have their own interests and they may be different or even in opposition with those of their principals. The misalignment of the principal's and the agent's interests is one of the key aspects of the agency problem (Bamberg and Spremann 1989). Agency theory, building on assumptions that both principals and agents are narrowly self-interested and boundedly rational and the agents are more risk averse than principals, asserts that extrinsic motivation is sufficient to convey the principal's interests and to effectively guide the agents.

While the deductive logic of the traditional agency theory is sound, its predictions have found limited empirical support (Bottom, Holloway et al. 2006). Behavioral agency theories, which recognize that agents are psychologically complex, socially rooted, and have rich relationships, provide an explanation for the limitations of the standard agency theory (Westphal and Zajac 2013). They acknowledge the agents' limited competences, their work under time pressure, obligations to others, and reliance on verbal and non-verbal communication (Bottom, Holloway et al. 2006; Pepper and Gore 2015).

The extrinsic motivation provided by incentives and rewards that corresponds to the agents' self-interests, was found insufficient in assuring the desired results. Oftentimes the principals were found to be worse off when providing incentive contracts to their agents than without such a contract (Kunz and Pfaff 2002). This is because extrinsic motivation may diminish or even erode intrinsic motivation (IM) rather than reinforce it (Ryan and Deci 2000a; b).

The impact of IM on negotiation behavior has been discussed in the literature. Experiments show the impact of motivation regarding information processing (cognitive complexity and problem-solving orientation) on negotiation outcomes (Pruitt and Lewis 1975). Information processing motivation also influences social value orientation (De Dreu, Koole et al. 1999). More recently, two studies have shown that epistemic motivation affects the negotiator's ability to use information effectively (Ten Velden, Beersma et al. 2010) and her ability to reach integrative agreements in complex negotiations (Van der Schalk, Beersma et al. 2010).

A number of studies show that different social motivations affect the negotiation process and its outcomes. For example, social motivation result in different size of concessions (De Dreu and van Lange 1995). Also, the perception of the counterpart affects the size of first offers and the final agreements (Olekalns, Smith et al. 1996) and affects the negotiators' satisfaction (Gillespie, Brett et al. 2000).

Studies on motivation in negotiations focus on the process and its outcomes. We know of no study that considers motivation in the context of negotiation preparation, which is particularly important in the negotiation conducted by agents.

To negotiate effectively and to obtain the best possible agreement the agent must prepare for the negotiation, which includes learning about the issues and their values as well as the principal's preferences over issues and issue options (values). It may be difficult or even impossible to accurately represent principals' preferences when they are ill-defined, heterogeneous, or conflicting. However, if they are clearly and unambiguously formulated and presented, then the agent may be able to establish a good representation of the principal's preferences. The early examples of studies in this field show that the effects of prenegotiation preparation may affect the process and outcomes (Bass 1966), as well as the agent's misrepresentation of the principal's preferences affect the outcomes and their understanding by the former (Wachowicz, Kersten et al. 2019).

This work presents the results of experiments in which students acted as agents and negotiated on behalf of principals. The purpose of the experiments was to study the relationship between the agents' internal motivation and the quality of their work, given the same extrinsic motivation. The agents were given both verbal and graphical information about negotiation issues and their principal's preferences and they were asked to represent this information in numerical form. We used this cognitively simple task to determine the accuracy of the representation. The accuracy was determined with cardinal and ordinal measures. Cardinal accuracy is difficult to achieve as it requires measuring every issue's value, i.e., every option. However, the agents should be able to be ordinal-accurate, i.e., know that the principal prefers one issue over another and one option over another and not vice versa. The preference strength is noted in the text and indicated by the height of bars.

The analysis of data obtained from the online bilateral negotiation experiments shows that 66% of agents make errors in the reconstruction of the principal's preferences. Their epistemic, social, and identity motivations affect their errors. The model obtained from the structural equation modeling (SEM) shows that the rational motivation system has a significant direct effect on social motivation (accommodating, competing, and collaborating intentions) and identity motivation (learning and substantive goals). The experiential motivational system has no significant effect on social motivation but it affects two goals of identity motivation (substantive and relational). The model also shows that only identity motivation has a direct effect on the accuracy of the preferences formulated by the agents in comparison with their principal's preferences.

We are unaware of prior studies that show how one type of motivation affect other types and relate motivation to the accuracy of the representation of the principals' preferences and their implications. The contribution of this work is at least three-fold.

Firstly, it proposes and verifies a three-level motivational framework that focusses on intrinsic motivation. The causal relationship between different types of motivation and their direct and indirect impact on preference representation are verified. In addition to negotiation preparation, the framework can be further studied in principal-agent interactions, negotiation conduct, and other forms of social interactions.

Secondly, it contributes to behavioral agency theory and behavioral negotiation analysis in presenting a rich and complex picture of the agents and their multiple interacting motivations. It shows the difficulties in imparting the principals' preferences to their agents. It also shows that while both intrinsic and extrinsic motivations impact the agent's behavior, some types of IM may be beyond the principal's control.

Thirdly, it shows the importance of preparing for negotiations. While negotiation literature stresses the significance of the preparation, there is a dearth of field and experimental studies on errors made during this phase of the negotiation. One may expected that such errors would carry over on the conduct of negotiations and thus their outcomes.

The paper has five more sections. The next section discusses three types of motivation and their measures. Then, a research model is presented in Sect. 3. To verify the hypotheses two online negotiation experiments were designed and implemented; they are discussed in Sect. 3. The results and their implications are discussed in Sect. 4. Conclusions and future work are given in Sect. 5.

2 Intrinsic Motivation in Negotiation

2.1 Intrinsic Motivation and Accuracy

There have been many studies on the different types of IM but a few on their structural representation. Vallerand (2000) presents a model of different types of motivation, their antecedents and consequents. In the model, motivation takes place at three levels: global, contextual, and situational. The global level refers to a broad disposition to engage in activities or to approach decision problems. Contextual motivation refers to motivational orientations that are specific to various contexts, e.g., negotiations. Lastly, situational motivation refers to concrete motivational determinants relevant to a situation at hand.

Recently, an IM classification has been proposed in the context of negotiations (De Dreu 2004; Carnevale and De Dreu 2006). It comprises: (1) *epistemic motivation*; (2) *social motivation*; and (3) *identity motivation*. The relationship between these motivations is not clear. De Dreu and Carnevale (2003) assume that social motivation and epistemic motivation are independent (op. cit., p. 269) and refer to earlier experimental studies confirming this assumption (e.g., De Dreu, Koole et al.

1999). Recent experiments, however, show that epistemic motivation affects social motivation (De Dreu, Beersma et al. 2006; Van der Schalk, Beersma et al. 2010).

Epistemic motivation refers to the innate desire for truthful representation of a state or an accurate solution of a problem (Fairweather 2001). This motivation involves the selection of information processing that is considered for the task-at-hand. High epistemic motivation relates to slow, systematic and rational processing. Low epistemic motivation is associated with reliance on heuristics, and fast experiential processing (Epstein, Pacini et al. 1996). In the context of negotiations, rational motivation (high motivation) contributed to higher joint outcomes and better ability to obtain and use relevant information (Ten Velden, Beersma et al. 2010).

Social motivation is the person's need to interact with other people, to participate and be involved in the allocation of inputs and distribution of outputs. It is captured by individual differences in social value orientation (Messick and McClintock 1968). In the context of negotiations, it refers to the preferred division of outcomes (De Dreu and van Lange 1995; Olekalns and Smith 2003). Epistemic motivation affects social motivation because it deals with information a person is processing and her willingness to deal with complexity.

Identity motivation refers to the achievement of well-being and the need for creating and maintaining an impression of self. It manifests itself through extrinsic and intrinsic goals (Kasser and Ryan 1996). The choice of goals, their specificity and difficulty affect task performance (Locke and Latham 2004). In the context of negotiations, extrinsic goals impact the outcomes (Zetik and Stuhlmacher 2002). Studies showed that the dyads who were given more specific and difficult goals achieved better results than those who were given no goal or a compromise goal (e.g., Huber and Neale 1987).

We use Vallerand's (2000) hierarchy and adapt De Dreu's classification to propose a motivational framework to analyze the relationship between motivation types and their effect on a negotiation preparation task. The framework corresponds to the goal hierarchy model in which the predisposition to organize and process information is at the highest level at the hierarchy followed by values that include social value orientations, and the task goals at the bottom of the hierarchy (Cropanzano, James et al. 1993).

To sum up, the focus of this paper is to determine the effect of three types of intrinsic motivation on the accuracy of the representation of the principal's preferences. Figure 1 presents the high-level structure of the research model.

In what follows we discuss the theories of the three types of motivation and their scales. We take a bottom-up approach, which makes the explanation of the linkages between the concepts easier. In Sect. 3, we discuss accuracy in detail.

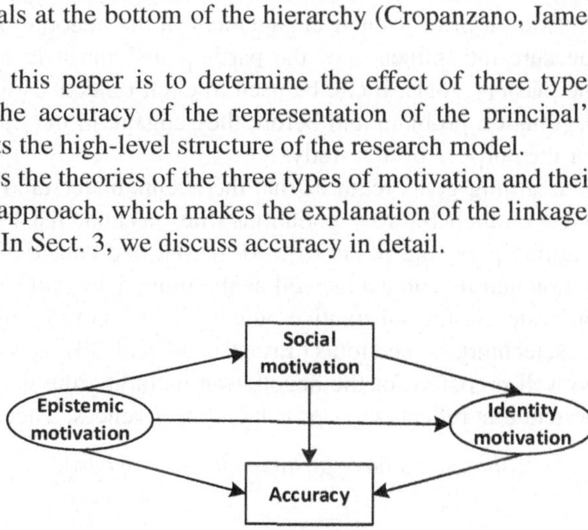

Fig. 1 A bird-view of the research model

2.2 Identity Motivation and Goals

Identity motivation may be extrinsic and focused on approval from others. This type of motivation involves performance and social goals, which refer to social reasons for engaging in a task and include pleasing and gaining approval and an opportunity for desired social interaction. Identity motivation may also comprise intrinsic goals, i.e., instrumental and relational (Wilson and Putnam 1990).

Intrinsic identity motivation involves satisfaction from learning and acquiring knowledge while performing and accomplishing a task (Ryan and Deci 2000a; b). This type of motivation involves the mastery of the goals related to self-regulated learning, the transfer of problems solving strategies, the achievement of the task, and a positive attitude towards others (Graham and Golan 1991; Levy, Kaplan et al. 2004).

The mastery and the performance goals concern self-perception of one's competence. These concepts have been developed within the *goal-orientation theory* (Dweck 1986). The theory has been confirmed in academic and workplace environments in which assigned tasks required both extrinsic and intrinsic motivations to be successfully performed (see a meta-analysis by Cellar, Stuhlmacher et al. 2011).

Negotiation goals may be associated with substantive and relational outcomes, the process, and the satisfaction with own performance. Curhan et al. (2006) observe that the negotiators assess the negotiation based on other achievements than substantive outcomes only. They propose to consider four types of feelings about: (1) the substantive outcomes; (2) the self; (3) the process; and (4) the relationship.

SVI is a post-negotiation questionnaire aimed at assessing the negotiation participants and the process. We are interested in the goals' influence on negotiation preparation. Furthermore, goal learning is important in academic and professional training contexts.

The substantive, relational, learning (SRL) scale is a modification of SVI. It includes substantive and relational goals of the SVI and adds goal learning (Kersten, Gimon et al. 2013). This makes SRL suitable for the assessment of the participants' motivation prior to their engagement in the negotiation activities. Its purpose is to measure the influence of the participants' intrinsic motivation operationalized in the mastery goals. It can be used to elicit objectives when students learn about the negotiation problem and before they engage in the interactions, making it suitable for the purpose of this study.

Learning goals requires that the agents understand the process and its potential effect on their future negotiations (Stevens and Gist 1997). Negotiators who pursue learning goals are likely to work harder and consider various aspects of the negotiation and its context as well as the immediate and future outcomes. They are less interested in the substantive outcomes and more interested in the effects of strategies, techniques, and tools (Tasa, Celani et al. 2013), which suggests that they should be well prepared for the negotiation including the understanding of the preference system that reflects their principals' preferences. Therefore, we hypothesize that:

Learning goals have positive effect on accuracy.

Elliot and Harackiewicz (1996) studied the effect of the extrinsic achievement goal on an intrinsic mastery goal. Their experiments show that when the achievement and the mastery goals are aligned, then individuals have the same level of intrinsic motivation as the participants who have only the mastery goal. This suggests that agents who are concerned with the substantive outcomes and know that the principal assesses these outcomes, need to make an effort to get their principals' preferences correctly, that is:

Substantive goals have positive effect on accuracy.

Bargh (2014, p. 37) who studied unconscious motivation and goals observes that "many people's goals are directed toward the welfare of others … these 'communally' oriented individuals should react by being more likely to help others and less apt to focus on themselves." Negotiators have power to accept or reject an agreement. If they are "communally" oriented, then they are likely to consider that atmosphere, emotions and relationship are more important than the agreement score. This suggests that they may be less concerned with the preferences than with creating a positive atmosphere from the outset. Therefore, we posit that:

Relational goals have negative effect on accuracy.

2.3 Social Motivation and Negotiation Approach

Theories and models on social motivation include social value orientation (McClintock and van Avermaet 1982). Theory of cooperation and competition (Rubin and Brown 1975) and dual concern theory (Pruitt and Rubin 1986) adapt the theory of social value orientation to study approaches to conflict resolution processes. Accommodating and collaborating approaches are associated with prosocial orientation and competitive (including individualistic) approaches correspond to proself orientation.

Thomas-Kilmann Inventory (TKI) formulated by Thomas and Kilmann (1974) uses a variant of the dual concern model with the dimensions describing assertiveness (effort to satisfy own concerns) and cooperativeness (effort to satisfy the counterpart's concern). The scale has been used to create a negotiation profile which is a vector of five values (from 0 to 12) associated with each approach (Meyer, McCormick et al. 2012; Fleck, Volkema et al. 2017). Negotiators typically have a medium-to-strong predisposition to two or three approaches rather than one (Shell 2001).

The TKI scale produces ipsative data limiting the types of statistical analyses to assess their validity. We decided to use TKI because it has been widely used in training, its results are relatively uncontaminated by social desirability effects, and it is very easy to use and interpret (Womack 1988; Shell 2001; Brown 2012). Moreover, the TKI categories represent behavioral intentions rather than conflict behaviors making it suitable for our research model (Thomas 1988).

TKI identifies five categories of intentions: collaborating, accommodating, competing, compromising, and avoiding. Because of the nature of the scale, i.e., the approaches are a linear combination; four rather than five categories should be used for statistical analysis. TKI does not clearly discriminate between avoiding and accommodating approaches (Kabanoff 1987; Van de Vliert and Kabanoff 1990). Therefore, avoiding intentions will not be included in our model.

Social Motivation and Accuracy. Collaborating and compromising approaches require considerable cognitive effort to consider the problem from different perspectives—their own, the principal's, and the counterparts. They search for solutions that are concordant with their values and the need to accommodate the requirements of the counterparts (Van Lange 1999). Prosocial agents were found to represent their principal's interests better than proself agents (Aaldering, Greer et al. 2013). Since it requires effort to understand the principal's preferences we posit the following:

Collaborating intentions have positive effect on accuracy.
Accommodating intentions have positive effect on accuracy.

Competitively predisposed agents are concerned with the achievement of the best substantive outcome. They serve the principal's interests only if these converge with their own interests (Jensen and Meckling 1976; Bazerman, Neale et al. 1992). This implies that given the opportunity to use a set of preferences that would allow achieving a higher score (utility) than the principal's preference set, they would do so. These agents are likely to increase the preference values for the options that have a good chance to be included in the agreement. Therefore, we hypothesize that:

Competing intentions have negative effect on accuracy.

Compromising is a "cooperative gesture" which, during the course of the negotiation, may involve a mix of cooperative and competitive intentions (Ruble and Thomas 1976). During the negotiation preparation, compromising predisposition resembles a collaborating predisposition. Therefore, we hypothesize that:

Compromising intentions have positive effect on accuracy.

Social Motivation and Identity. Cooperation requires the exchange of information focused on preferences rather than positions (Liu, Chua et al. 2010), which has a positive effect on the relationship (Gelfand, Major et al. 2006). Cooperative and compromising negotiators' are concerned with both substantive and relational outcomes (Curhan, Elfenbein et al. 2006). This suggests that:

Collaborating intentions have positive effect on relational goals.
Collaborating intentions have positive effect on substantive goals.
Compromising intentions have positive effect on relational goals.
Compromising intentions have positive effect on substantive goals.

Cooperation and compromising require an interest in learning of the counterpart's needs and preferences (Tasa, Celani et al. 2013).

Collaborating intentions have positive effect on learning goals.
Compromising intentions have positive effect on learning goals.

Competition is characterized by an emphasis on own interests irrespectively of the counterpart's needs (Putnam 1990). Competitive agents focus on their own substantive results and relative gains (Griessmair and Druckman 2018).

Competing intentions have positive effect on substantive goals.
Competing intentions have negative effect on relational goals.

The successful competition requires that the agent understands the counterpart's weaknesses and ways to exploit them (Tasa, Celani et al. 2013). This attention to the discovery of their counterparts' interests suggests that:

Competing intentions have positive effect on learning goals.

Accommodating means that the counterpart's interests are important for the agent, but her own are not. Such intentions may be due to the willingness to establish or maintain good relations, trust and goodwill. Therefore, we posit that:

Accommodating intentions have negative effect on substantive goals.
Accommodating intentions have positive effect on relational goals.

Accommodating agents are likely to let their counterparts control the agenda and the process. Because they do not need to learn about their counterparts and the problem, we hypothesize that:

Accommodating intentions have negative effect on learning goals.

2.4 Epistemic Motivation and Information Processing Scale

There have been numerous studies on the individual differences in perception and interpretation of information presented in different forms. *Dual process theory* that distinguishes between intuitive vs. rational, heuristic vs. systematic, and implicit vs. explicit modes of information processing, has been found particularly useful to gain insights into the underlying causes behind these differences (Epstein, Pacini et al. 1996; Chaiken and Trope 1999). The theory posits that people process information using two distinct systems: (1) *experiential system* in which assessment is holistic, intuitive, and fast; and (2) *rational system* in which assessment is reflective, analytical, and slow. The underlying assumption is that rational and experiential styles are stable; they predispose individuals to heed and process information in different ways (Berger 2005).

Our experiments involve students of business and economics with good analytical skills and the ability to process information in a systematic way. They may use heuristics and naïve methods if their motivation is low. This allows us to consider the two processing systems as intrinsic motivation rather than stable predispositions.

Rational-Experiential Inventory (REI) is grounded in dual process theory and it is one of the well-known instruments used to study perception and processing of information (Epstein, Pacini et al. 1996). This scale was used to show how reliance on one system affects preference for the type of information presentation: people who use the experiential system prefer images and graphs and people who use the rational system prefer text and numbers (Epstein 2003). A large and growing body of research supports the reliability of REI and validity of the two styles as two independent constructs (Sladek, Bond et al. 2010; Wang, Highhouse et al. 2017) leading us to rely on this scale.

Epistemic Motivation and Accuracy. People, who have low ability to expend cognitive effort or their extrinsic motivation is low, are more likely to rely on the experiential system (Shiloh, Salton et al. 2002). Reliance on the experiential system allows bypassing the task difficulty and producing results quickly, which suggests that those who are uninterested in the task may revert to the experiential system. Witteman et al. (2009) found a significant positive correlation of the relational system and a significant negative correlation between the experiential system and performance on tasks that required computations and deductive reasoning. Hence, we posit that:

Rational system has positive effect on accuracy.
Experiential system has negative effect on accuracy.

Epistemic and Identity Motivations. Mastery goals require greater effort and more systematic information processing than performance goals (Stevens and Gist 1997). The three goals associated with identity motivation are of the mastery type; hence, we hypothesize that:

Rational system has positive effect on substantive goals.
Rational system has positive effect on relational goals.
Rational system has positive effect on learning goals.

Negotiators use situated concepts of negotiation; their experience is typically limited to the situations in which it was acquired (Loewenstein and Thompson 2006). When they are inexperienced, they are likely to show more decision biases and make more errors, then the experienced negotiators. Therefore, we posit that:

Experiential system has negative effect on substantive goals.
Experiential system has negative effect on learning goals.

Negotiation experience may be either positive or negative. In the past, the negotiators could establish a good relationship which they may expect this to continue. Because the reverse is also possible, we do not posit what effect the experiential system has on relational goals.

Epistemic and Social Motivations. The hierarchical model shown in Fig. 1 posits a causal relationship between epistemic and social motivation. This relationship reflects the difference in the motivations: epistemic motivation refers to the system being used to process information while social motivation uses the information to support one or more social motives. The rational system, which involves systematic information processing and a deep understanding of the task, is required when the negotiators seek integrative outcomes (Carnevale and De Dreu 2006). The experiential system relies on heuristics and intuition is likely to focus on one's own interests. Three experiments confirmed that prosocial motivation is associated with the rational system, while proself with the experiential system (De Dreu, Beersma et al. 2006). Given these results, we hypothesize the following:

Rational system has positive effect on collaborating intentions.
Rational system has negative effect on competing intentions.
Experiential system has positive effect on competing intentions.
Experiential system has negative effect on collaborating intentions.

Accommodating intentions are prosocial but they do not require significant cognitive efforts; therefore, they are often associated with high experiential and low analytical systems. Accommodating agents tend to be emotional, agreeable, and sensitive (Tucker and Ferson 2008). They are likely to rely more on their experiential than a rational system.

Experiential system has positive effect on accommodating intentions.
Rational system has negative effect on accommodating intentions.

Compromising agents search for solutions that are concordant with their values and accommodate the requirements of the counterparts (Van Lange 1999). These agents rely on the analytical processing system and they are predisposed to cooperate (Shabbir, Atta et al. 2014).

Rational system has positive effect on compromising intentions.
Experiential system has negative effect on compromising intentions.

3 Research Model and Experiments

3.1 Research Model

In Sect. 2, we hypothesized that there are causal relationships between the three motivations and between the motivations and accuracy. These concrete relationships allow us to formulate a more detailed model shown in Fig. 2. Continuous and dashed arrows represent positive and negative effects, respectively. We do hypothesize a relationship between experiential systems and relational goals but for completeness, we indicate such a possibility with a gray line.

Rational and experiential information processing, as well as the substantive, relational and learning goals, are represented with latent variables. They are inferred from, respectively, the REI and SRL scales. Accommodating, compromising, competitive and cooperative intentions as well as ordinal and cardinal accuracies are directly observable variables. Because of the specifics of the TKI scale, discussed in Sect. 2.3, social motivation, i.e., negotiation intentions, are represented as directly calculated variables.

It is possible that social and identity motivation mediate the relationship between epistemic motivation and accuracy and identity motivation mediate the relationship between social motivation and accuracy. We were not able to find earlier results

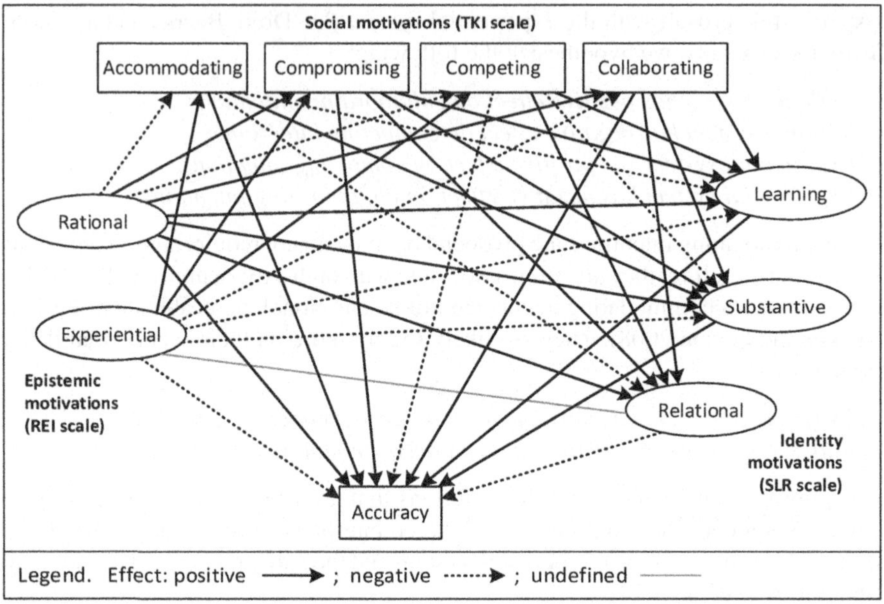

Fig. 2 Overview of the research model

that would allow us to hypothesize mediated relationships; therefore, in Sect. 2 we did not discuss mediation. The model analysis will allow us to determine if such relationships are significant.

3.2 *Online Negotiation Experiment and Business Case*

To verify the research model online bilateral negotiations were organized. The negotiations were conducted via the Inspire negotiation support system, which has both communication and decision support aids (Kersten and Noronha 1999).

Inspire negotiations are anonymous. They are organized for groups of students (see: http://invite.concordia.ca/inspire/). They are requested by course instructors who receive readings, instructions, slides, and tests. After the instructors sign up their groups, students register individually using the group-specific URL. Both instructors and students are informed that the registration is voluntary and the students may have their information deleted upon their completion of the negotiations. Upon request, researchers receive the anonymized dataset three months after the negotiations.

Inspire negotiations comprise three phases. In the preparation phase, each participant is required to read the case and the handouts. Groups are also shown slides and given a short in-class quiz. The preparation phase, which is discussed in this work, is completed when the participants, depending on the case, either accept or formulate the preference system they will use in the negotiations.

Negotiations are asynchronous. The students log in to the Inspire system, prepare their offers and when these are submitted, Inspire informs their counterparts by sending emails. The offers may be prepared by selecting one option per issue or by specifying the offer utility value; in the latter case, the system displays seven alternatives for the user to choose from. Participants may send offers with or without messages or send only messages. The negotiation transcripts are recorded and stored in the database; they are used to show negotiation progress in both tabular and graphical formats.

The Inspire system offers several business cases. For this study, we requested that the negotiations take eight days and the case be a contract negotiation between an entertainment company and an artist. Note, however, that the participants were asked to complete the preparation phase no later than by the end of the third day. The principals that are the company and the artists are represented by their agents, Mosico and Fado, respectively.

One negotiation participant plays the role of Mosico and the others play the role of Fado. The participants are asked to negotiate a three-year contract in which the following four negotiation issues are specified: no. of concerts, no. of new songs, royalties for CDs and the signing bonus. Issues, their values (i.e., options), as well as background information about the artist and the company are public (http://invite.concordia.ca/cases/inspireYowlPop.html). There is also private information, which includes the specifics of the interests of each principal.

3.3 Accuracy of the Agents' Preferences

During the negotiation preparation, the agents are given verbal and graphical information about the preferences of their principal. Then, they are asked to construct a rating system (using the scale 1–100) that represents the principal's preferences. The principal's underlying preferences, which are not shown to the agents, and the preference constructed by the agents are both represented with a vector of natural numbers.

The preferences the agents construct may be inaccurate because the imparted preferences are not expressed in numerical but in verbal and graphical format. The accuracy of the two systems is measured using ordinal and cardinal scales. Hamming distance (D_H) is used to measure ordinal accuracy scales and block (Manhattan) distance (L_1) is used to measure cardinal accuracy. The two scales are discussed in the Appendix.

The ordinal accuracy is a numerical representation of a qualitative difference—it measures if one preference is stronger (is assigned greater value) than another. Both verbal and graphical representation provide unequivocal information regarding the comparative strength of the preferences. Therefore, one may expect that the agents constructed ordinal-accurate preferences, i.e., $D_H = 0$.

The cardinal accuracy measures the numerical difference between the two preferences. It could be determined precisely if a ruler was used to measure the size of the graphs representing preferences. Because of this additional effort and inaccurate measurement, the agents are likely to be inaccurate in the cardinal sense, i.e., $L_1 > 0$.

The principal's preference system is additive and represented by integers. The negotiated issues are allocated scores that total 100. Issue scores are the same as the scores allocated to the best option in these issues. The worst option of each issue has a score of 0 and the remaining option's scores are between the best and the worst score for this issue. Thus, the best alternative has a score of 100 and the worst one—0.

The preferences of the two principals (the singer and the entertainment company) differ but are not strictly opposite. An example of Mosico's private information with a principal's preference description for issues is shown in Fig. 3. The priorities for various options within each issue are described in a similar way.

To determine the scoring system precise measurement of the bar heights is not necessary because the preferences are measured with integers. The approximation is sufficient albeit it may result in some cardinal inaccuracy. However, we can see in Fig. 3 that ordinal accuracy can easily be attained.

The scoring system of the entertainment company, i.e., Mosico's principal, is given in Table 1. The scores allocated to the issues (totaling 100) are the same as the

Before meeting Fado, you discussed the Agency's priorities and requirements with senior management. Senior managers could not give you very detailed information regarding the importance of the negotiated issues and options, but during a few short meetings they gave you many indications as to the relative importance of the issues and the agency's preferences. To help visualize the relationship between the issues you drew bars with their height indicating the issues' importance. You did the same for the options of each issue.

Importance of the four issues:

- It is clear that the most important issue is the **number of promotional concerts.** This is because successful concerts are critical to the artists' popularity and approval ratings. Without the concerts the agency cannot establish the artist in a particular market.
- The second most important issue is the **number of new songs.** Obviously, the artist has to produce new songs to be recognized and accepted.
- **Royalties for CDs** are less important than the number of songs. The management considers the royalties to be a motivating factor for the artist to produce good CDs.
- The **contract signing bonus** is the least important issue. It is less important than the royalties for CDs. This is because the agency views a contract as an investment opportunity that can bring in many of millions of dollars. The bonus size is seen as a token of appreciation, but obviously within limits.

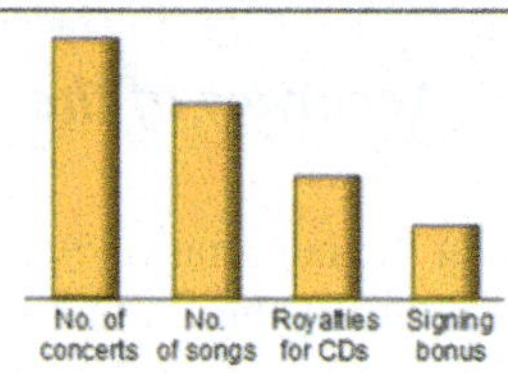

Fig. 3 Preferences for four issues of Mosico's principal

Table 1 Scoring system S_P of the principal (Entertainment Company) and S_M—formulated by Mosico

Issues	No. of concerts (C)				No. of songs (S)					Royalties for CDs (R)				Signing bonus (B)		
Options	5	6	7	8	11	12	13	14	15	1.5	2.0	2.5	3.0	125	150	200
System S_P	0	23	31	39	0	6	16	30	24	10	20	15	0	11	5	0
System S_M	0	18	31	39	0	11	**26**	30	**14**	10	**15**	**20**	0	11	5	0
$\left\|w_{ij}^P - w_{ij}^M\right\|$	0	5	0	0	0	5	10	0	10	0	5	5	0	0	0	0

scores allocated to the best option in these issues (e.g., issue "No. of concerts" and the best option "8 concerts" have a score of 40). The worst option has a score of 0 and the scores for the remaining options are between the best and the worst score. Note that scores for issues "No. of songs" and "Royalties for CDs" are non-monotonic.

The agent, i.e., Mosico, is required to convert the text and graphs given in private information (see Fig. 3) to numerical scores. Let's assume that Mosico formulated her scoring system S_M as shown in row "System S_M" in Table 1. We can see that there are both ordinal and cardinal inaccuracies. The best option for Royalties is R(2.0) for the principal but it is R(2.5) for the agent, the second best is R(2.5) and R(2.0) for the principal and the agent, respectively. Also, the order for the "No. of songs" is switched for the options S(13) and S(15). Therefore, the ordinal inaccuracy (highlighted in Table 1) is $D_H(S_P, S_M) = 2$ (see Appendix, formula (3)).

To determine the cardinal accuracy, we need to calculate the absolute differences between the option scores for S_P and S_M; these differences are given in the last row of Table 1. Using the formula (4) given in the Appendix, we obtain cardinal inaccuracy $L_1(S_P, S_M) = 40$.

3.4 Participants and Evaluation of Their Performance

The negotiation experiments were organized in 2015 and 2016 with students from seven universities from Europe, North America and Asia. In total, there were 442 active participants: 221 Mosicos and 221 Fados. To achieve a high level of homogeneity and avoid any unintended effects of cultural and role-playing factors on the results, we decided to limit our research sample to the Mosicos coming from one culture only. After eliminating some participants because of incomplete data, we obtained a group of 181 agents, uniform in both their culture and role-playing. The general description of participants' demographics and skills is presented in Table 2.

Note that because the participants of the experiment were students, hence the age distribution is skewed and the sample consists of rather young negotiators (85% of them not older than 25) with particularly low negotiation experience, with 75% of participants with the rating below average (4.0). Yet, the reported case understanding is moderately high (4.97), which allows us to posit that the potential errors in the representation of the principal's preferences by the agents are not related to any misunderstandings of the assignment.

Reducing the sample allowed us to assure uniformity of the performance evaluation; all 181 students participated in courses given by two instructors who used a single course assignment with the same performance evaluation format. Students' performance was assessed using four criteria: (1) diligence in pre-negotiation activities; (2) engagement during the negotiation process; (3) contract profitability; and (4) participants' reports justifying their negotiation approach.

The evaluation of the pre-negotiation activities included the online quiz administered by Inspire and the similarity of the principal's and the agents' preferences. Students could lose their marks for violating the principal's instruction regarding the

Table 2 Participants

Characteristics		
Demographics and age data		*No. (%)*
Gender	Female	119 (65.7)
Age (years)	20 and less	10 (5.5)
	21–25	145 (80.1)
	26–30	20 (11.0)
	31–40	5 (2.8)
	41–50	1 (0.6)
Skills and ability		*Average (SD)*
English proficiency#		4.00 (1.33)
Case understanding#		4.97 (1.23)
Negotiation experience#		2.51 (1.31)

#Scale 1–7 (1-low; 7-high)

order of preferences as well as for assigning ratings significantly different from those given in the case. To get the maximum number of grading points their preferences had to be ordinal-accurate (i.e., $D_H(S_P, S_A) = 0$, see Appendix). Cardinal inaccuracy was acceptable.

The engagement was evaluated by measuring the students' engagement in the actual conduct of negotiation, i.e., early engagement in talks, prompt responses to the counterpart's offers and messages, and argumentation.

The contract, if reached, was evaluated based on its value to the principal that the agent had represented. Students who negotiated a contract that was not lower than the principal's reservation level (RL) and those who could not negotiate it because they would have to accept a worse contract than RL were rewarded. Students who did not get a contract even though they could have accepted an offer that exceeded their RL were penalized.

Finally, students were required to write reports which were evaluated based on the reasoned and justified description of their activities, including the pre-negotiation activities. Students were asked to illustrate their negotiation actions and assessments with the negotiation transcript and graphs. The assignment had a significant effect on the final grade in the course.

4 Data analysis and results

The results discussed below were obtained for a group of 181 participants who played the role of Mosico (representing an entertainment company) and who came from the same country. To assess both the two measurement models and the structural model we used structural equation modeling (SEM) implemented in IBM Amos 26. To analyze the data also IBM SPSS Statistics 26 was used.

Preliminary results show that the participants made both ordinal errors and cardinal errors, i.e., their assessments of the principal's preferences were not accurate. The average ordinal error for all participants was 2.34 (i.e., $D_H(S_P, S_A) = 2.34$), that is, over two pairwise comparisons were opposite to the principal's comparisons. We expected small cardinal errors but the errors were quite high; the average error was 58.85 (i.e., $L_1(S_P, S_A) = 58.85$). This result indicates that different motivations may contribute to inaccuracy.

4.1 Measurement model

Before verifying the structural model shown in Fig. 2, we validated both measurement models for epistemic and identity motivation. The participants' motivation was measured with three scales: REI was used to measure epistemic motivation (see Sect. 2.4); TKI measured social motivation (see Sect. 2.3); and SLR measured identity motivation (see Sect. 2.2).

The TKI scale identifies five categories of intentions with one of them being a linear combination of the remaining four. Therefore, we used four categories of intentions. Factor analysis is not appropriate on forced-choice measures, as each item loads on two scales. There are only four degrees of freedom among the five scale scores so that Cronbach's alpha is not an appropriate index of scale reliability. Instead, the reliability of TKI was evaluated with a number of instances of test–retest (Johnson, Thompson et al. 2014). We rely on these results and model the categories of intentions as variables rather than latent constructs.

The REI and SRL scales were analyzed in two steps E/CFA analysis. Implementing both exploratory and confirmatory approaches can be justified as the modified instruments were used, and additionally, non-native English speakers were answering the test written in English. Their comprehension of similar wordings used in some questions could result in correlated residuals and also the potential cross-loadings needed to be verified (Marsh, Balla et al. 1996).

EFA shows the general structure of two factors, as expected in the REI model, and the absence of cross-loadings. Five items were removed for having the loadings below the rejection threshold, set at the level of 0.5. CFA analysis revealed a correlation of residuals for three pairs of items. The correlated items with lower loadings within each pair were removed. This was due to a relatively small sample of the dataset; introducing another relationship in the model would increase the number of parameters to be estimated.

Multivariate non-normality of data led us to conduct the bootstrap with 2000 samples. The Bollen-Stine test confirmed the model fit ($p = 0.164$) and the fit indices are: $\chi^2(63) = 77.64$; CFI = 0.968; RMSEA = 0.051. All items in a measurement model REI are statistically significant ($p < 0.001$), so the convergent validity is achieved. The values of Cronbach alpha are greater than 0.70 (0.804 for *Rational* and 0.839 for *Experiential* modes), hence the internal reliability is also achieved.

A similar analysis was done for the SRL scale. All items have loadings significant and higher than 0.5 with no cross-loadings and no residual correlations. The data again appeared to be multivariate non-normal, hence bootstrapping was used with 2000 samples and the Bollen-Stine test for model acceptance. The model is acceptable ($p = 0.087$ in Bollen-Stine test) and its fit indices are: $\chi^2(51) = 90.88$; CFI $= 0.955$; RMSEA $= 0.066$. The value for Cronbach Alpha for constructs is 0.789 for variable *Learning*, 0.862 for *Substantive*, and 0.832 *Relational*, thus the internal reliability is achieved. Convergent validity is achieved because all items in a measurement model SRL are statistically significant ($p < 0.001$).

4.2 Structural Model

The analysis of the measurement models resulted in the removal of 10 observed variables. The resulting model has 28 observed variables, exceeded the suggested relation of the number of variables to the number of observations in the dataset (i.e., 181). To obtain a model with a more reasonable number of observations of variables we combined highly correlated items in the REI scale (Bagozzi and Edwards 1998) so that *Rational* and *Experiential* latent constructs were operationalized with five items each.

The TKI scale used to measure negotiation intentions does not clearly discriminate between accommodating and avoiding intentions (Van de Vliert and Kabanoff 1990) therefore, avoiding intentions were removed from the model (see Sect. 2.3). However, the remaining four intentions share at least one common unmeasured cause mentioned by some alternative tests, e.g., extroversion-introversion dichotomy from the MBTI instrument (Johnson, Thompson et al. 2014). This means that the specification of the residual correlation of the intention variables should be included in the structural model.

The AMOS test for the input data normality shows that some variables have non-normal distribution because their critical ratios for skew and kurtosis statistics are either smaller than -1.96 or greater than 1.96 (Kline 2015). Similarly, we observed multivariate non-normality. However, using rules of thumb suggested by Kline (2015) and Byrne (2010) that departure from normality is more substantial when kurtosis is greater than 7 and skewness greater than 3, we may say that departure is not very substantial because kurtosis is smaller than 3.5 and skewness is smaller than 1.9. Therefore, we used the maximum likelihood method with bootstrapping with 2000 samples. The Bollen-Stine bootstrap produced a *p*-value of the overall model fit with $p = 0.15$, which does not allow to reject the null hypothesis that the model fits the data well at the level $p = 0.05$.

ML estimation "yields unbiased, consistent, and efficient parameter estimates and provides a model test statistic or assessing the adequacy of a hypothesized model" (Nevitt and Hancock 2001). ML assumes multivariate normality and large sample but it is a common practice to use ML with bootstrap for relatively small samples (op. cit., 354). We estimated the model with ML with the bootstrap and 0.95

bias-corrected confidence level. The fit indices (χ^2/df = 1.361; RMSEA = 0.045; CFI = 0.940; NFI = 0.810; and GFI = 0.850) suggest good fit of the model.

Figure 4 shows the results of the model estimation; the values are standardized coefficients with two-tailed bias-corrected significance at a confidence level of 0.95. Because of the small sample size and a relatively complex model, we decided to show paths with 0.1 significance (acceptable in social sciences), in addition to more traditional paths with $p < 0.01$ and $p < 0.05$.

Note that we changed the arrow depicting the effect on *Accuracy* so that it corresponds to the regression coefficient's sign. The positive effect on *Accuracy* formulated in the hypotheses (Sects. 2.3–2.5) and shown in Fig. 2 means that the accuracy increases. However, accuracy increase is achieved when the value of D_H decreases (full accuracy is achieved when $D_H = 0$). Therefore, in Fig. 4 the depiction of the arrows leading to *Accuracy* is reversed, i.e., the positive (negative) coefficient value associated with a continuous (dotted) arrow indicates a decrease (increase) in accuracy. We also added the cardinal measure of accuracy L_1; it strongly and directly depends on the ordinal inaccuracy.

During the testing of the structural model, we observed that goal *Learning* affects goals *Substantive* and *Relational* (see Fig. 4). This indicates a hierarchical relationship; *Learning* has a strong (0.40) and significant ($p < 0.01$) effect on goal *Substantive* and it has a weaker (0.16) and less significant ($p < 0.05$) effect on goal *Relational.*

A more complete analysis requires including also indirect and total effects. Although these effects can be computed using the results given in Fig. 4, to make the discussion in the next section easier, the total effects (i.e., the sum of all direct and indirect effects, of one model variable on another) are given in Table 3.

Total effects are sums of direct effect (e.g., $A \rightarrow C$) and indirect effect (e.g., $A \rightarrow B \rightarrow C$). If direct effect ($A \rightarrow B$) is insignificant, then the indirect effects ($A \rightarrow B \rightarrow C$) may be insignificant even when two effects ($A \rightarrow B$ and $B \rightarrow C$) are significant. In Table 3 five total effect are insignificant because they comprise only indirect insignificant effects, which are not significant.

4.3 Results

A cursory assessment of the results presented in Fig. 4 indicates that many of the hypotheses are not confirmed. However, 14 out of 34 hypotheses presented in Sect. 2 and indicated in Fig. 2 are confirmed. In addition, we observed the effects not hypothesized before. Figure 4 shows only significant direct effects. In the following, three tables (Tables 4, 5 and 6), we show all significant effects for the three types of motivations.

Epistemic Motivation. The results shown in Tables 3 and 4 confirm that motivation *Rational* has a positive effect on *Collaborating* and a negative effect on *Accommodating* intentions. Contrary to our hypothesis, it has a positive effect on *Competing.* The postulated positive effect on competitive intentions was based on the association of proself motivation on experiential motivation (De Dreu, Beersma et al. 2006).

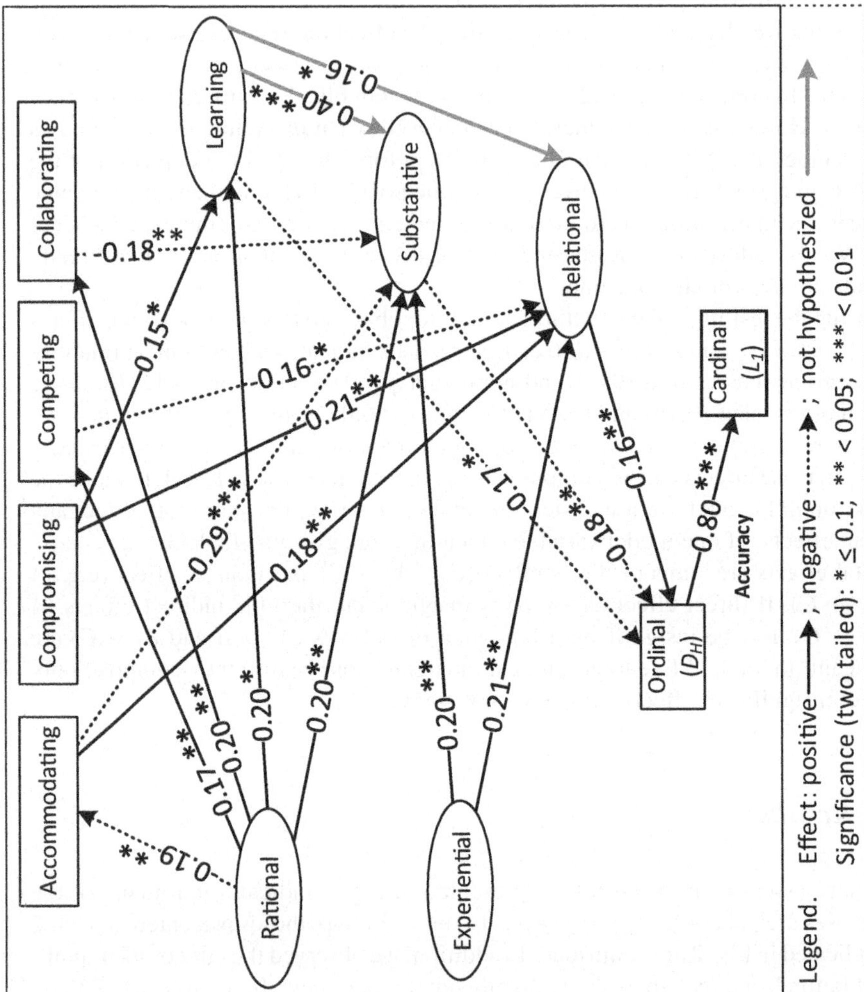

Fig. 4 Results of the structural model: standardized coefficients and significance

Table 3 Standardized total effect

	Epistemic motivation		Social motivation				Identity motivation			Accuracy
	Rational	Experient	Accommod	Compr	Compet	Collab	Learn	Subst	Relat	D_H
Accommod	−.190	.000	.000	.000	.000	.000	.000	.000	.000	.000
Competing	.174	.000	.000	.000	.000	.000	.000	.000	.000	.000
Collaborating	.204	.000	.000	.000	.000	.000	.000	.000	.000	.000
Learning	.202	.000	.000	.146	.000	.000	.000	.000	.000	.000
Substantive	.300	.198	−.286	**.059**	.000	−.181	.405	.000	.000	.000
Relational	**−.028**#	.213	.179	.230	−.156	.000	.163	.000	.000	.000
DH	**−.094**	**−.002**#	**.081**	**.001**#	**−.025**	**.033**	−.221	−.181	.160	.000
L1	**−.075**	**−.001**#	**.064**	**.001**#	**−.020**	**.026**	**−.176**	**−.145**	**.128**	.799

– total effect insignificant, i.e., $p > 0.1$; bolded indirect ef fect only

Table 4 Effects of epistemic motivation on social and identity motivations and on accuracy

Hypothesis	Effect	Confirmed
Epistemic and social motivation		
Rational system has positive effect on collaborating intentions	Direct, positive***	Yes
Rational system has negative effect on competing intentions	Direct, positive**	No
Rational system has negative effect on accommodating intentions	Direct, negative**	Yes
Epistemic and identity motivation		
Rational system has positive effect on substantive goals	Indirect, positive* Direct, positive**	Yes
Rational system has positive effect on learning goals	Direct, positive*	Yes
Experiential system has negative effect on substantive goals	Direct, positive**	No
Experiential system has positive effect on relational goals. ^	Direct, positive**	–
Epistemic motivation and accuracy		
Rational system has positive effect on accuracy	Indirect, negative on D_H*** and L_1***	Yes

*$p < 0.1$; **$p < 0.05$; ***$p < 0.01$; ^ – not hypothesized

Another possibility is that competing, like collaborating, requires the involvement of systematic information processing and a good understanding of the task (Carnevale and De Dreu 2006), which would explain the positive effect of rational motivation on competing.

Neither *Rational* nor *Experiential* motivation affect *Compromising* intentions (Table 4). This may be due to the particularity of the TKI scale we used to measure

Table 5 Effects of social motivation on identity motivation and on accuracy

Hypothesis	Effect	Support
Social and identity motivations		
Compromising intentions have positive effect on substantive goals	Indirect, positive*	Yes
Compromising intentions have positive effect on relational goals	Direct, positive** Indirect, positive* Total, positive**	Yes
Compromising intentions have positive effect on learning goals	Direct, positive*	Yes
Accommodating intentions have negative effect on substantive goals	Direct, negative***	Yes
Accommodating intentions have positive effect on relational goals	Direct, positive*	Yes
Competing intentions have negative effect on relational goals	Direct, negative*	Yes
Collaborating intentions have positive effect on substantive goals	Direct, negative***	No
Social motivation and accuracy		
Collaborating intentions have positive effect on accuracy	Indirect, positive on D_H** and L_1**	No
Accommodating intentions have positive effect on accuracy	Indirect, positive on D_H*** and L_1***	No
Competing intentions have negative effect on accuracy	Indirect, negative on D_H* and L_1*	No

*$p < 0.1$; **$p < 0.05$; ***$p < 0.01$

Table 6 Effects of identity motivation on accuracy

Hypothesis	Effect	Support
Social and identity motivations		
Learning goals has positive effect on accuracy	Direct, negative on D_H* Total, negative on D_H** Indirect, negative on L_1**	Yes
Substantive goals has positive effect on accuracy	Direct, negative on D_H** Indirect, negative on L_1**	Yes
Relational goals has negative effect on accuracy	Direct, positive on D_H** Indirect, positive on L_1**	Yes

*$p < 0.1$; **$p < 0.05$; ***$p < 0.01$

social motivation (Sect. 2.3) and the fact that compromising is a mixture of competing, collaborating, and accommodating actions.

We found that *Experiential* does not affect any social motivation. This result may be explained with the experiential system not requiring planning and strategizing. Social motivation refers here to the negotiation intentions, which form strategies and tactics requiring forward-thinking and analysis. Experiential motivation allows for cognitive misery and focuses on tasks that must be done, i.e., participation in the negotiation, and compromise achievement. These tasks are related to goals *Relational* and *Substantive*; indeed, we found a significant effect of *Experiential* on *Relational* and *Substantive*. The same cognitive misery rationale suggests that *Experiential* has no effect on goal *Learning*.

Rational motivation has a direct significant positive effect on two goals: *Learning* and *Substantive* (Fig. 4). For *Substantive*, the mediation of *Accommodating* and *Collaborating* increases the effect from 0.20 (Fig. 4) to the total effect of 0.30 (Table 3). Although *Rational* has no direct effect on *Relational* it has an indirect effect through *Accommodating* and *Competing* motivations; the total effect is small (−0.03) and insignificant.

Neither *Rational* nor *Experiential* motivation have a direct impact on *Accuracy*, but both have an indirect effect. *The rational* motivation effect on *Accuracy* comprises paths with one and with two mediators (Fig. 4). There are two one-mediator paths with mediators describing identity motivation (e.g., *Rational* →*Substantive* → *Ordinal*); three two-mediator paths with pairs of mediators describing social and identity motivation (e.g., *Rational* →*Accommodating* → *Relational* → *Ordinal*); and two two-mediator paths with one *Learning* mediator. Because some mediators have a positive effect and others – negative, the total effect of *Rational* on *Accuracy* is, as shown in Table 3, negative and small (-0.094).

A simpler relationship is between *Experiential* and *Ordinal*. In this case, there are only two mediators: *Substantive* has a negative effect (−0.18) and *Relational* has a positive effect (0.16); the total effect is small and it is insignificant. Consequently, the effect on *Cardinal* is small and insignificant. Therefore, the hypothesis on the negative effect of *Experiential* motivation on *Accuracy* is not confirmed.

Social Motivation. The results summarized in Table 5 show that six hypotheses (out of 12) confirm the causal effect between social and identity motivation.

As hypothesized, the effect of *Compromising* intentions on *Learning* (0.15) and on *Relational* (0.21) is direct and significant.

The indirect effect on *Relational* via *Learning* mediator (i.e., *Compromising* → *Learning* → *Relational*) is equal to 0.02; it increases the total effect to 0.23 (Table 3). There is only an indirect and weakly significant positive (0.06) effect of *Compromising* on *Substantive* goals. This way, *Compromising* is the only intention, which affects positively all three goals defined in our model.

As hypothesized, the effect of *Accommodating* intentions on *Substantive* is negative (−0.29) and positive (0.18) on *Relational* goals. *Competing*, as hypothesized, has a negative (−0.16) effect on *Relational* but the significance is weak. There is no significant impact of *Competing* on *Learning* and on*Substantive*. This result suggests that competing intentions have only a negative impact on relational goals and they are not associated with willingness to achieve a highly rated agreement. This departure from the results reported in the literature needs to be further investigated.

The significant negative (−0.18) impact of *Collaborating* intentions on *Substantive* goals is opposite to the hypothesized one. This may suggest that the focus on the substantive goals is one-sided, i.e., only own rather than joint results and cooperative intentions indicate a willingness for joint work. On the other hand, no significant effect of *Collaborating* on *Relational* puts this result into question. The anonymity and the fact that the possibility of the participants meeting each other in the future is remote may play a role.

There are no direct effects of social motivation on the accuracy of preference representation. There are significant indirect effects of some intentions that, surprisingly, occurred to confirm the opposite relationship. These effects were, however, much smaller than other effects observed in the model (see Table 3). For *Accommodating*, there are two mediation paths with *Relational* and *Substantive* goals of mediators; the total indirect effect on *Ordinal* accuracy is 0.08. We also have a positive (0.03) indirect effect of *Collaborating* on *Ordinal* accuracy via *Substantive*.

Competing intentions have a small indirect negative (−0.03) effect on the *Ordinal* via *Relational* goals. We hypothesized that *Competing* decreases accuracy (increases *Ordinal*) to reflect the agents' interest in assigning ratings that would maximize their own rather than the principal's substantive outcomes. The relationship *Competing* → *Substantive* is not significant but *Competing* → *Relational* is significant with, as expected, negative effect. Because, the relationship *Relational* → *Ordinal* has, as expected, positive effect (Table 3), the indirect effect of *Competing* → *Relational* → *Ordinal* has to be negative.

Identity Motivation. Table 6 shows the effects of identity motivation on accuracy. All hypotheses related to the effect of the goals on accuracy (both *Ordinal* and *Cardinal*) are supported. However, in the case of the *Cardinal* measure of accuracy, the effects were only indirect.

Substantive and *Learning* goals have a positive effect on accuracy (decrease value of *Ordinal* D_H and *Cardinal* L_1 measures), while *Relational* have negative effects on both types of *Accuracy* (increase value of D_H and L_1). *Learning* has also indirect small negative effect (−0.05) on *Ordinal* (*Substantive* and *Relational* are mediators) increasing total ordinal accuracy by −0.22.

4.4 Discussion

In Sect. 4.3, we discussed the causality between rational and social motivation but found no significant relationship between experiential and social motivation (see Table 4). De Dreu and Carnevale (2003) argue that social motivation drives the kind of information (cooperative or competitive) negotiators seek while epistemic motivation determines the extent to which this information is being processed. Furthermore, they assume that these two motivations are independent and that this assumption holds empirically (op. cit., p. 269). In contrast, our results show that social motivation is a consequence of one type of rational epistemic motivation and it is independent of experiential epistemic motivation. The difference is in the perspective on and the measurement of motivation. De Dreu and Carnevale (2003) refer to studies in which epistemic motivation is a one-dimensional variable measured using the "Need for closure" scale. The REI scale used here produces two-value epistemic motivation. The independence of social motivation that we observe for experiential motivation confirms but the dependence of this motivation on rational motivation contradicts the earlier results. It does, however, confirm experimental results in which epistemic motivation was manipulated by making participants accountable or not. Van Der Schalk et al. (2010) report negotiators with low epistemic motivation, i.e., highly experiential, do not change their social motivation but those with high epistemic motivation, i.e., highly rational, do (see also, De Dreu, Beersma et al. 2006).

The purpose of this study was to analyze how—in the context of preparation for negotiations—motivation affects accuracy. The model presented in Sect. 4.2 (Fig. 4) shows that every type of motivation affects accuracy either directly or indirectly (Table 3). Only identity motivation (*Learning*, *Substantive* and *Relational*) has direct effects on *Accuracy* but only on ordinal accuracy (measured with D_H). They are fairly large (between 16 and 18%) and significant. *Learning* has the largest effect on ordinal accuracy (D_H) confirming that the negotiators who wish to learn how to negotiate and improve their negotiating skills make efforts to perform the preference specification task diligently. These efforts are successful.

Negotiation literature suggests that focus on *Substantive* goals is typical for competitive negotiators and, to a lesser degree, those who compromise. *Competitive* motivation has no effect and *Compromising* has a negative effect on *Substantive*. Other reasons for the negotiators' focus on *Substantive* may be the natural association of analytical approach characteristic for *Rational* and interest in *Learning*. We observe both and they may be behind the significant effect of *Substantive* on D_H.

Relational goals, in the context of online anonymous negotiations, may be indicative of the participants' interest in having nice, easygoing interactions with their counterparts or interests in learning how to build a relationship. We observe that both *Accommodating* and *Learning* have a significant and positive effect on *Relational*, creating conditions for little interest in the quantitative aspect of the negotiation that includes the formulation of preferences. Thus, *Relational* has a negative effect on D_H.

The effect of two motivations, *Experiential* and *Compromising*, on accuracy D_H is indirect, very small, and insignificant. Also small (between 2 and 9%) but significant are the indirect effects of *Rational* and social motivation (*Accommodating*, *Compromising*, *Competing* and *Collaborating*) on D_H.

Cardinal accuracy measure L_1 is strongly and significantly dependent on the ordinal measure D_H. This means that if participants made an ordinal error, then they had to make also cardinal errors (Sect. 3.3, Table 1). While it was possible to make cardinal but not ordinal errors, such errors would have to be relatively small.

In summary, the motivation model that fits the experimental data shows that rational information processing significantly affects most of the negotiation intentions and identity goals, except for the relational goals. Negotiation intentions significantly affect identity goals, which, in turn, affect accuracy.

5 Conclusions

Discussions on motivation recognize its different types but the opinions differ regarding the relationship between the types. One reason for these differences is the fact that experiments were done in different contexts and involved different populations and tasks. Negotiation experiments take time during which the participants first learn about the case and their role and then interact with the purpose to achieve an agreement. It is possible that their initial intrinsic motivation changes because they get interested in the process, surprised by their counterpart, or realize that no agreement is likely. These changes, observed in real life and in experiments, lead researchers to notice causal and non-causal relations as well as different directions of causality. For example, prior social motivation affects epistemic motivation and identity motivation affects social motivation in the same context (Harackiewicz and Elliot 1993; Dupont, Carlier et al. 2009). This dependence on prior experiences and the ability to weaken one type of motivation and strengthen another may question the reliance on theories and associated measures, which posit stable, time-independent motivation.

This study adds to the discussion about the stability of the rational and experiential systems assumed by (Berger 2005) and others. It is possible that some people rely on only one system. Most people have one system stronger, the other weaker or their strength is similar. These people can choose; depending on the circumstances, they can then rely on one or another system, e.g., if one uses a rational system, then time pressure or information overload may cause switching to the experiential system. They also can move between the systems. Our students learn and practice analytical and critical thinking hence they can employ the rational system. On occasions, however, they skip classes, do not work on assignments, or, even if they work on them, they do poorly. The REI scale used here does not allow distinguishing between the experiential system, cognitive misery, and laziness, which—depending on the circumstances—may produce different results (Chater, Felin et al. 2018).

The results discussed here confirm that people, who have low ability to expend cognitive effort or their extrinsic motivation is low, are more likely to rely on the experiential system (Shiloh, Salton et al. 2002). Reliance on the experiential system allows bypassing the task difficulty and producing results quickly, which suggests that those who are uninterested in the task may revert to the experiential system. Witteman et al. (2009) found a significant positive correlation between the relational system and a significant negative correlation between the experiential system and performance on tasks that required computations and deductive reasoning.

The ability to choose information processing mode raises the issue of the relationship between mode and task. We used REI-20 because it was found effective in earlier studies. Other instruments may be better suited for this type of experiments. Novak and Hoffman (2008) observe that the relationship between epistemic motivation and the task assumed in REI may be reversed, i.e., "the rational or experiential nature of a task, as well as the motivation for performing a task, influence situation-specific thinking style." (op. cit., p. 57). They propose SSTS scale designed to measure the impact of the situation-specific information processing system and the nature of the task on motivation and performance. Comparisons of REI and SSTS in experiential and analytical types of tasks led the authors to conclude (op. cit., p. 46) that in some cases REI acts as a suppressor variable. The task used in this experiment is strongly analytical and it is possible that it influences the participants' use of their processing system. This suggests that SSTS may be a more appropriate instrument used after the participants constructed their preference system rather than before.

Another scale, which we used but proved to be problematic, is TKI. This instrument is well known and widely used but its characteristics make the statistical analysis of its results difficult (see Sect. 2.3 and 4.1). We used it because it was already implemented in Inspire to provide users with a post-negotiation assessment of their negotiation predispositions. Some negotiation studies use decomposed games to determine prosocial and proself motivation. ROCI-II designed by Rahim (1983) has also been used and its results can be analyzed but the questions would have to be modified to make them relevant to students. NEO-PI(R) designed by (Costa Jr, McCrae et al. 1991) provides another possibility to assess how negotiation preparation is affected by distinct facets of conscientiousness.

The shortcomings of both scales suggest that SST and ROCI-II and/or NEO-PI(R) should be used in future experiments. In addition, larger samples would help in using SEM without bootstrapping.

De Dreu (2004, p. 129) concludes his discussion on epistemic, social and impression motivation with an observation that impression motivation is not a well-developed area of negotiation research. Identity motivation is the theory that encompasses impression motivation (Oyserman, Lewis Jr et al. 2017) and De Dreu comments may be extended to identity motivation. While we considered identity motivation in a narrow sense and limited it to three goals, the results show that these goals play a very significant role both as mediators between epistemic and social motivation and as predictors of accuracy. We also showed that goals may form their own hierarchy; there are higher-level goals (e.g., learning) that affect lower-level goals (substantive and relational).

We noted that the participants made both ordinal errors and cardinal errors. The fact that they made ordinal errors (2.34 on average) requires further investigation. Further analysis shows that 62 out of 181 participants were ordinal-accurate, the remaining 119 participants made 3.55 ordinal errors on average. The average cardinal inaccuracy value was 35.18 for ordinary-accurate and 71.18 for ordinary-inaccurate participants. The accuracy of the two groups is significantly different, using the Mann–Whitney *U* test yields $z = -11.333$, $p < 0.001$ for ordinal and $z = -7.255$, $p < 0.001$ for cardinal accuracy. The sample size for these two groups was too small to determine differences in their motivation using SEM. We also found differences in their social and identity motivation. The accommodating and competitive intentions are significantly different; accommodating is weaker (TKI: 4.10 vs. 4.93; $p = 0.021$) and competing is stronger for the ordinal-accurate than for the inaccurate groups (5.63 vs. 4.61; $p = 0.023$). Mann–Whitney U test also shows that the two groups significantly differ in their relational and learning motivation. Relational goals are much less important for the ordinal-accurate group than for the ordinal-inaccurate group (0.87 vs. 1.26; $p = 0.006$) but the reverse is true of the learning goals (1.46 vs. 1.17; $p = 0.013$). Larger data sets are required to perform a more complex analysis that would help distinguish between agents who are likely to be well prepared and those who are not.

Positive agency theory assumes that an extrinsic motivation system can be devised to motivate agents to negotiate on behalf of their principals to the best of their abilities. Many negotiation experiments also rely on the assumption that extrinsic motivation can direct the participants to behave in a particular way. Arguably, the extrinsic motivation formulated in the course requirements was quite strong. The results achieved by the majority (66%) of the participants are not, however, satisfactory. Although they completed all negotiation activities and they discussed their activities in their reports, they had made easily avoidable errors. This confirms the limitations of the positive agency theory and indicates that the principals need to know their potential agents' intrinsic motivation.

Extrinsic motivation may increase the use of the rational system and reduce the use of the experiential system. The proposed model of intrinsic motivation shows that this change may not be sufficient because rational motivation does not affect accuracy directly. The model also indicates that the participants of this experiment might not have had a good understanding of the requirements associated with compromising, competing, and collaborating negotiation strategies. The compromising positive and the competing negative affect of the relational goals are correct, but they should also affect the substantive goals, which they do not. Collaborating, being a strategy that promotes substantive outcomes of both negotiation sides, should not have a negative effect on the substantive goals and it should have a positive effect on the relational goals.

We included many relationships in the research model because they have been theorized in literature coming from management, psychology, and sociology. Rather than make assumptions that some of these relationships are irrelevant, our purpose was to determine their relevancy through data analysis. As far as we know, this is the first attempt to construct a framework of intrinsic motivation, which puts together the different types of motivation discussed in the literature. It is likely that the particular relationships and their strength differ in different settings and with different participants, which lead us to the conclusion that the presented approach and model may provide directions for future studies.

Acknowledgements This work was supported by the grant from Polish National Science Center (2016/21/B/HS4/01583) and Natural Sciences and Engineering Research Council Canada.

Appendix. Similarity of Two Preference Systems

The principal's and the agents' preference systems are both represented with a vector of natural numbers. The similarity of these two systems can be measured on *ordinal* and *cardinal* scales. Ordinal comparisons can be done with Hamming distance because it counts the number of differences made in pairwise comparisons and is easy to interpret (Hamming 1950).

Let $S_P = \left\{w_{ij}^P, i = 1, \ldots, m_j, j = 1, \ldots, n\right\}$ and $S_A = \left\{w_{ij}^A, i = 1, \ldots, m_j, j = 1, \ldots, n\right\}$ denote the preference systems of respectively the principal and the agent, where w_{ij} is the score of the i th option of the j th negotiation issue, m_j is the number of options in issue j, and n is the number of issues. Without loss of generalization, we assume that the score of the best option for an issue is equal to the score (rating) of this issue, i.e. $\max_i\left(w_{ij}^P\right) = w_j^P$ and $\max_i\left(w_{ij}^A\right) = w_j^A$.

Hamming distance measures the similarity of S_P and S_A; it is the number of times the agent made an ordinal error in comparing two issues or two options of an issue. If an issue (option) is more important for the principal than another issue (option) but it is less or equally important for the agent, then 1 is added to the distance. If it is also more important for the agent, then 0 is added. Let $r_{jj'}$ define the agent's error in scoring issues j and j'. The values $r_{jj'}$ are obtained from:

$$r_{jj'} = \begin{cases} 1 \text{ if } \operatorname{sign}\left(w_j^P - w_{j'}^P\right) \neq \operatorname{sign}\left(w_j^A - w_{j'}^A\right) \\ 0 \text{ otherwise} \end{cases} \tag{1}$$

where: $j = 1, \ldots, n-1$; $j' = j+1, \ldots, n$, and sign(x) is a function that extracts the sign of w.

Error for comparison for options i and i' of j th issue is $r_{jj'}^j$ and its values are obtained from:

$$r_{ii'}^j = \begin{cases} 1 \text{ if } \operatorname{sign}\left(w_{ij}^P - w_{i'j}^P\right) \neq \operatorname{sign}\left(w_{ij}^A - w_{i'j}^A\right) \\ 0 \text{ otherwise} \end{cases} \tag{2}$$

for each $j = 1, \ldots, n$; $i = 1, \ldots, m_j - 1$ and $i' = i + 1, \ldots, m_j$.

Hamming distance between S_P and S_A is

$$D_H(S_P, S_A) = \sum_{j=1}^{n-1} \sum_{j'=j+1}^{n} r_{jj'} + \sum_{j=1}^{n} \sum_{i=1}^{m_j-1} \sum_{i'=i+1}^{m_j} r_{ii'}^j. \tag{3}$$

Agents who make no pairwise comparison errors are ordinal-accurate, i.e., $D_H(S_P, S_A) = 0$.

The second measure of accuracy is cardinal and we use block (Manhattan) distance L_1. The cardinal distance between two preference systems S_P and S_A is the sum of the absolute differences of the principal's and the agent's scores, that is:

$$L_1(S_P, S_A) = \sum_{j=1}^{n} \sum_{i=1}^{m_j} \left| w_{ij}^P - w_{ij}^A \right|. \tag{4}$$

The agent is cardinal-accurate if $L_1(S_P, S_A) = 0$, for inaccurate agents $L_1(S_P, S_A) > 0$.

References

Aaldering, H., Greer, L.L., Van Kleef, G.A., De Dreu, C.K.: Interest (mis)alignments in representative negotiations: do pro-social agents fuel or reduce inter-group conflict? Organ. Behav. Hum. Decis. Process. **120**(2), 240–250 (2013)

Bagozzi, R.P., Edwards, J.R.: A general approach for representing constructs in organizational research. Organ. Res. Methods **1**(1), 45–87 (1998)

Bamberg, G., Spremann, K. (eds.): Agency Theory, Information, and Incentives. Springer, Heidelberg (1989)

Bargh, J.A.: Our unconscious mind. Sci. Am. **310**(1), 30–37 (2014)

Bass, B.M.: Effects on the subsequent performance of negotiators of studying issues or planning strategies alone or in groups. Psychol. Monogr. Gen. Appl. **80**(6), 1 (1966)

Bazerman, M.H., Neale, M.A., Valley, K.L., Zajac, E.J., Kim, Y.M.: The effect of agents and mediators on negotiation outcomes. Organ. Behav. Hum. Decis. Process. **53**(1), 55–73 (1992)

Berger, C.R.: Slippery slopes to apprehension: rationality and graphical depictions of increasingly threatening trends. Commun. Res. **32**(1), 3–28 (2005)

Bottom, W.P., Holloway, J., Miller, G.J., Mislin, A., Whitford, A.: Building a pathway to cooperation: negotiation and social exchange between principal and agent. Adm. Sci. q. **51**(1), 29 (2006)

Brown, J.G.: Empowering students to create and claim value through the Thomas-Kilmann conflict mode instrument. Negot. J. **28**(1), 79–91 (2012)

Byrne, B.M.: Structural Equation Modeling with AMOS: Basic Concepts, Applications, and Programming. Routledge, New York (2010)

Carnevale, P.J., De Dreu, C.K.W.: Motive: the negotiator's raison d'être. In: Thompson, L. (ed.) Frontiers of Social Psychology: Negotiation Theory and Research, pp. 55–56. Psychosocial Press, New York (2006)

Cellar, D.F., Stuhlmacher, A.F., Young, S.K., Fisher, D.M., Adair, C.K., Haynes, S., Twichell, E., Arnold, K.A., Royer, K., Denning, B.L.: Trait goal orientation, self-regulation, and performance: a meta-analysis. J. Bus. Psychol. **26**(4), 467–483 (2011)

Chaiken, S., Trope, Y.: Dual-Process Theories in Social Psychology. Guilford Press, New York (1999)

Chater, N., Felin, T., Funder, D.C., Gigerenzer, G., Koenderink, J.J., Krueger, J.I., Noble, D., Nordli, S.A., Oaksford, M., Schwartz, B., Stanovich, K.E., Todd, P.M.: Mind, rationality, and cognition: an interdisciplinary debate. Psychon. Bull. Rev. **25**(2), 793–826 (2018)

Costa, P.T., Jr., McCrae, R.R., Dye, D.A.: Facet scales for agreeableness and conscientiousness: a revision of the NEO personality inventory. Pers. Individ. Differ. **12**(9), 887–898 (1991)

Cropanzano, R., James, K., Citera, M.: A goal hierarchy model of personality, motivation, and leadership. Res. Organ. Behav. **15**, 267–267 (1993)

Curhan, J.R., Elfenbein, H.A., Xu, H.: What do people value when they negotiate? Mapping the domain of subjective value in negotiation. J. Pers. Soc. Psychol. **91**(3), 493–512 (2006)

De Dreu, C.K.W., van Lange, P.A.M.: The impact of social value orientations on negotiator cognition and behavior. Pers. Soc. Psychol. Bull. **21**(11), 1178–1188 (1995)

De Dreu, C.K., Koole, S.L., Oldersma, F.L.: On the seizing and freezing of negotiator inferences: need for cognitive closure moderates the use of heuristics in negotiation. Pers. Soc. Psychol. Bull. **25**(3), 348–362 (1999)

De Dreu, C.K., Beersma, B., Stroebe, K., Euwema, M.C.: Motivated information processing, strategic choice, and the quality of negotiated agreement. J. Pers. Soc. Psychol. **90**(6), 927 (2006)

De Dreu, C.K.: Motivation in negotiation: a social psychological analysis. In: Gelfand, M.J., Brett, J.M. (eds.) The Handbook of Negotiation and Culture, pp. 114–135. Stanford Business Books, Stanford (2004)

DuDreu, C., Carnevale, P.: Motivational bases for information processing and strategic choice in conflict and negotiation. Adv. Exp. Soc. Psychol. **35**, 235–291 (2003)

Dupont, J.-P., Carlier, G., Gerard, P., Delens, C.: Teacher-student negotiations and its relation to physical education students' motivational processes: an approach based on self-determination theory. Eur. Phys. Educ. Rev. **15**(1), 21–46 (2009)

Dweck, C.S.: Motivational processes affecting learning. Am. Psychol. **41**(10), 1040 (1986)

Eisenhardt, K.M.: Agency theory: an assessment and review. Acad. Manag. Rev. **14**(1), 57–74 (1989)

Elliot, A.J., Harackiewicz, J.M.: Approach and avoidance achievement goals and intrinsic motivation: a mediational analysis. J. Pers. Soc. Psychol. **70**(3), 461 (1996)

Epstein, S.: Cognitive-experiential self-theory of personality. In: Millon, T., Lerner, M.J. (eds.) Handbook of Psychology, pp. 159–184. Wiley, Hoboken (2003)

Epstein, S., Pacini, R., Denes-Raj, V., Heier, H.: Individual differences in intuitive–experiential and analytical–rational thinking styles. J. Pers. Soc. Psychol. **71**(2), 390 (1996)

Fairweather, A.: Epistemic motivation. In: Fairweather, A., Zagzebski, L. (eds.) Virtua Epistemology. Oxford University Press, Oxford (2001)

Fleck, D., Volkema, R., Pereira, S., Vaccari, L.: Factors affecting desire to negotiate again. J. Manag. Psychol. **32**(1), 16–29 (2017)

Gelfand, M.J., Major, V.S., Raver, J.L., Nishii, L.H., O'Brien, K.: Negotiating relationally: the dynamics of the relational self in negotiations. Acad. Manag. Rev. **31**(2), 427–451 (2006)

Gillespie, J.J., Brett, J.M., Weingart, L.R.: Interdependence, social motives, and outcome satisfaction in multiparty negotiation. Eur. J. Soc. Psychol. **30**(6), 779–797 (2000)

Graham, S., Golan, S.: Motivational influences on cognition: task involvement, ego involvement, and depth of information processing. J. Educ. Psychol. **83**(2), 187 (1991)

Griessmair, M., Druckman, D.: To match or not to match? Reactions to turning points in negotiation. Group Decis. Negot. **27**(1), 61–83 (2018)

Hamming, R.W.: Error detecting and error correcting codes. Bell Syst. Tech. J. **29**(2), 147–160 (1950)

Harackiewicz, J.M., Elliot, A.J.: Achievement goals and intrinsic motivation. J. Pers. Soc. Psychol. **65**(5), 904 (1993)

Huber, V.L., Neale, M.A.: Effects of self- and competitor goals on performance in an interdependent bargaining task. J. Appl. Psychol. **72**(2), 197–203 (1987)

Jensen, M.C., Meckling, W.H.: Theory of the firm: managerial behavior, agency costs and ownership structure. J. Financ. Econ. **3**(4), 305–360 (1976)

Johnson, C.A., Thompson, R.C., Anderson, M.G.: Technical brief for the Thomas-Kilmann conflict mode instrument. CPP (2014). https://kilmanndiagnostics.com/wpcontent/uploads/2018/04/TKI_TranslationsStudy.pdf. Accessed 20 May 2020

Kabanoff, B.: Predictive validity of the MODE conflict instrument. J. Appl. Psychol. **72**(1), 160 (1987)

Kasser, T., Ryan, R.M.: Further examining the American dream: differential correlates of intrinsic and extrinsic goals. Pers. Soc. Psychol. Bull. **22**(3), 280–287 (1996)

Kersten, G.E., Gimon, D., Wu, S.: Sellers' objectives in E-procurement auction and negotiation experiments. In: 46th Hawaii International Conference on System Sciences (HICSS). IEEE, pp. 538–547 (2013)

Kersten, G.E., Noronha, S.J.: WWW-based negotiation support: design, implementation, and use. Decis. Support Syst. **25**, 135–154 (1999)

Kline, R.B.: Principles and Practice of Structural Equation Modeling. Guilford Publications, New York (2015)

Kunz, A.H., Pfaff, D.: Agency theory, performance evaluation, and the hypothetical construct of intrinsic motivation. Acc. Organ. Soc. **27**(3), 275–295 (2002)

Levy, I., Kaplan, A., Patrick, H.: Early adolescents' achievement goals, social status, and attitudes towards cooperation with peers. Soc. Psychol. Educ. **7**(2), 127–159 (2004)

Liu, L.A., Chua, C.H., Stahl, G.K.: Quality of communication experience: definition, measurement, and implications for intercultural negotiations. J. Appl. Psychol. **95**(3), 469 (2010)

Locke, E.A., Latham, G.P.: What should we do about motivation theory? Six recommendations for the twenty-first century. Acad. Manag. Rev. **29**(3), 388–403 (2004)

Loewenstein, J., Thompson, L.L.: Learning to negotiate: novice and experienced negotiators. In: Thompson, L.L (ed.) Negotiation Theory and Research, pp. 77–97. Psychology Press, New York (2006)

Marsh, H.W., Balla, J.R., Hau, K.-T.: An evaluation of incremental fit indices: a clarification of mathematical and empirical properties. In: Marcoulides, G.A., Schumacker, R.E. (eds.) Advanced Structural Equation Modeling: Issues and Techniques, pp. 315–353. Lawrence Erlbaum Associates, Mahwah (1996)

McClintock, C.G., van Avermaet, E.: Social values and rules of fairness: a theoretical perspective. In: Derlega, V.J., Grezlak, J. (eds.) Cooperation and Helping Behavior, pp. 43–71. Academic Press, New York (1982)

Messick, D.M., McClintock, C.G.: Motivational basis of choice in experimental games. J. Exp. Soc. Psychol. **4**(1), 1–25 (1968)

Meyer, C.J., McCormick, B., Clement, A., Woods, R., Fifield, C.: Scissors cut paper: purposive and contingent strategies in a conflict situation. Int. J. Confl. Manag. **23**(4), 344–361 (2012)

Nevitt, J., Hancock, G.R.: Performance of bootstrapping approaches to model test statistics and parameter standard error estimation in structural equation modeling. Struct. Equ. Model. **8**(3), 353–377 (2001)

Novak, T.P., Hoffman, D.L.: The fit of thinking style and situation: new measures of situation-specific experiential and rational cognition. J. Consum. Res. **36**(1), 56–72 (2008)

Olekalns, M., Smith, P.L.: Social motives in negotiation: the relationships between dyad composition, negotiation processes and outcomes. Int. J. Confl. Manag. **14**(3/4), 233–254 (2003)

Olekalns, M., Smith, P.L., Kibby, R.: Social value orientations and negotiator outcomes. Eur. J. Soc. Psychol. **26**, 299–313 (1996)

Oyserman, D., Lewis, N.A., Jr., Yan, V.X., Fisher, O., O'Donnell, S.C., Horowitz, E.: An identity-based motivation framework for self-regulation. Psychol. Inq. **28**(2–3), 139–147 (2017)

Pepper, A., Gore, J.: Behavioral agency theory new foundations for theorizing about executive compensation. J. Manag. **41**(4), 1045–1068 (2015)

Pruitt, D.G., Lewis, S.A.: Development of integrative solutions in bilateral negotiation. J. Pers. Soc. Psychol. **31**(4), 621–633 (1975)

Pruitt, D.G., Rubin, J.Z.: Social Conflict: Escalation, Stalemate, and Settlement. Random House, New York (1986)

Putnam, L.L.: Reframing integrative and distributive bargaining: a process perspective. In: Sheppard, B.H., Bazerman, M.H., Lewicki, R.J. (eds.) Research on Negotiation in Organization, vol. 2, pp. 3–30. JAI Press, Greenwich (1990)

Rahim, M.A.: A measure of styles of handling interpersonal conflict. Acad. Manag. J. **26**(2), 368–376 (1983)

Rubin, J.Z., Brown, B.R.: The Social Psychology of Bargaining and Negotiation. Academic Press, New York (1975)

Ruble, T.L., Thomas, K.W.: Support for a two-dimensional model of conflict behavior. Organ. Behav. Hum. Perform. **16**(1), 143–155 (1976)

Ryan, R.M., Deci, E.L.: Intrinsic and extrinsic motivations: classic definitions and new directions. Contemp. Educ. Psychol. **25**(1), 54–67 (2000a)

Ryan, R.M., Deci, E.L.: When rewards compete with nature: the undermining of intrinsic motivation and self-regulation. In: Sansone, C., Harackiewicz, J.M. (eds.) Intrinsic and Extrinsic Motivation: The search for optimal motivation and perfrmance, pp. 13–54. Academic Press (2000b)

Shabbir, N., Atta, M., Adil, A.: Conflict management and decision making styles in college management. J. Behav. Sci. **24**(2), 52 (2014)

Shell, G.R.: Bargaining styles and negotiation: the thomas-kilmann conflict mode instrument in negotiation training. Negot. J. **17**(2), 155–164 (2001)

Shiloh, S., Salton, E., Sharabi, D.: Individual differences in rational and intuitive thinking styles as predictors of heuristic responses and framing effects. Pers. Individ. Differ. **32**(3), 415–429 (2002)

Sladek, R.M., Bond, M.J., Phillips, P.A.: Age and gender differences in preferences for rational and experiential thinking. Pers. Individ. Differ. **49**(8), 907–911 (2010)

Stevens, C.K., Gist, M.E.: Effects of self-efficacy and goal-orientation training on negotiation skill maintenance: what are the mechanisms? Pers. Psychol. **50**(4), 955–978 (1997)

Tasa, K., Celani, A., Bell, C.M.: Goals in negotiation revisited: the impact of goal setting and implicit negotiation beliefs. Negot. Confl. Manag. Res. **6**(2), 114–132 (2013)

Ten Velden, F.S., Beersma, B., De Dreu, C.K.: It takes one to tango: the effects of dyads' epistemic motivation composition in negotiation. Pers. Soc. Psychol. Bull. **36**(11), 1454–1466 (2010)

Thomas, K.W.: The conflict-handling modes: toward more precise theory. Manag. Commun. Q. **1**(3), 430–436 (1988)

Thomas, K.W., Kilmann, R.H.: Thomas-Kilmann Conflict Mode Instrument. Xicom, Tuxedo (1974)

Tucker, W.T., Ferson, S.: Evolved altruism, strong reciprocity, and perception of risk. Ann. N. Y. Acad. Sci. **1128**(1), 111–120 (2008)

Vallerand, R.J.: Deci and Ryan's self-determination theory: a view from the hierarchical model of intrinsic and extrinsic motivation. Psychol. Inq. **11**(4), 312–318 (2000)

Van de Vliert, E., Kabanoff, B.: Toward theory-based measures of conflict management. Acad. Manag. J. **33**(1), 199–209 (1990)

Van der Schalk, J., Beersma, B., Van Kleef, G.A., De Dreu, C.K.: The more (complex), the better? The influence of epistemic motivation on integrative bargaining in complex negotiation. Eur. J. Soc. Psychol. **40**(2), 355–365 (2010)

Van Lange, P.A.: The pursuit of joint outcomes and equality in outcomes: an integrative model of social value orientation. J. Pers. Soc. Psychol. **77**(2), 337 (1999)

Wachowicz, T., Kersten, G.E., Roszkowska, E.: How do I tell you what I want? Agent's interpretation of principal's preferences and its impact on understanding the negotiation process and outcomes. Oper. Res. **19**(4), 993–1032 (2019)

Wang, Y., Highhouse, S., Lake, C.J., Petersen, N.L., Rada, T.B.: Meta-analytic investigations of the relation between intuition and analysis. J. Behav. Decis. Making **30**(1), 15–25 (2017)

Westphal, J.D., Zajac, E.J.: A behavioral theory of corporate governance: explicating the mechanisms of socially situated and socially constituted agency. Acad. Manag. Ann. **7**(1), 607–661 (2013)

Wilson, S.R., Putnam, L.L.: Interaction goals in negotiation. Ann. Int. Commun. Assoc. **13**(1), 374–406 (1990)

Witteman, C., van den Bercken, J., Claes, L., Godoy, A.: Assessing rational and intuitive thinking styles. Eur. J. Psychol. Assess. **25**(1), 39–47 (2009)

Womack, D.F.: Assessing the Thomas-Kilmann conflict mode survey. Manag. Commun. Q. **1**(3), 321–349 (1988)

Zetik, D.C., Stuhlmacher, A.F.: Goal setting and negotiation performance: a meta-analysis. Group Process. Intergroup Relat. **5**(1), 35–52 (2002)

Decision and Negotiation Systems

Automated Decision Systems: Why Human Autonomy is at Stake

Sabine T. Koeszegi

Abstract For more than 50 years, humans have been using model- and data-based support systems in decision making with the hope that system supported decisions are not only better, but also more objective and fairer (i.e. more efficient and less biased). With data-based Artificial Intelligence (AI) systems, this hope is revived. AI systems are to support physicians in diagnosing illnesses, or assist managers when recruiting personnel. Furthermore, AI systems could even replace humans when deciding whether or not applicants receive a loan, an insurance policy or can lease a car. Even governments and public institutions are already making use of such automated decision systems including such diverse applications as setting bail in legal proceedings, risk assessment in youth welfare services or eligibility for state assistance to the unemployed and access to health services. The results of the current applications of automated decision systems are, however problematic. In this essay I examine the question of how automated decision systems impact the autonomy of humans and what requirements are to be placed on automated decision systems in order to protect individuals and society.

1 Introduction

Humans have been using model-based and data-based support systems in decision making for approximately 50 years—primarily with complex decision-making problems. Not only do humans have limited cognitive capabilities and capacities, but they also have conscious and unconscious biases which can reduce the quality of their decision-making. The use of decision systems is therefore justified by the argument

This essay has been published in an earlier version in: 1. Koeszegi, Sabine T. (2020): The Autonomous Human in the Age of Digital Transformation, in: Markus Hengstschläger/Austrian Council for Research and Technology Development (Eds.): Digital Transformation and Ethics. Salzburg, München: Ecowin Verlag, pp. 60–84.

S. T. Koeszegi (✉)
TU Wien, Institute of Management Science, Theresianumgasse 24, 1040 Vienna, Austria
e-mail: Sabine.Koeszegi@tuwien.ac.at

T. Szapiro and J. Kacprzyk (eds.), *Collective Decisions: Theory, Algorithms And Decision Support Systems*, Studies in Systems, Decision and Control 392,
https://doi.org/10.1007/978-3-030-84997-9_7

that, with their support, decisions can be made not only more efficiently, but also more objectively, and thus more fairness and justice can be achieved (Wihlborg et al. 2016).

To date, however, these expectations have only been partially fulfilled in practice. Today such hopes are once again gaining steam as a result of the technological advances in data-based Artificial Intelligence (AI). AI systems are designed to support physicians in diagnosing illnesses, assist managers when recruiting personnel, and are to replace employees when deciding whether or not applicants receive a loan, an insurance policy, or can lease a car. Even governments and public institutions are using automated decision systems. In its report, Algorithm Watch (2019) lists twelve European countries that are already using automated decision systems today. The widely diverse areas of application include setting bail in legal proceedings, risk assessment in youth welfare services as well as eligibility for state assistance to the unemployed, and access to health services. Current applications of automated decision systems, however, show a very problematic track record with a large number of proven forms of discrimination (cf. for example Molnar and Gill 2018).

Humans are eager to use new decision-making aids in their private lives as well. The application of decision support systems have expanded from complex difficult decisions to simple, frequently occurring everyday decisions, in which we can be supported by preselection of suitable alternatives or which we delegate to the system entirely: Personal AI assistants filter products and services, songs and romantic partners based on individual click behaviours signalling our preferences. While information that we were not originally looking for when paging through a newspaper draws our attention, nevertheless, algorithms are preselecting news content which is presumably interesting to us. Since we do not know what decision alternatives, information or news automated decision systems have kept from reaching us, we also do not know what we are missing. When everything is filtered for us, according to past behaviour, on what is familiar to us, and by similar criteria, the spectrum of our preferences and our freedom of action gradually shrink, since we hardly have a chance to discover things that are new, different, contradictory or surprising.

In many situations, we are not aware of the fact that automated decision systems are being used at all. Generally, we are not informed of the data being used by such systems neither about how they work nor about our rights relating to automated decisions (cf. the EU General Data Protection Regulation, articles 14 and 22, on the right to transparency and right to be informed). Empirical studies have shown that people, therefore, underestimate the areas in which automated decision systems are used as well as the dangers associated with such systems for their individual freedom to make decisions and their personal rights (Wouters et al. 2019). Even without our knowledge, as of today personal and biometric data is collected, and personality profiles are generated which expose deeply personal, private and even intimate information and are used for a variety of purposes, some of them manipulative (Kosinski et al. 2013).

A lack of transparency and restrictions on autonomy is only one side of the coin. Delegation of decision-making to automated decision systems creates a vacuum regarding responsibility for the decisions made. The question remains who is ultimately responsible for incorrect predictions and wrong decisions, especially if the users neither know, let alone can understand or explain, the decision-making criteria and models that are used. In the case of many machine learning strategies and technologies, not even the system designers can provide satisfying explanations of the forecasts generated and decisions made.

In this essay, I address the question of how automated decision systems affect the autonomy of humans and what requirements automated decision systems have to fulfil to protect human autonomy. After defining the terms, I will go into detail in the following section on the anatomy of decision-making processes to subsequently show how automated decision systems can influence our autonomy and self-determination. Finally, I will conclude with the requirements for automated decision systems that can ensure that we will be able to continue to live self-determined, autonomous lives in the future.

2 The Anatomy of Decision-Making Processes

In essence, data-based AI technology is software with four characteristics: 1) It works autonomously (i.e. without the direct control of the user), 2) its results are statistical, which means it does not connect cause and effect, 3) it is adaptable, that is, it adapts its behaviour when it learns more about context, and 4) it is interactive, which means that its actions and its results influence humans and are also influenced by us and our social and physical environments. AI systems can interpret structured or unstructured data. Various machine learning approaches and techniques are used for this (cf. Dignum et al. 2020; HLEG, 2019). The term algorithm refers to the logical sequence of steps for organizing, processing and analysing these data volumes (cf. Gillespie 2016) and is equivalent to a formula. The decisive factor, however, is that AI systems require clear target functions in order to learn and act autonomously. **Algorithms are thus the result of modelling that includes both the formalization of a problem and an objective.**

The term **automated decision systems**[1] is used when algorithms execute decision-making models and human assessment is completely or partially replaced by the system. The term system defines the socio-technical framework and, in addition to the technical components includes the political, social and economic environment in which algorithms for decision-making are implemented. This term also takes into account the context in which an automated decision system is to be used.

Figure 1 schematically illustrates an automateed decision process using a simple process model.

[1] In public documents and the media, many other terms are used synonymously for automated decision systems, such as »predictive analytics«, »algorithmic decision-making«, »automated decision-making« and »machine learning«. There are differences between the individual terms, but these are also controversial among experts (see e.g. Molnar/Gill, 2018).

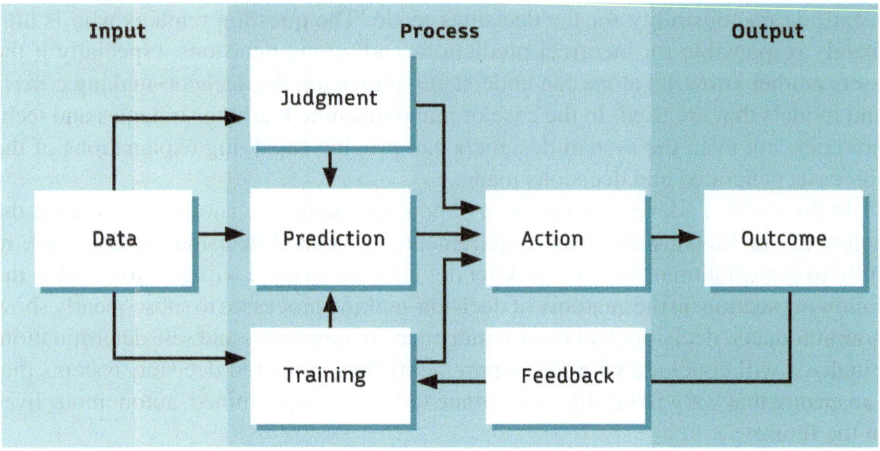

Fig. 1 Anatomy of Decision-Making Processes, adapted from Agrawal et al. 2018

The quality of automated decision-making processes (output) fundamentally depends on the quality of the training data (input), the modelling as well as on the implementation of the decision by the decision-makers (process).

Automated decision-making processes are based on data from the past. The values and assumptions that are incorporated in the selection and preparation of the training data are just as important as the algorithm which learns from them. First and foremost, the training data must be a reasonable approximation of the data that an algorithm will be applied to later, since machine learning works on the assumption that new data will have the same pattern as the training data or, in other words, that the future will look like the past. This is why AI systems are also referred to as »prediction machines«.

In order that algorithms can predict future states—or to classify current states—a model must first be developed that determines which parameters are relevant for a given decision problem in the first place (see Fig. 1: Evaluation). Humans provide a definition of the decision problem and determine which data may be relevant to solving the problem. Usually, the future users work together with AI experts here to evaluate the decision model. The challenge here is to clearly formulate the problem and the goal so that an algorithm can actually derive meaningful and precise classifications or predictions from the data.

For example, if we want to develop an automated decision system that analyses job applications in order to find suitable candidates for a company, we must first determine what the actual decision problem is. What does »suitable« mean in this context? Suitability for a particular given job can be determined by various criteria. A suitable candidate can bring the right qualifications, the right motivation, the right values, the right expectations regarding pay, the right experience or all of these criteria. The loyalty of future employees to the company could be defined as successful recruitment. By defining what success is and selecting the target variables which result in this success, humans determine what an algorithm actually tries to predict. If success in the recruiting process is specified as »employee loyalty«, then those people who are likely to stay with the company for a long time are the right people, so the target variable would be retention. The target variables are broken down further in a process known as *feature selection (c.f. Molnar and Gill 2018)*. Here the programmers decide on the specific primary criteria they prioritize, evaluate or classify individual cases. The qualities that determine whether an employee will stay with a company in the long term may include, for example, the time that a person spent at their previous workplace.

The prediction model or classification model can interpret structured or unstructured data using various machine learning approaches and techniques (cf. Dignum et al. 2020; HLEG 2019). The decisions about suitable methods for analysing existing data are also made by AI engineers. Ultimately the decision system learns to recognize patterns and relationships based on training data. The result of this learning process is the algorithm, which can then be applied to new data for prediction. The result of an algorithm is either presented to a decision-maker in the form of decision support or the system executes decisions directly based on the decision model. Therefore, it is crucial to determine for which problems automated decision models are relevant at all. Do I want to use such a system, for instance, for all positions in the company, for example also at the top management level, or only for certain positions? Which decisions should be fully automated, and which decisions must a human make? Thus, human judgment is essential for successful implementation, as the quality of automated decision processes ultimately depends on how they are implemented in practice.

Ultimately, only the execution of decision leads to intended and potentially unintended consequences (output). Thus, also the expertise of users plays a major role. Only when they understand the decision-making model, they can assess the quality of decisions and their effects and learn from them. However, it is precisely the use of such systems that leads to the fact that humans no longer acquire such important expertise or lose it over time (Bainbridge 1983). In the case of a system failure or system errors, it then becomes increasingly difficult for humans to control these systems or to take corrective action.

An algorithm is thus only the procedure with which a task can be solved in an operationalized form. »What was a social judgment—" What's relevant?"—gets modeled: posited and measurable relationships, actionable and strategic targets, and threshold indicators of success.« (Gillespie 2016, 20). When we speak of algorithms, we are thus actually referring to an undefined network of technical arrangements in which the participation of humans remains hidden in every process step: It is humans who decide on the methods and model design, it is humans who curate and correct the training data, it is humans who design the algorithms by deciding which parameters are relevant in which contexts and which target function is to be achieved (cf. Figure 1). Thus, we have to understand that … »these algorithmic systems are not standalone little boxes, but massive, networked ones with hundreds of hands reaching into them, tweaking and tuning, swapping out parts and experimenting with new arrangements. […] We need to examine the logic that guides the hands.« (Seaver, quoted in Gillespie 2016).

Corinna Balkow and Irina Eckhardt have worked out a schematic overview the development process of algorithmic systems and pointed out where errors and systematic distortions (»bias«) can negatively affect the quality of automated decisions, thus resulting in unjustified discrimination, disadvantage and wrong decisions. Figure 2 makes clear that various forms of distortions can occur at each stage of development. Regarding unjustified discrimination, factors which can lead to direct discrimination are even easier to identify than implicit or indirect discrimination. For example, an algorithm that has been trained on historical hiring decisions of a company in order to recommend suitable candidates may reproduce patterns of ethnic or gender-based discrimination from the past. By knowing about such bias in the past system, designers can take appropriate precautions and counteract. This is not possible for forms of indirect discrimination that are represented in data on the institutional or structural level, because they usually remain unknown. If, for

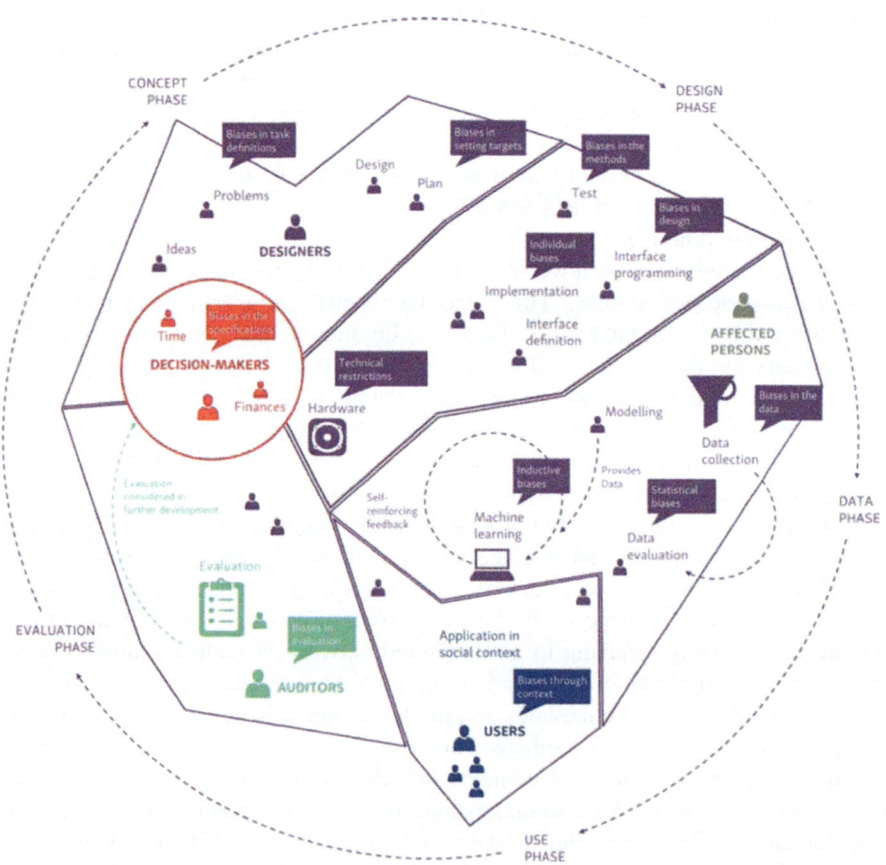

Fig. 2 Bias in algorithmic systems. *Source* Balkow et al. 2019

example, a company has in the past predominantly recruited graduates of an elite university in which people from minorities or women were underrepresented, this pattern will also be represented subsequently in the algorithm. The disclosure of indirect discrimination is made more difficult since the critical variable is »university« rather than »gender« or »ethnicity«. The problem of indirect and implicit discrimination is also a critical factor for human decision-makers. The hope that algorithms could eliminate human inadequacies in decision-making processes is in vain, since »algorithms that make decisions open up the same host of questions we have for humans making decisions: transparency, accountability, discrimination, error, and so forth.« (Tufekci 2015 216).

Algorithms are therefore a socio-technical ensemble. Labelling these complex socio-technical networks »algorithms« in a way disguises that cultural, societal and political values are integrated by designers. The designers decide about how certain activities, objectives or parameters are operationalized in a model and are thus represented in individual mathematical allocations of preference values and threshold values (Molnar and Gill 2018).

3 The Autonomy of the Human Being at Stake

The paradigmatic change triggered by technological progress is based on the ever-increasing autonomy and the resulting agency of AI systems. Decisions which we in the past made ourselves are partially or completely transferred to these systems. In many applications of algorithmic decision-making, the boundaries between automated decision-making and decision-making support are blurred. The extent of how people's options for action, control and self-determination are restricted as a result, usually remains opaque.

The word *autonomy* comes from the Greek and is made up of the words *autos* (»self«) and *nomos* (»law«). It literally means self-regulation respectively self-determination. Autonomous humans can make decisions independent of outside influences or the will of others. The contrast, there is heteronomy, i.e. regulation by others or direction of actions by others. Self-determined actions and responsibility for individual actions are two sides of the same coin and are therefore closely interwoven. The following examines different aspects of automated decision systems in more detail. On the one hand, it is analysed what consequences the attribution of machine agency has for us as humans on the interaction with these machines, and on the other hand, behavioural engineering mechanisms which can be used to directly influence our behaviour are illustrated.

4 The Attribution of Agency to Automated Decision Systems

In a sketch, the British TV comedy show »Little Britain« draws attention to the agency of machines: A mother and her five-year-old daughter come to the hospital for an agreed appointment for a tonsillectomy. Having entered the daughter's data, the receptionist says that the child is scheduled for a bilateral hip replacement surgery. Despite the objections from the mother, which the receptionist first types into her computer, she keeps responding with the answer: »Computer says no!« In the meantime, the phrase »Computer says no!« has become iconic for the customer-hostile attitude on the part of companies and service facilities. However, the sketch also illustrates in a rather drastic way how the use of information technologies can shift the roles of humans and machines absurdly. Regardless of how reasonable the customer's objections are, and regardless of how obviously wrong the computer's statements are, the machine's proposed decision ultimately triumphs.

The delegation of decision-making to machines is accompanied by the attribution of agency in a social context. Humans develop assumptions on the capabilities and skills of »intelligent« machines and place certain expectations upon them. In the context of social interaction, this subsequently leads to shifts in the roles of humans and machines. So who is ultimately responsible for a decision in a socio-technical ensemble?

Humans who use AI systems and computers to support their decision-making usually have great trust in the technology. In a recently conducted survey in which a program uses facial recognition software to identify not only test subjects' gender, ethnicity and age but allegedly also their emotional state and personality traits, scientists show how impressed laymen can be even by obviously false results. »[The readout] must be right. Because the assessment is made by a computer, and computers are better than people at drawing such conclusions.« (test subject, as quoted in Wouters et al. 2019, 454). This can even go so far that subjects question their own self-image due to a wrong classification by the computer (ibid.).

In another example from Sweden, in which an algorithm was used for decision support in a public agency, officers reveal their own roles in interviews: »In the end, the automated system suggested a decision based on all the information that has been put in, provides a decision. [...] There is no room for doubting the decision« (Wihlborg et al. 2016, 2908). The use of algorithmic decision support changes the people's roles and tasks in their work processes, and thus their understanding of their own roles and ultimately their understanding of their responsibility for the work process and the result. The conclusion drawn in the study by Wihlborg et al. (2016, 2911) is quite clear here: »The officer becomes a mediator rather than decision-maker, keeping the system in operation throughout the process.« While autonomy and self-determined action require a certain degree of personal responsibility, delegating decisions to automated systems not only limits the agency of humans but also their perceived control over the decision-making process. The sole fact that a result is the result of an automated process can give it a greater legitimacy, increase the perceived 'neutrality' of a result and weaken the accountability of the human agents and institutions (Gillespie 2016). According to a study by (Logga et al. 2019), this is particularly problematic because, in particular, people who have little or no knowledge in a certain field prefer to trust an algorithm rather than to rely on the specialized expertise of humans, while experts are significantly less likely to rely on the credibility of algorithmic predictions.

The attribution of agency and associated expectations on roles placed on automated decision systems reduces human agency due to changes in the roles in the socio-technical system. It becomes much more difficult and unlikely for humans to act based on personal accountability and to accept responsibility and oversight over the process and its results.

5 The Threat of Manipulation by Automated Decision Systems

Not only do we attribute agency to automated decision systems, but they also have it effectively. This means that they have the potential to directly influence human behaviour. Critical voices warn that AI technologies can invade our privacy and reveal our most personal and intimate areas to others. By disclosing data, we make ourselves vulnerable because we become predictable on our desires, needs and longings and thus we become manipulable. Targeted »behavioural engineering« using manipulative tactics will gradually restrict individual freedom and self-determination. According to Shoshana Zuboff (2018, 335), the true power of »profiling« user data is modifying the behaviour of humans in the real world. The information which humans disclose voluntarily or involuntarily (through their observed behaviour) in the virtual world is used to change their behaviour for specific purposes. The manipulative approaches for modifying human behaviour are referred to as »behavioural engineering« or »tuning« and are based on behavioural science approaches (cf. Zuboff 2018).

In the case of conditioning, desired user behaviours are reinforced through reward, recognition and praise. Shoshana Zuboff quotes an app's programmer as follows: »When people use our app, we can record their behaviours and identify the good and the bad. Then we develop ›treatment methods‹ or ›data pills‹ that select the good behaviours. We can test how useful our cues are for them and how profitable a certain behaviour is for us.« (Zuboff 2018, 339). While in the »Social Credit System«, in use in China since 2014, the government defines what »good« behaviour is, this direct quote shows that in the western world companies determine which behaviours are »good« or desirable—because they are profitable—based on their own economic interests. Accordingly, they implement positive conditioning and more or less obvious cues for certain behaviours in their application. It is only a tiny step from the fitness app or diet app used by consumers, in the hope of achieving a healthier lifestyle and better self-discipline through positive conditioning, towards manipulating user behaviour on behalf of the company's own economic interests. However, this manipulation usually remains hidden from the user. Although conventional marketing has always placed targeted, group-specific advertising, companies have so far not been able to focus on individuals with their highly personal vulnerabilities in specific situations. It is furthermore possible to directly influence the users' moods and their behaviour in social networks. The Facebook study by Kramer et al. (2014), shows that we can be manipulated through the conscious filtering of our news feed for the commercial or political purposes of others, without being aware of it.

Nudging, another form of behavioural manipulation, is based on a behavioural economics method from Thaler and Sunstein (2008): Decision-making architectures in systems are structured and deployed in such a way that they predictably influence human behaviours. This form of manipulation also subtly takes place, does not require economic incentives or direct influence on decision-making behaviours, and remains hidden from the consciousness and will of the manipulated person. The inventors attempt to legitimize the method by claiming it influences the behaviour of individuals only to make their lives longer, healthier and better «(Thaler and Sunstein 2008). Although people have a highly individual notion of »healthier and better«, behavioural manipulation through *nudging* can also be used without the commendable goal of improving quality of life. A live experiment with 61 million Facebook users for electoral mobilization, impressively demonstrated how nudging works. Without the knowledge or consent of its users, Facebook manipulated their personal news feed and the information content of election-related news. The prompt to go to the polls, an »I voted« button, as well as a link to election information, changed the voter turnout in the US congressional elections in 2010 by hundreds of thousands of votes (Bond et al. 2012). Although the experiment caused a collective outcry and Facebook assured that it did not influence the election results, it becomes clear that selective manipulation of certain user groups would make that possible.

There are many other examples of manipulative decision-making architectures: General Terms of Business are presented in such a cumbersome way in the literal »fine print«, that users quickly give up understanding them. It is also common practice to pre-select cookies for users to agree. This manipulative decision-making architecture prompts users to click on the clearly visible »Okay« or »Continue« buttons to allow tracking cookies to be set since the decision not to allow cookies requires several more clicks and is thus less convenient.[2] Another nudging strategy is to make access to communication channels extremely difficult for users with complaints, problems or other issues. Thus, it is often impossible to ascertain a telephone number or e-mail address for a specific contact person to raise a concern personally.

The third form of behavioural manipulation is »herding«, an approach in which people are remotely »orchestrated« in certain situations: Shoshana Zuboff quotes a software developer as saying »We learn to write the music and then make sure that it makes people dance,« (2018, 338). The game »Pokémon Go« is a good example

[2] According to the e-privacy guideline of the European Commission, referred to as the »Cookie-Directive«, setting cookies is only allowed if users expressly and actively consent. The European Court of Justice confirmed this in a recent judgment (cf. Ionos, Digital Guide, 2020).

of how the behaviour of crowds can be controlled and orchestrated in the real world using a digital app. With »Internet of Things« technologies it is possible to directly influence networked devices connected to the Internet remotely and the behavioural options of their users can be restricted. Thus, for example, if there are outstanding lease payments, the operation of a vehicle can be prevented. Devices can also take into account context-specific user data: A car cannot be started because the driver is under the influence of alcohol. Even if the legal blood alcohol level is not exceeded, a leasing company could enforce stricter rules.

Every decision that machines make for us humans, is a decision we do not make ourselves. In many contexts and many decision-making problems, this can be helpful and practical, however, it can be problematic and wrong in others. We must therefore ensure that certain requirements and quality criteria are fulfilled when using automated decision systems so that we can have confidence in their benefits.

6 Requirements for Trustworthy AI Systems

Goal transparency, clear decision-making criteria and a transparent decision-making process are the basic conditions for self-determined, responsible action. The discussion in the preceding sections outlines the impacts which automated decision systems can have on our self-determination, our privacy, the democracy and society. We are currently facing a decisive phase: In many areas, technology is not yet so advanced that we can no longer influence its design. However, we do not have much time to take advantage of this timeframe. The ethical guidelines of the European Commission's expert group on AI, which I also worked on, call for a human-centred technology that protects and respects the rights of the individual. The guidelines once again emphasize the fact that AI technologies are not an end in themselves, but rather should serve to increase our well-being and our self-determination and freedom. Here, however, human-centredness does not mean drawing the attention only on the individual, but also on the well-being of society as a whole and to the environment in which we live.

In the expert group, we formulated a total of seven requirements[3] for trustworthy AI technology that go beyond compliance with existing relevant legislation. Following Fig. 1 »Anatomy of Decision-Making Processes«, these requirements can be structured as shown in Fig. 3.

The quality of automated decisions essentially depends on the quality of the data that is used for modelling and for developing algorithms. Here, ensuring the quality and integrity of the data as well as access to relevant data is often a major challenge. Furthermore, system developers must ensure the protection of privacy whenever personal data is used.

[3] Cf. AI HLEG 2019, https://ec.europa.eu/digital-single-market/en/news/ethicsguidelines-trustworthy-ai.

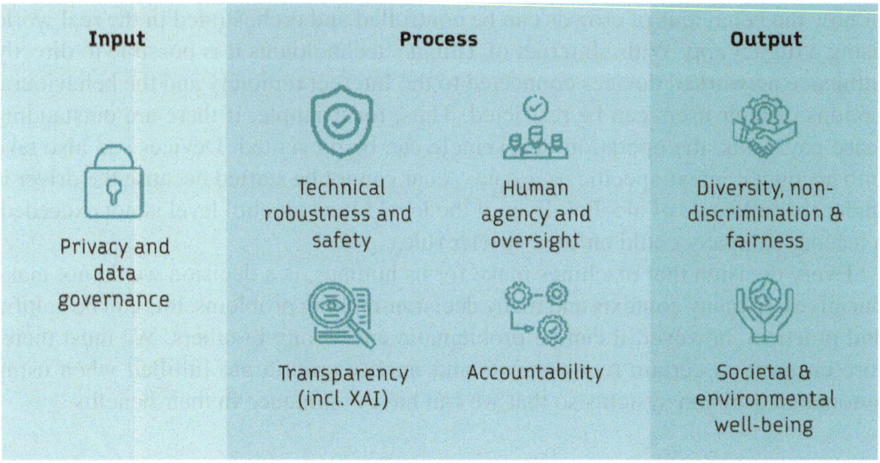

Fig. 3 Requirements for trustworthy AI systems

Before systems can be implemented in practice, their technical robustness and safety must also be guaranteed. Automated decision systems are considered technically robust when they are accurate, reliable and their results can be reproduced. In addition to data security, these systems have to be secure against external attacks and resistant to manipulation of data or algorithms.

The criteria of transparency, accountability and human agency and oversight are closely linked with one another. Transparency (including explainability, i.e. XAI) of modelling, decision-making criteria and methods is an important quality criterium which is currently not guaranteed, in particular in automated decision systems with implemented machine learning strategies. The explainability of a decision or forecast is not only necessary to understand it at the level of the overall model, but also on the level of the algorithm and with regard to individual components (Lipton 2016). The problem is that the explainability of a system does not necessarily mean understanding. Intuitive explanations are particularly necessary for non-experts in order not to overwhelm them (cf. Papagni and Koeszegi 2020). However, with AI systems there is a trade-off between explainability and accuracy of predictions (cf. Hagras 2018). It may be necessary to consider which of the requirements—explainability or accuracy—should be given precedence in a specific application case.

Users are only able to decide whether or not they want to use these systems at all and take the responsibility for possible positive or negative consequences when they know the objectives of the AI systems and can understand the decision-making processes the systems use. As this essay shows, the attribution of agency to automated decision systems and the resulting threat of behavioural manipulation by these systems jeopardizes this basic right of autonomy and self-determination. Therefore, AI systems are required to forgo all manipulative behavioural engineering tactics and are to take into account the modification of the roles of humans and machines from a socio-technical perspective. Even if various forms of human oversight over

AI systems can be implemented in specific application cases, it must always be ensured that humans have control over the use of these systems at all times. This includes the right to opt-out, i.e. the right to refuse the use of these systems. Ultimately, the accountability for the consequences of using AI systems has to be clearly regulated. Before AI systems can be used, it must be ensured that possible negative consequences for individuals and society as a whole can be ruled out or at least minimized. In addition, appropriate legal remedies to enforce the rights of those affected must also be guaranteed, in the event of negative consequences.

The two last criteria relate to the objectives of AI systems. Diversity, non-discrimination and fairness, as well as social well-being and the environment, must also have high priority when implementing AI systems.

The implementation of these requirements, which can only be presented very superficially here, is a great challenge. The urgency of these requirements, however, has only partially made its way into public discourse. While the need to safeguard privacy and data protection through the General Data Protection Regulation is an important first regulatory step, there is unfortunately still little discussion about protecting our basic right of freedom and self-determination. The consequences of an increasing delegation of decisions to machines are certainly context-dependent, but lead, as the present analysis shows, to a sensitive limitation of our self-determination as well as to an increased dependence on technologies which in the current state of research are not very transparent and therefore difficult for users to comprehend. For this reason, there is often a lack of fundamental awareness of the problem. Statements like »I have no problem with them collecting data on me, I have no secrets«, or »Which music I listen to, which news I read and which products I buy is up to me! The algorithm merely points out things I might be interested in«, make it evident that there is an enormous need for clarification regarding the way automated decision systems work.

7 Conclusion

In this essay, I have shown that automated decision systems are not efficient, objective computing systems, but rather socio-technical ensembles which are susceptible to the same problems as humans when it comes to decision making. If we intend to develop decision support systems which respect the dignity of humans and strengthen their autonomy and well-being—both as individuals and as a society—we must not consider these systems in isolation, but must understand them in their respective contexts of application as a socio-technical system, and design them carefully. No less than our self-determination, our democracy and our freedom is at stake!

References

Agrawal, J., Gans, J., Goldfarb, A.: Prediction Machines: The Simple Economics of Artificial Intelligence. Harvard Business Review Press, Boston (2018)

AW Algorithm Watch: Automating Society: Taking Stock of Automated Decision-Making in the EU (2019). www.algorithmwatch.org/automating-society. (zuletzt abgerufen am 28.02.2020)

Bainbridge, L.: Ironies of automation. Automatica **19**(6), 775–779 (1983). https://doi.org/10.1016/0005-1098(83)90046-8

Balkow, C., Eckardt, I.: Bias in Algorithmischen Systemen – Erläuterun-gen, Beispiele und Thesen. In: Denkimpulse digitale Ethik, Initiative D21 (2019). Quelle: https://initiatived21.de/app/uploads/2019/03/algomon_denkimpuls_bias_190318.pdf.

Bond, R., Fariss, C.J., Jones, J.J., Kramer, A.D.I., Marlow, C., Settle, J.E., Fowler, J.H.: A 61-million-person experiment in social influence and political mobilization. Nature **489**, 295–298 (2012)

Collingridge, D.: The Social Control of Technology. Open University Press, Milton Keynes (1981)

Council of Europe: Algorithms and Human Rights: Study on the Human Rights Dimensions of Automated Data Processing Techniques and Possible Regulatory Implications DGI (2017)12 (2018). Quelle: https://rm.coe.int/algorithms-and-human-rights-en-rev/16807956b5. (zuletzt abgerufen am 28.02.2020)

Dignum, V., Muller, C., Theodorou, A.: First Analysis of the EU Whitepaper on AI (2020). http://allai.nl/first-analysis-of-the-eu-whitepaper-on-ai/. (zuletzt abgerufen am 26.02.2020)

Gillespie, T.: Algorithm. In: Peters, B. (ed.) Digital Keywords: A Vocabulary of Information Society and Culture, pp. 18–30. Princeton University Press, Princeton (2016)

Hagras, H.: Toward human-understandable, explainable AI. Computer **51**(9), 28–36 (2018)

High-Level Expert Group on Artificial Intelligence: Ethics Guidelines for Trustworthy AI (2019). https://ec.europa.eu/digital-single-market/en/news/ethics-guidelines-trustworthy-ai. (zuletzt abgerufen am 28.02.2020)

Ionos, Digital Guide: Cookie Richtlinie (2020). https://www.ionos.at/digital-guide/websites/online-recht/cookie-richtlinie/. (letzter Zugriff 25.02.2020)

Kosinski, M., Stillwell, D., Graepel, T.: Private traits and attributes are predictable from digital records of human behavior. PNAS **110**(15), 5801–5805 (2013)

Kramer, A.D.E., Guillory, J.E., Hancock, J.T.: Experimental evidence of massive scale emotional contagion through social networks. PNAS **111**(24), 8788–8790 (2014)

Lipton, Z. C.: The mythos of model interpretability. In: ICML Workshop on Human Interpretability in Machine Learning (WHI 2016), New York (2016). arXiv:1606.03490 1470 arXiv:cs, stat/1606.03490

Logga, J.M., Minsona, J.A., Moore, D.A.: Algorithm appreciation: people prefer algorithmic to human judgment. Org. Behav. Hum. Decis. Processes **15**, 90–103 (2019)

Molnar, P., Gill, L.: Bots at the gate: a human rights analysis of automated decision-making in Canada's immigration and refugee system. In: University of Toronto, Research, p. 114 (2018)

Papagni, G., Koeszegi, S. T.: Understandable and trustworthy explainable robots: a sensemaking perspective. Paladyn. J. Behav. Robot. **12** (2020). https://www.degruyter.com/view/journals/pjbr/12/1/article-p13.xml. https://doi.org/10.1515/pjbr-2021-0002

Thaler, R., Sunstein, C. R.: Nudge: Improving Decisions about Health, Wealth, and Happiness. Yale University Press (2008)

Tufekci, Z.: Algorithmic harms beyond facebook and google: emergent challenges of computational agency. In: Colo. Tech. L.J 13, pp. 203–218 (2015)

Wihlborg, E., Larsson, H., Hedström, K.: The computer says no!– a case study on automated decision-making in public authorities. In: IEEE Computer Society, Proceedings of HICCS 2016, pp. 2903–2912 (2016). https://doi.org/10.1109/HICSS.2016.364

Wouters, N., Kelly, R., Velloso, E., Wolf, K., Ferdous, H. S., Newn, J., Joukhadar, Z., Vetere, F.: Biometric mirror: exploring values and attitudes towards facial analysis and automated decision-making. In: Conference on Designing Interactive Systems, vol. 1145 (2019)

Zuboff, S.: Das Zeitalter des Überwachungskapitalismus. Campus Verlag, Frankfurt, New York (2018)

Supporting Users' Utilitarian Needs for Systems in Online Negotiations

Bo Yu, Rustam Vahidov, and Gregory E. Kersten

Abstract From users' perspective, an e-negotiation system should help them achieve their objective (e.g., an agreement) to the best possible extent. This perspective aligns well with the view of utilitarian system use that conceptualizes the system use as embedded in a process within which a user attempts to achieve an exogenous goal. Given this conceptualization, users' assessment of system use can take place in three tiers, i.e., their assessment of goal achievement, technology-interaction, and the utilitarian value of the system. These theoretical propositions were tested by conducting an experiment involving e-negotiations. The results support the proposed theoretical model. The results also showed that social psychological impacts on users of a system can influence their assessment of system use.

1 Introduction

E-negotiations can be simply defined as negotiations conducted through electronic means. A family of systems has been developed to facilitate and support negotiations involving multiple parties. These systems are generally classified as electronic negotiation systems (ENSs) (Kersten and Lai 2010). ENSs evolved mainly from two areas of research. The first includes research on decision support systems and negotiation support systems (Kersten and Lai 2007), while the second comprises research for the design and development of groupware. The key focus of the second area was to facilitate group activities and assist members, with or without decision support. ENSs, as a joint product of the above two areas of research, are concerned not only with support for individual negotiators, but also for collective interaction and decision making. Negotiation processes can be facilitated, managed, and supported with ENSs. Lim and Benbasat (1992) noted the minimum requirement by drawing

B. Yu (✉)
Dalhousie University, Halifax, Canada
e-mail: bo.yu@dal.ca

R. Vahidov · G. E. Kersten
Concordia University, Montreal, Canada

T. Szapiro and J. Kacprzyk (eds.), *Collective Decisions: Theory, Algorithms And Decision Support Systems*, Studies in Systems, Decision and Control 392,
https://doi.org/10.1007/978-3-030-84997-9_8

insights into both areas of research and stated that an ENS requires the capabilities of both negotiation support systems and decision support systems. ENS research also needs to address the issue of how negotiation can be effectively conducted through electronic communication channels.

Negotiation is a practice wherein dependencies exist. Negotiators cannot resolve the target issues unilaterally. Negotiations often involve various types of social entities, such as individuals, groups, and organizations. In general, negotiators are looking for an agreement that will help to resolve conflicts or set the stage for further cooperation or coordination. The achieved agreement can be assessed using various criteria (e.g., utilities). The process and outcomes of negotiation can influence negotiators' assessment of ENSs. For instance, their experiences when interacting with other negotiators can affect their opinion of the systems they used.

Information systems (IS) research identifies two types of system use: hedonic and utilitarian (Van der Heijden 2004). In the former, the hedonic experience (e.g., enjoyment) is the end of using the technology. In contrast, utilitarian use implies that the objective of using a system is exogenous to the use. Here, using a system is instrumental to achieve an exogenous objective rather than merely using the technology as the end.

When users employ ENSs to conduct their negotiations, it meets the definition of utilitarian system use, as users have an exogenous goal, e.g., to achieve an agreement. Use of the system is embedded in the goal achievement processes of negotiators. Users must work with other negotiators to explore and exploit the possibilities. When users negotiate with each other through an ENS, the system is a channel for their communication. It can also provide structured process management and decision support features to facilitate negotiations or help users to analyze issues and make decisions.

Focusing on users' perspective, the current study argues that the use of ENSs is utilitarian. Given this conceptualization, users' assessment of an ENS will take place in three sequential tiers, i.e., assessment of achievement, assessment of technology-interaction, and assessment of system. The theoretical propositions were tested by conducting an e-negotiation experiment. The results support the proposed theoretical model and show that users' perception of their negotiation partner may influence their assessment of system use, which implies that the latter can be situationally distorted. Situational bias is likely when users assess the utilitarian value of ENSs. This finding has implications for system designers, helping them to understand how users will assess ENSs and how to better address their needs.

The remainder of the paper is organized as follows. The next section will provide the research background of the current study. The following section will present a research model featuring users' use and assessment of e-negotiation systems. Based on the research model, two sections, i.e., methodology and results, will present the details of the current study and its main findings. The discussions section will elaborate on these findings. The paper will conclude with a summary and implications of the study.

2 Research Background

This section will first briefly review competing perspectives in the IS field regarding user assessment of system use. Next, the concept of utilitarian system use will be introduced and its implication for user assessment discussed. Drawing on this background, user assessment of ENSs will be presented, laying out the foundation of our research hypotheses in the next section.

2.1 IS Research About User Assessment and System Use

Businesses keep investing resources in IS that can help users and organizations achieve better decision making, increased productivity, and improved organizational performance. IS play a critical role in organizational success (Orlikowski and Lacono 2001). Among various ways of assessing systems, user evaluation is an important one. Early practices of user evaluation (e.g., in the 1980s) focused on user satisfaction, which was used as a surrogate for IS effectiveness. Similar practices using satisfaction as an indicator of effectiveness were adopted in other fields such as organization theory and marketing (Melone 1990). Later, more and more theoretical models and approaches were proposed in the IS field. Among them, the most influential ones include the Technology Acceptance Model (TAM) of Davis (1989), the Task Technology Fit model (TTF) of Goodhue and Thompson (1995), and the IS success model of DeLone and McLean (1992).

The TAM, one of the most intensively investigated models in IS, adopts users' behavioral intention to use a system as the key predictor of their actual use behavior when they are exposed to the system (Davis et al. 1989). An implicit assumption of the TAM is that users are self-interested when considering whether they should use a technology or not, i.e., the technology needs to be useful to the users. Latterly, multiple extensions to the TAM have been proposed since it gained recognition and popularity. The recent unified or revised TAM integrated many factors (e.g., performance expectancy, effort expectancy, social influence and facilitating conditions) that may influence users' behavioral intention to use a system. While the unified or revised TAM includes more factors that may help explain variation in users' behavioral intention to use the system, it still implicitly assumes users are self-interested: the systems are being used to contribute to users' individual performance.

The TTF postulates that individual performance depends on the fit between the tasks undertaken by individual users and the technology used for the tasks (Goodhue and Thompson 1995). The better the fit, the higher the user performance and thus the more valuable the technology. A distinct feature of TTF is its way to measure fit, which is based on the user's evaluative perceptions. Due to the de facto difficulty of assessing the actual fit, the TTF uses user beliefs and attitudes towards a system as a surrogate. Such a surrogate approach makes the validity of the construct questionable. Since the actual fit is difficult to determine, all the reported performance gain or loss can be interpreted in terms of better or worse fit. Moreover, the appropriateness of using users' attitude and beliefs to assess the success of IS has been challenged (e.g., Melone 1990). In addition, dynamic fit may be a more common phenomenon. In actual applications, the fit can be gained by either modifying tasks when new technology is introduced (e.g., business process re-engineering) or selecting alternative technologies for available tasks. In recent years, organizations and IT projects placed more importance on the capability of being agile. TTF may not be reliable or robust enough to capture such a dynamic fit because both technologies and tasks may change quickly.

While these two models shed a lot of light on user evaluation of IS, the requirement to assess a system is often more complex and broader than these two models can describe and predict. Assessing system use in organizations usually involves multiple level relations between systems, individual users, and organizations. The IS success model (DeLone and McLean 1992) seeks to explain a chain of impacts that connects systems, individual users, and the organization. The benefit of using a system for an organization depends on the characteristics of system (i.e., system and information quality) and usage (i.e., use and user satisfaction). Although the IS success model helps to bridge different levels, the use of the whole model is often difficult. The impacts of system on organizational level are often multiple and complex. For instance, organizational effectiveness as a construct can be defined and operationalized in various ways (Thong and Yap 1996). Consequently, it is difficult to clearly ascertain the relation between organization performance and the systems used. As an integral part of management, system evaluation highlights the complex nature of any organization. To be part of the management control process, an evaluated system must meet the organizational needs, comply with the organizational policies, and be of good quality. An IS project often has multiple stakeholders whose interests are different and often conflicting. The evaluation of a system must examine, encourage, and promote collective satisfaction as a legitimacy process within the relational network. To appropriately evaluate a system, approaches need to be applied carefully with respect to both the identified organizational objectives and the actual difficulties (Hamilton and Chervany 1981a; b).

User evaluation relies on users' subjective beliefs about and attitudes towards a system. A premise underlying this approach is these subjective perceptions will influence their use of the system. However, beliefs and attitudes cannot fully explain and reliably predict actual behaviors. Both criticism (e.g., Melone 1990) and support (e.g., Gatian 1994; Gelderman 1998) of this approach have been voiced. Thus, examining users' actual system use behaviors may be helpful to gain a complete picture.

System use is a notion that focuses on the observable behaviors of system users. It is believed to be a critical factor influencing the success of information systems. As Orlikowski (2000, p. 425) notes, "Technology per se can't increase or decrease the productivity of workers' performance, only use of it can." This idea is simple, yet powerful. System usage has been among the most frequently used measures of assessing IS success (DeLone and McLean 2003). It may also underpin other important processes of information technology development and implementation, such as IS adoption, acceptance, and diffusion. The IS field has dedicated a lot of effort to develop and understand this notion. According to Barki et al. (2007), the journals of MIS Quarterly and Information System Research published 60 papers between 1992 and 2007 in which IS use was a key construct. In total, in 120 issues of these two influential journals, half of these issues have an article that is related to IS use.

Despite being an important construct in IS research, IS use is still problematic in both its conceptualization and measurement. For instance, IS use may be measured very differently in terms of frequency, duration, or portfolios of used system features. Empirical studies have shown mixed results for the influence of this construct on individual performance. Positive, negative, and insignificant effects have been reported. Barki et al. (2007) identified six problems with this construct:

- *Ignoring how IT is actually used in organization,*
- *Failing to consider the multidimensional nature of IS use,*
- *Overlooking the richness of organizational contexts,*
- *Lack of relevance in mandatory used context,*
- *Ignorance of what level of use should be considered sufficient for successful IS*
- *Inadequacy for capturing the relationship between usage and the realization of expected results.*

Both the importance and problem of the IS use construct raises calls for improving its conceptualization and measurement. Barki et al. (2007) proposed to expand the focus of existing conceptualizations of IS use. As they commented, the current measurement of IS use exclusively focuses on user technology interaction behaviors. This practice excludes many other use-related activities that may contribute to IS use. Based on task-technology fit and activity theory, they extended traditional IS use to a new construct of IS use-related activity, which is conceptualized as a second-order aggregate construct that comprises both user technology interaction behaviors and activities that users undertake to adapt the task-technology-individual system. They also provide empirical results that support the proposed extension.

Burton-Jones and Straub (2006) explicitly recognized the difficulty of obtaining a unified definition of IS use. They proposed a systematic approach for re-conceptualizing the system usage construct in the particular nomological contexts that are of interest for researchers. The approach includes two stages: definition and selection. In the definition stage, researchers need to define system usage and clarify its underlying assumptions. In the selection stage, IS use needs to be conceptualized in terms of its structure and function. IS use has three key components, i.e., a user, system, and task. Researchers need to justify the relations of these three components and their relevance for studies. In terms of function, researchers need to choose measures for each component. A desirable rich measure for IS use would tie these three components together and connect with other closely related constructs. However, obtaining such a measure in practice is challenging. Burton-Jones and Straub's approach provide a useful methodological guide if IS use is adopted as a construct in studies.

In contrast to the approach of Barki et al. (2007) which expands the focus of IS use, Burton-Jones and Grange (2013) proposed to move from use to effective use.

The authors define effective use as "using a system in a way that helps attain the goals for using the system." The authors draw on the representation theory to state that IS are not like a regular tool. According to them, a system consists of structures that serve to represent parts of the actual world. The effective use of the system derives from users' and other stakeholders' understanding of the system. They also presented a framework to explain how effective use and performance are related. Intuitively, this approach may narrow the operationalization scope of IS use, which in turn may help to establish a consistent relation between this and other constructs.

2.2 Utilitarian System Use and User Assessment

The studies by Barki et al. (2007) and Burton-Jones and Grange (2013) identified three key elements in IS use, i.e., user, system, and task. In the view of Barki et al. (2007), IS use-related activity includes two types of behaviors, i.e., technology interaction and task-technology-individual adaptation. Both types are related to individual users, the used system, and tasks undertaken by users. The approach proposed by Burton-Jones and Grange (2013) identified six ways to measure IS use, which differ in terms of how user, system, and task are related. In their view, a rich and ideal measure of IS use needs to consider all the three elements. They also point out that IS use often involves multiple stakeholders, who may expect different results from IS use. Although directing research attention to users' behaviors in using a system has the potential to better explain system success, there are challenges to this approach as well. Human behaviors are complicated, and they cannot be understood or interpreted without referring to their contexts. The studies of Barki et al. (2007) and Burton-Jones and Grange (2013) showed the importance of a coherent theoretical account of which user, system, and task can be related.

Drawing on the consumer behavior literature, Van der Heijden (2004) argued that technology use can be classified into two categories: utilitarian or hedonic. Hedonic technologies provide self-fulfilling values to users, such as enjoyment and playfulness. The value of the technologies is derived from and encapsulated in hedonic use itself. In contrast, utilitarian technologies provide instrumental value to users to achieve objectives that are outside the technology use, such as job performance or profit. In utilitarian technology use, the derivation of technology value is exposed to factors exogenous to technology use. Users' assessments of their utilitarian technologies depend on how well their objectives are met. For instance, if a user's objective is not achieved, he or she may downplay the value of the used utilitarian technology.

Van der Heijden's study (2004) stimulated research on the hedonic aspect of systems influencing users' acceptance. For instance, constructs representing hedonic aspect have been integrated in both the TAM3 (e.g., computer playfulness and perceived enjoyment) and the UTAUT (Unified Theory of Acceptance and Use of Technology) 2 (e.g., hedonic motivation). Following a strong momentum, utilitarian technology features, use, and assessments continuously remain at the core of technology acceptance research. Davis (1989) published the famous paper, proposing that perceived usefulness and perceived ease of use are the key factors influencing users' acceptance of technology. Perceived usefulness captures users' assessment of the utilitarian value of technology. The TAM has been tested and applied to various technologies and in different contexts. It has been extended to the TAM2 and the TAM3. More recently, it has been refined and converted into new models, such as the UTAUT model and the UTAUT2 models (see, Venkatesh et al. 2016).

In utilitarian use, users employ a system because they have a goal to achieve. Barki et al. (2007) and Burton-Jones and Grange (2013) view the task as a goal direct activity: a goal that drives a user and the course of actions the user undertakes to achieve his or her goal. Detaching goal from actions in utilitarian use makes it clear that the use of a system is embedded in a process of the user's goal achievement. When this is the case, the achieved performance of the user will become an important factor influencing their assessment of the system. This is in contrast to the traditional research of IS use that treats individual performance as a dependent variable that is influenced by both users' use of the system and their subjective attitudes and beliefs about their use.

2.3 E-negotiation as a Case of Utilitarian System Use

Negotiations often involve various types of social entities, such as individuals, groups, and organizations. In general, negotiators are looking for a potential agreement, which will help to calm conflicts or further cooperation or coordination. The achieved agreement can be assessed using some criteria (e.g., utilities).

E-negotiations are those conducted through electronic means. ENSs have been developed to facilitate and support negotiators (Kersten and Lai 2010). Two major lines of research contribute to the evolution of the ENS concept. The first includes decision support systems and negotiation support systems (Kersten and Lai 2007). This line attempts to help users identify, construct, and solve negotiation problems. The second line includes research for the design and development of groupware, e.g., computer-supported cooperative work, group decision support systems, group support systems, and meeting support systems. This line attempts to facilitate group activities and then help group members undertake some joint tasks, with or without support functions. Evolved from both lines of research, ENSs can support not only individual negotiators, but also their collective interaction and decision making (Lim and Benbasat 1992).

With e-negotiations, the participants want to produce a potential agreement. This characterizes ENSs as a utilitarian technology in general, although users may have hedonic experiences when negotiating. This conceptualization shares common ground with the socio-technical theory, which was initially proposed to help understand the relation between IS and organizations. Figure 1 presents an adapted model from Bostrom and Heinen (1977) that demonstrates the relation of user, task, and technology in a scenario of e-negotiation. Users are social entities who reside in a social system and have some goals to achieve. Their goal achievement processes may

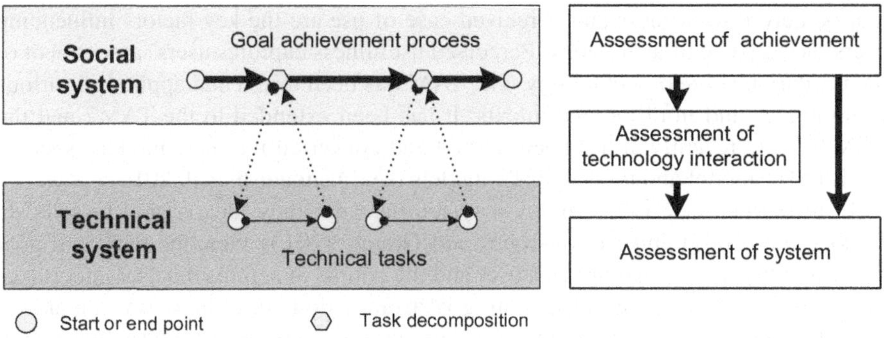

Fig. 1 Utilitarian system use and user assessments

be assessed by some means (e.g., how good a deal is made). Users' goal achievement processes involve the use of ENSs or related technology. To achieve their goals, users may divide their overall tasks into more manageable sub-tasks, some of which are technical tasks that require users to interact with their systems: user-technology interaction. User-technology interaction depends on the requirements of users' goals and the functionalities of the used systems, i.e., what users want to do with the system and what the system can offer.

Figure 1 illustrates that individual users' assessment can take place in three sequential tiers. First, users will assess the achievement of their goals, which is a critical factor influencing many other perceptions and attitudes of the users. Second, users have utilitarian needs for a system and when those are fulfilled, they will tend to be satisfied with the system. The users' needs for the system can be multiple. Thus, they will assess the system from multiple aspects in terms of their experiences when interacting with the system. Third, users' experiences of interacting with the system will influence their assessment of the contribution made by the system, e.g., perceived usefulness.

For instance, individual users may need some information from the system and thus they would submit an inquiry. The system provides information upon inquiry and the users may evaluate the quality of the information. Users' perception of information quality is not solely determined by either the users or the system. It depends on both what users want and what the system provides, except when the users are fully passive information subscribers to whom the system pushes all the information. By borrowing the term 'technology interaction' from Barki et al. (2007), users' evaluation of their interaction with the system can be defined as their assessment of technology interaction. Based on users' goal achievement and experiences interacting with the system, they will formulate their beliefs and attitudes about the system. which leads to the formation of perceived usefulness: the contribution made by the system. In summary, users will assess an embedded system use in three tiers. Their assessment of goal achievement will be the first tier, which will influence the assessment of technology interaction. The first two tiers will influence users' assessment of the system.

In an ENS use scenario, the actual use of the system is jointly influenced by how users want to achieve their negotiation goal (e.g., an agreement) and how the system may help. The goals of users are constructed or prescribed in a social system. Users' interaction with ENSs, such as proposing an offer is partially prescribed by the social system. At the same time, the actions undertaken by a user are shaped by the system when he or she uses it, e.g., how an offer can be formulated and sent through the system. Usually, a system will have a set of functionalities and features provided to the user. It is also possible for the system to implement and enforce part of social rules or norms to regulate the behaviors of the users making them comply with some institutions and policies.

3 The Research Model and Hypotheses

The objective of the current study is to understand how users will assess ENSs they use. The above sections argue that ENS use in general is a case of utilitarian system use. During negotiations, users exercise their freedom to make choices and interacting with each other using an ENS. Their use of the system is a means, and not the end. They may or may not get an agreement.

A research model that aligns with the theoretical foundation laid out in the prior section is presented in Fig. 2. Three constructs, i.e., users' satisfaction with achievement (SA), satisfaction with technology interaction (STI), and perceived usefulness of the system (PU), constitute the main structure of the model. The construct of satisfaction with outcome represents users' assessment of goal achievement. The construct of satisfaction with technology interaction captures users' assessment of their experiences of interacting with the system and how well their utilitarian needs for system are fulfilled. The construct of perceived usefulness represents users' assessment of the contribution made by the system to their goal achievement.

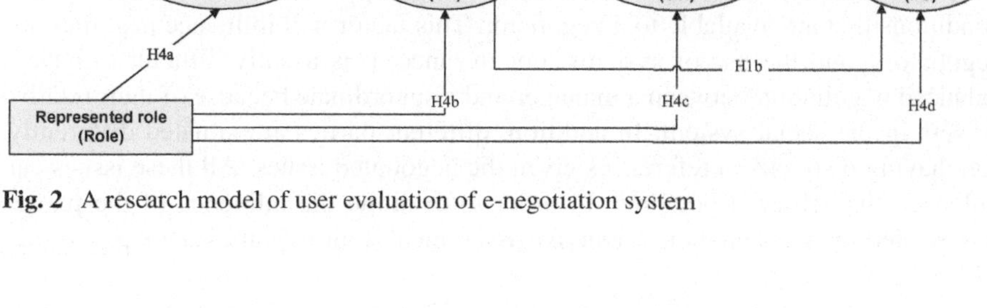

Fig. 2 A research model of user evaluation of e-negotiation system

Three hypotheses have been developed according to the expectations that ENS use is utilitarian and users' assessments will take place in three sequential tiers.

H1a: Users' satisfaction with achievement will positively influence their satisfaction with technology interaction.

H1b: Users' satisfaction with achievement will positively influence their perceived usefulness of the system.

H1c: Users' satisfaction with technology interaction will positively influence their perceived usefulness of the system.

Since negotiators need at least one other negotiator, their experiences in the social system can influence their assessment from multiple aspects. In the e-negotiation scenario, the experience of individual users interacting with their negotiation counterparts are important. Users' perceptions of their negotiation counterparts have a social psychological impact when they are engaged in a social activity involving other social actors. To examine the role of social psychological impacts on users, three hypotheses are proposed to examine the role of perception about the negotiation partner.

H2a: Users' perception about negotiation partner will positively influence their satisfaction with outcome.

H2b: Users' perception about negotiation partner will positively influence their satisfaction with technology interaction.

H2c: Users' perception about negotiation partner will positively influence their perceived usefulness of system.

A distinctive feature of the conceptualization of utilitarian system use is that the performance achieved by users will have significant influence on their assessments of systems. In contrast to traditional research of IS use that makes performance a dependent variable, the current study posits performance on the independent variable side. Four hypotheses are proposed to examine the influence of achieved outcome (e.g., the value of agreement).

H3a: Users' achieved outcome will positively influence their perception of negotiation partner.

H3b: Users' achieved outcome will positively influence their satisfaction with outcome.

H3c: Users' achieved outcome will positively influence their satisfaction with technology interaction.

H3d: Users' achieved outcome will positively influence their perceived usefulness of system.

In a negotiation, users represent a role. A role maps to a collection of situational conditions that are available to a negotiator. This factor will influence negotiations, negotiators, and the use of systems. For instance, it is usually difficult to have a balanced negotiation between a manager and a subordinate because of their relative powers in the social system. In addition, different parties are situated differently, e.g., having distinctive preferences given the negotiated issues. All these issues can influence the achieved outcome. In order to examine the influence of the parties represented by users on their subjective assessment, four hypotheses are proposed.

H4a: The role represented by users will have an influence on their perception of negotiation partner.

H4b: The role represented by users will have an influence on their satisfaction with outcome.

H4c: The role represented by users will have an influence on their satisfaction with technology interaction.

H4d: The role represented by users will have an influence on their perceived usefulness of system.

It needs to be noted that the influence of the role represented by users on their outcomes is profound. However, the influence strength and direction of this factor in the real world are not easy to determine for at least two reasons. First, social actors may have relative power over each other that will influence their choices and outcomes. Usually, the more power a role has, the better the agreement he or she can have. Second, the situational conditions possessed by each role may influence their choices and the value of the agreement as well. The current study tried to provide to each negotiator an equal social position, thus reducing the influence of power. Second, users can determine their own preferences given the negotiation issues. Thus, no pre-existing social conditions were expected before the negotiation. More details are provided in the methodology section. Given the experimental settings, it is expected that no significant effect will exist between the role represented by users and their achieved outcome.

4 Methodology

4.1 Experiment and Data Collection

To test the research model, an online experiment was conducted. The experiment adopted the e-negotiation system Inspire (http://interneg.concordia.ca/inspire), which supports bilateral negotiations. A business case was used to provide the negotiation context and task. The case involves a contract negotiation between an agent representing an artist (i.e., the seller side) and a manager representing an entertainment company (i.e., the buyer side). The contract comprises four issues: (1) number of new songs, (2) royalties for CDs, (3) number of promotional concerts, and (4) contract signing bonus. Each issue has three to five options to choose from. Every contract package to be negotiated is a particular combination of one option from each issue. No role was allowed to propose new issues or options. The information of the case is decomposed into two parts, i.e., public and confidential. All participants were provided with the same public information, which described who is involved in the negotiation and what the negotiation issues are. The confidential information conveyed the interests of the artist or the entertainment company both graphically and textually. Based on the confidential information, users could specify their preferences regarding the issues using the system. These included the relative importance

Table 1 The descriptive statistics of samples

Sample feature	Statistics
Total number of participants	278
Number of effective samples	207
Gender	Male: 96; female: 104, missing data: 7
Number of participants in groups	Austria-1: 65; Austria-2: 69; Portland: 13; Taiwan: 48; U.S.A: 3; Canada: 9
Prior experience of using negotiation support system	Yes: 9; no: 198
Self-ranked knowledge of negotiation (1 to 7 to indicate the levels from novice to expert)	Level-1:38; level-2:37; level-3:36; level-4:42; level-5:23; level-6:14; missing data: 7
Prior experience with negotiation experiment	Yes: 28; no:172; missing data: 7

of issues and options in forms of both text and graph. The relative importance of issues was not fully opposite on the sides of buyer and seller, i.e., buyers and sellers had opportunities to obtain joint gains.

Users' preferences were represented in the form of utilities, whereby users specified a numerical value for each issue and its options. The system used these values to calculate ratings of each issue-option package to support users. Users were paired into dyads and could communicate with each other with text messages sent through the system. The progress of negotiations and the exchanged offers were provided to participants, along with a graphical display support. Users may have had two stages to reach agreement. In the first stage, users looked for an initial agreement. After an initial agreement was reached, users could choose whether to enter a post-settlement stage to improve their initial agreements that were not Parento optimal.

The participants in the study were undergraduate students, 278 participants in total. They were from six classes in four universities located in Europe, North America, and Asia. The participants were given three weeks to complete the task but could finish earlier if they reached an agreement. After negotiations were ended, the participants were invited to fill in an online questionnaire measured the four subjective constructs in the research model. On a voluntary basis, 207 complete sets of responses to the questionnaire were obtained. Some descriptive statistics of the samples are provided in Table1. The age of most participants was in the 20s and 30s.

4.2 Measurements

The achieved negotiation outcome for each individual negotiator was measured in terms of the utility value which was calculated by an additive compensatory function based on preferences specified by the users. It should be noted that the achieved utilities for both users in a dyad would be zero if no agreement was made.

In total, seven sets of questions were used to capture users' subjective assessment. Five sets were used to measure users' satisfaction with outcome, self-performance, information, communication, and process. One set measured users' perceived usefulness of the system and one set measured users' perception about their negotiation partner in the dyad. The instrument was developed according to guidelines suggested in Straub et al. (2004). The factor structure and the items contained in questionnaires administered in the experiment are presented in Table 2. The five sets of items measuring users' satisfaction are constructed into two secondary reflective factors. Users' satisfaction with achievement used users' satisfaction with outcome and self-performance as first order factor indicators. Users' satisfaction with technology interaction used users' satisfaction with communication, information and process as first order factor indicators. More details are provided in the scale validation section (5.2).

5 Results

5.1 Manipulation Checks

Manipulation checks were conducted for the experiment. The first examined whether the achieved outcome by users will differ given the parties they represented. Since users determined their own preferences, no significant effect was expected although the relative importance of issues and options were provided to users. A comparison between two groups divided by the two parties confirmed this expectation. Second, a question was asked about users' perceived ease of understanding their tasks before negotiation. The comparison between two parties showed no significant difference. Therefore, the results show that the manipulation effects were obtained.

5.2 Scale Validation

The potential effects of demographic variables on the subjective measures were examined at the very beginning. Besides the variables presented in the methodology section, demographic variables included participants' spoken languages, English proficiency, the country of residence and others. Multivariate analysis of variance revealed no significant effect of each demographic variable on the set of subjective measures. After this initial examination, a confirmatory factor analysis was conducted. An initial model contained 27 indicators, grouped into seven first order factors and two second order factors. The two second order factors included users' satisfaction with achievement and technology interaction. Details of the model structure can be found in Table 2. There were 207 observations resulting from the experiment. The software EQS was used to test the factor model, with the maximum-likelihood method plus the robust option. The fit indices of the initial model showed

Table 2 The factor loading, reliability, and validity of subjective constructs

	Items	Factor Loading 2nd order	Factor Loading 1st order	Rho	AVE
Satisfaction with achievement	***Satisfaction with outcome:*** Now that the negotiation has concluded, I can say that...			0.88	0.59
	I am satisfied with the results that I achieved.	0.80	0.90		
	I am satisfied with the results as compared to my expectations.		0.91		
	I am satisfied that the results were favorable for me.		0.87		
	Satisfaction with self-performance:				
	I am satisfied with my own performance in this negotiation.	0.73	0.84		
	I was confident in performing the negotiation tasks.		0.71		
	I adequately represented my client.		0.80		
Satisfaction with technology interaction	***Satisfaction with communication:***			0.88	0.63
	*I felt that it was easy to communicate with my counterpart.	0.82			
	I could write freely to my counterpart.		0.84		
	I was able to express myself effectively.		0.90		
	Satisfaction with information:				
	I felt that the information provided by the system was sufficient to conduct the negotiation.	0.81	0.73		
	I felt that the information was represented in a format that I felt comfortable with.		0.83		
	I felt that the information exchanged through the system could be relied upon.		0.73		
	Satisfaction with process:				
	I feel the negotiation process was efficient in performing my tasks.	0.75	0.94		
	I am pleased with the effectiveness of the negotiation process.		0.92		
	I feel the negotiation process was adequate for this business scenario.		0.72		
Perceived usefulness of the system				0.85	0.65
I felt that the system helped to achieve my objectives.		0.84			
I felt that the system helped to improve my performance.		0.84			

(continued)

Table 2 (continued)

I felt that the system helped to obtain faster results.	0.75		
Perception about negotiation partner: nine questions of 7-point bi-polar scale What can you say about your counterpart in the negotiations?		0.86	0.61
* Uninformative _ _ _ _ _ _ _ Informative			
* Push-over _ _ _ _ _ _ _ Persuasive			
Deceptive _ _ _ _ _ _ _ Honest	0.76		
* Exploitative _ _ _ _ _ _ _ Accommodating			
Competitive _ _ _ _ _ _ _ Cooperative	0.68		
Untrustworthy _ _ _ _ _ _ _ Trustworthy	0.86		
* Unlikable _ _ _ _ _ _ _ Likable			
Unfair _ _ _ _ _ _ _ Fair	0.82		
* Rigid _ _ _ _ _ _ _ Flexible			

Table 3 The goodness of fit of the factor model

Indices	Initial Model	Final Model	Cut-off Point
NFI (Bentler–Bonett Normed Fit Index)	0.772	0.832	NFI > 0.90 good fit (Salisbury et al. 2002); NFI > 0.8 reasonable fit (Hair et al. 1998; Hadjistavropoulos et al. 1999)
CFI (Comparative Fit Index)	0.888	0.936	CFI > 0.90 (Salisbury et al.2002; Bentler and Bonett 1980)
IFI	0.890	0.938	IFI > 0.90 good fit (Salisbury et al. 2002; Bollen 1989)
RMSEA (Root Mean Square Error of Approximation)	0.059	0.049	RMSEA < 0.01 excellent, < 0.05 good, and < 0.08 reasonable fit (MacCallum et al. 1996)
90% confidence interval of RMSEA	0.050–0.067	0.036–0.060	

a poor fit of the model (see Table 3) and the factor model was further refined. In total, six items were removed, one from satisfaction with communication and five from the perception about negotiation partner. During the refining process, both the statistic indices and face validity of the instruments were simultaneously checked. The final factor model showed an improvement on the indices of goodness of fit and those of residuals of errors (see Table 3). All indices meet the recommended cut-off points.

Table 4 The correlations of variables in the research model

	Outcome	PNP	SA	STI
PNP	0.346**			
SA	0.465**	0.448**		
STI	0.144*	0.517**	0.594**	
PU	0.149*	0.320**	0.424**	0.605**

**Correlation is significant at the 0.01 level (2-tailed)
*Correlation is significant at the 0.05 level (2-tailed)

The reliability of the scales was assessed by examining the reliability coefficient Rho (Raykov 1997). All values of the coefficient were greater than 0.7 which indicates a good coherent reliability of the measurement of factors. The discriminant validity of the scales was assessed by examining AVEs (Average Variance Extracted) of the factors. The values of AVE and shared variance are reported in Table 2. All the values of AVE are above the recommended value of 0.5 (MacKenzie et al.2011).

5.3 Research Model Testing

A path analysis method was adopted to test the research model. There are two reasons for such a choice. First, the model needs to be identifiable. A path analysis is helpful to reduce the overall number of paths, as the full structure equation model with all factorial constructs having multiple indicators contains many more paths that need to test. Second, the data feature of achieved outcome needs to be coped in the model testing. Since all negotiators were supposed to obtain higher utility value, this variable cannot be in normal distribution at all. To test the research model, the sums of the values of items representing each factor were used. In total, the tested model included six variables, which is consistent with the research model. In variance analysis the use of sum score to represent factors is deemed to be an appropriate or even preferred technique (DiStefano et al. 2009). The tested research model contains a categorical variable, i.e., the role represented by users. The model treated the variable as continuous (i.e., coded as either 0 or 1) in order to examine the effect of this variable together with other variables. Categorical variables can also be used to conduct between-group tests mainly for moderating effect. But, it will be more useful for the current study to examine the effect of this variable along with others. The correlations of the five variables are presented in Table 4. The variable of role is excluded because it is a categorical variable.

The software EQS 6.1 was used for the path analysis of the research model testing. The maximum-likelihood method with robust option was used. The results showed very good fit (Chi-square = 0.421 with 1 degree of freedom, probability value for chi-square = 0.514, NFI = 0.998, CFI = 1, IFI = 1.003, RMSEA = 0, and 90% confidence interval of RMSEA is between 0 and 0.159). The coefficients for the

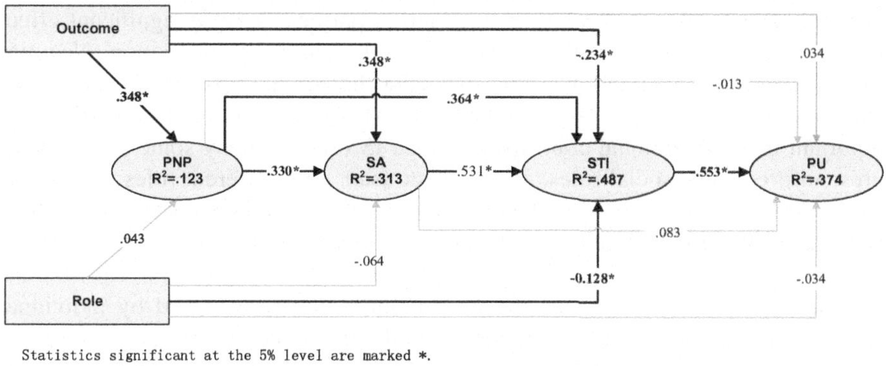

Fig. 3 The results of testing the research model

paths contained in the model are presented in Fig. 3. The significant paths (at 5% significance level) are highlighted.

6 Discussion

The results of testing the research model show several interesting findings. First, the relation of the three main constructs are consistent with the prior theoretical analysis. Users' satisfaction with their achievement positively influences their satisfaction with technology interaction, which in turn influences users' perceived usefulness of the system. The direct effect from users' satisfaction with achievement on their perceived usefulness of the system is not significant. These results show a clear sequence of how users assess the system that is involved in their goal achieving processes. When the use of a system is utilitarian, the use is embedded in the users' goal achievement process, in which the attempt of using the system is instrumental to the goals. The goal achievement by users will be the primary factor influencing the users' other assessments. Therefore, the sequential assessments in three tiers postulated in the prior analysis are well supported.

Second, users' satisfaction with technology interaction has a full mediation effect on their perceived usefulness of the system. This finding is unexpected, but also reasonable according to the prior analysis of the utilitarian system use. When users are pursuing their goals by using a system, they have utilitarian needs for the system helping them to achieve their negotiation goals. The effects resulted from the user-system interaction are not solely determined by either the users or the system. It is a joint product of what users need to achieve their goals and what the system is able to provide. Users' assessment of technology interaction captures their experiences of interacting with the system when the users were pursuing their goals. Without interaction, the assessment of the contribution of the system is meaningless. Thus, the full mediation effect is very possible.

Third, users' perception of their negotiation partner shows a significant effect on both their satisfaction with achievement and technology interaction. Negotiation involves social interaction. Some social psychological impacts may result from using a system if multiple users are involved. This issue may become particularly important in organizational contexts, in which IS often embody some social structures, enforce some social rules, and connect users with different roles (DeSanctis and Poole 1994). Given the complexity of organizational and system structures, how users may influence each other can be an important issue in their adoption and use of a system.

Fourth, the results show that the negotiation outcome achieved by individual users has profound influences. It significantly influences users' perceptions of the negotiation partner, satisfaction with achievement, and satisfaction with technology interaction. In a negotiation scenario, negotiators have a goal to obtain the best outcome possible. According to goal achievement theory (Pintrich 2000), the better the outcome, the better the users may perceive or assess the negotiation from multiple aspects. From the psychological perspective, it is reasonable that the achieved outcome influences users' perceptions and evaluations on multiple aspects in negotiations.

Fifth, the role represented by users shows a significant effect on their satisfaction with technology interaction. The negative effect indicates that the seller side evaluated the technology interaction better than the buyer side. The causes of this effect may be multiple and difficult to pin down. First, the private information given to the seller is slightly more complex than that provided to the buyer. For sellers, the values of two issues are not straightforward given the option values of each issue. Thus, the sellers may need more support from the system and the system may better satisfy them with its decision support features. Second, it may be possible that the public information of the case describes the seller as a successful artist, which gave the seller slightly more power over the buyer. In the current study, sellers obtained slightly higher values from the negotiation. By considering all these possible reasons, further investigation is needed to better understand the influence of this variable.

Overall, the results yielded from the testing of the research model well support the theoretical analysis of utilitarian system use and users' assessment. However, the model testing is based on users' use of the Inspire system, which is a particular type of ENS implementation. It would be interesting to see future research extends the testing to other systems.

7 Conclusion and Implications

By taking the users' perspective, the current study conceptualized that the use of a ENS is utilitarian and embedded in negotiations. The use of the system is instrumental to an exogenous goal, i.e., reaching a potential agreement. Given this conceptualization, the negotiation outcomes achieved by users will influence their assessments. In addition, users will have a set of utilitarian needs for the system that need to be satisfied when interacting with the system. Users' achievement of their goals and experiences of interacting with system will jointly influence users' assessment of the value contributed by the used system. The current study proposed a three-tier model of users' assessments. The theoretical model is tested by conducting an experiment involving e-negotiation. The results of the current study well support the proposed theoretical model. The results also show that social psychological impacts on users may influence users' assessment.

There are several implications of the study. First, it proposed a competing way of conceptualizing system use and user assessment of system. The concept of utilitarian system use necessitates a re-examination of the relation between users and IS. We need to develop a clearer picture of user-system relations to better understand how users will use and assess systems. Second, it may provide an alternative way for IS research to view the casual relation between system use and IS success. Traditionally, a perspective of the designers or managers has been preferred in IS research to understand system use and user evaluation. Efforts have been made to find the relation between system use and individual impacts or organizational impacts. However, system use may be better understood if we adopt the users' perspective by examining how they may look at and respond to system use. Third, the management may be better informed if system use is viewed as being utilitarian in most organizational contexts.

The results of the current study show that users are influenced by both their achieved outcome and other participating parties. This implies that the success of ENS depends on the social system, e.g., how users' jobs are designed, and their performance evaluated, which will lead to how they would use and then assess the system.

Although the study only tested its theoretical analysis using a single experiment on a particular type of system, its conceptualization makes a lot of sense from the users' perspective. The study also provides to IS research some insights about system use and user evaluation. Our results show that users' assessments were sequential. On one hand, this implies that what users' want to achieve will affect their needs for a used system or partially determine how the system will be used. On the other hand, the used system needs to satisfy users' needs derived from their objectives, which will influence users' opinion of the value of the system. Users' objectives are not always self-determined but, are often formed in the social context. Thus, the formation of users' goals and the way their performance will be evaluated in the social system can play an important role of influencing their assessments of their used systems.

References

Barki, H., Titah, R., Boffo, C.: Information system use-related activity: an expanded behavioral conceptualization of individual-level information system use. Inf. Syst. Res. **18**(2), 173–192 (2007)

Bentler, P.M., Bonett, D.G.: Significance tests and goodness of fit in the analysis of covariance structures. Psychol. Bull. **88**, 588–606 (1980)
Bollen, K.A.: Structural Equations with Latent Variables. Wiley, New York (1989)
Bostrom, R.P., Heinen, J.S.: MIS problems and failures: a socio-technical perspective Part I: the causes. MIS Q. **1**(3), 17–32 (1977)
Burton-Jones, A., Grange, C.: From use to effective use: a representation theory perspective. Inf. Syst. Res. **24**(3), 632–658 (2013)
Burton-Jones, A., Straub, D.W.: Reconceptualizing system usage: an approach and empirical test. Inf. Syst. Res. **17**(3), 228–246 (2006)
Davis, F., Bagozzi, R., Warshaw, P.: User acceptance of computer technology: a comparison of two theoretical models. Manage. Sci. **35**(8), 982–1003 (1989)
Davis, F.D.: perceived usefulness, perceived ease of use and user acceptance of information technology. MIS Q. **13**(3), 319–340 (1989)
DeLone, W.H., McLean, E.R.: Information systems success: the quest for the dependent variable. Inf. Syst. Res. **3**(1), 60–95 (1992)
DeLone, W.H., McLean, E.R.: The DeLone and McLean Model of Information Systems success: a ten years update. J. Manage. Inf. Syst. **19**(4), 9–30 (2003)
DeSanctis, G., Poole, M.S.: Capturing the complexity in advanced technology use: adaptive structuration theory. Organ. Sci. **5**(2), 121–147 (1994)
DiStefano, C., Zhu, M., Mindrila, D.: Understanding and using factor scores: considerations for the applied researcher. Pract. Assess. Res. Eval. **14**(1), 20 (2009)
Gatian, A.W.: Is user satisfaction a valid measure of system effectiveness? Inf. Manage. **26**(3), 119–131 (1994)
Gelderman, M.: The relation between user satisfaction, usage of information systems and performance. Inf. Manage. **34**(1), 11–18 (1998)
Goodhue, D.L., Thompson, R.L.: Task-technology fit and individual performance. MIS Q. **19**(2), 213 (1995)
Hadjistavropoulos, H.D., Frombach, I.K., Asmundson, G.J.G.: Exploratory and confirmatory factor analytic investigations of the Illness Attitudes Scale in a nonclinical sample. Behav. Res. Ther. **37**(7), 671–684 (1999)
Hair, J.F., Anderson, R.E., Tatham, R.L., Black, W.C.: Multivariate Data Analysis with Readings, 5th edn. Prentice Hall, Hoboken (1998)
Hamilton, S., Chervany, N.L.: Evaluating information system effectiveness - Part I: comparing evaluation approaches. MIS Q. **5**(4), 55–69 (1981)
Hamilton, S., Chervany, N.L.: Evaluating information system effectiveness - Part II: comparing evaluator viewpoints. MIS Q. **5**(4), 79–86 (1981)
Kersten, G., Lai, H.: Negotiation support and e-negotiation systems: an overview. Group Decis. Negot. **16**(6), 553–586 (2007)
Kersten, G., Lai, H.: Electronic negotiations: foundations, systems, and processes. In: Kilgour, D.M., Eden, C.: (eds.) Handbook of Group Decision and Negotiation. Advances in Group Decision and Negotiation, vol. 4, pp. 361–392 Springer, Dordrecht (2010). https://doi.org/10.1007/978-90-481-9097-3_22
Lim, L.-H., Benbasat, I.: A theoretical perspective of negotiation support systems. J. Manage. Inf. Syst. **9**(3), 27–44 (1992)
MacCallum, R.C., Browne, M.W., Sugawara, H.M.: Power analysis and determination of sample size for covariance structure modeling. Psychol. Methods **1**(2), 130–149 (1996)
MacKenzie, S.B., Podsakoff, P.M., Podsakoff, N.P.: Construct measurement and validation procedures in MIS and behavioral research: integrating new and existing techniques. MIS Q. **35**(2), 293–334 (2011)
Melone, N.P.: A theoretical assessment of the user-satisfaction construct in information systems research. Manage. Sci. **36**(1), 76–91 (1990)
Orlikowski, W.J.: Using technology and constituting structures: a practice lens for studying technology in organizations. Organ. Sci. **11**(4), 404–428 (2000)

Orlikowski, W.J., Lacono, C.S.: Research Commentary: desperately seeking the 'IT' in IT research – a call to theorizing the IT artifact. Inf. Syst. Res. **12**(2), 121–134 (2001)

Pintrich, P.R.: An achievement goal theory perspective on issues in motivation terminology, theory, and research. Contemp. Educ. Psychol. **25**(1), 92–104 (2000)

Raykov, T.: Estimation of composite reliability for congeneric measures. Appl. Psychol. Meas. **21**(2), 173–184 (1997)

Salisbury, W.D., Chin, W.W., Gopal, A., Newsted, P.R.: Better theory through measurement-developing a scale to capture consensus on appropriation. Inf. Syst. Res. **13**(1), 91–103 (2002)

Straub, D., Boudreau, M.-C., Gefen, D.: Validation guidelines for IS positivist research. Commun. Assoc. Inf. Syst. 3(1) (2004)

Thong, J.Y.L., Yap, C.-S.: Information systems effectiveness: a user satisfaction approach. Inf. Process. Manage. **32**(5), 601–610 (1996)

Van der Heijden, H.: User acceptance of hedonic information systems. MIS Q. 695–704 (2004)

Venkatesh, V., Thong, J.Y., Xu, X.: Unified theory of acceptance and use of technology: a synthesis and the road ahead. J. Assoc. Inf. Syst. **17**(5), 328–376 (2016)

Setting the Right Tone: The Role of Language Sentiment in E-negotiations

Simon Alfano, Bo Yu, Gregory E. Kersten, Dirk Neumann, and Nil-Jana Akpinar

Abstract E-negotiations play an integral role in interconnected global markets. Previous research has studied in great detail various aspects of social interaction in e-negotiations, including the linguistic features of messages exchanged. However, insights on the importance of language dynamics are scarce. Thus, we evaluate the role of the language tone (i.e., sentiment) in messages on outcomes in bilateral e-negotiations. Our study uses the messages extracted from 1,092 bilateral e-negotiations of the Inspire experiment. Our results suggest positive language is helpful in achieving an agreement. Avoiding negative language is a stronger driver of agreement value than increased levels of positive language. Intriguingly, successful e-negotiations exhibit a more positive sentiment during the opening, relationship-building stage, and a reduced utilization of both positive and negative words during the core e-negotiation phase. Our findings suggest that interpersonal skills, such as a careful language choice, remain crucial despite the more transactional nature of e-negotiations.

1 Introduction

Negotiations are a critical mechanism within economic systems; they regulate the interactions of participants seeking to exchange goods and services. Well-designed negotiation processes can facilitate a potential agreement. In addition to agreements, negotiators would like to achieve a better deal (e.g., either joint or individual

S. Alfano · D. Neumann
Albert-Ludwigs-Universität Freiburg, Freiburg im Breisgau, Germany

B. Yu (✉)
Dalhousie University, Halifax, Canada
e-mail: bo.yu@dal.ca

G. E. Kersten
Concordia University, Montreal, Canada

N.-J. Akpinar
Carnegie Mellon University, Pittsburgh, USA

T. Szapiro and J. Kacprzyk (eds.), *Collective Decisions: Theory, Algorithms And Decision Support Systems*, Studies in Systems, Decision and Control 392,
https://doi.org/10.1007/978-3-030-84997-9_9

outcomes) through their interaction than they would without negotiating (Lax and Sebenius 1986).

Modern economic transactions are often digital, using the Internet and other computer-supported services, which leads to negotiations being increasingly conducted online. Exchanged information, either personal or impersonal, often relies on electronically submitted linguistic content. In their messages, negotiators encode qualitative information so that they can express their opinions and emotions. The positivity and negativity embedded in such linguistic communication can be assessed using sentiment analysis (Tetlock et al. 2008; Loughran and McDonald 2016).

Research on emotion in negotiations has been done in the context of face-to-face interactions. These studies draw on the premise that emotions influence decision-making of individuals. It can be both internal motive (e.g., emotional state) driving individuals and an instrument influencing other negotiators (e.g., expressed emotion). More recently, emotions are studied in e-negotiations (e.g.,Griessmair et al. 2015; Griessmair and Koeszegi 2009; Hine et al. 2009; Kurtzberg et al. 2005; Parlamis and Geiger 2015; Sheehy and Palanovics 2006). For instance, Hine et al. (2009) find that more positive emotions and less negative language have a favorable effect on the success rate of e-negotiations.

Earlier research about language-focused negotiations relied on machine learning to predict negotiation outcomes based on certain linguistic elements, of which expression of emotion can be one. For instance, Sokolova et al. (2008) and Sokolova and Szpakowicz (2007) used machine learning approaches to identify a group of language expressions, including emotion. They showed that the use of these expressions in the closing part of e- negotiations can better predict the outcome than in the opening part.

Recently, sentiment analysis has gained broader adoption. It is a computing approach, in which dictionaries are usually used to classify words as either positive or negative. Positive and negative words are counted, and a formula is used to obtain a sentiment score. The sentiment scores obtained from linguistic content align with the valence of expressed emotion, indicating the strength of positive and negative tones. Dictionary-based approaches are used in various fields of business. For example, Tetlock (2007) shows that sentiment in financial media is positively related to stock returns. Antweiler and Frank (2004) find that stock price volatility can be more accurately predicted when accounting for online stock message boards.

This paper applies sentiment analysis to assess the effect of language sentiment on negotiation outcomes. It assesses messages exchanged by participants of the bilateral Inspire negotiations (Kersten and Noronha 1999). In total, the study assesses 10,929 messages from 1,304 negotiations involving a so-called "Yowl-Pop" scenario.

We find that negotiators who employ less negative language sentiment are more likely to achieve an agreement with a higher value. The negative language sentiment component exerts a stronger effect on negotiation outcomes than positive language. Using less negatively connoted words increases the probability of achieving an agreement and a better outcome, while using more positively connoted words has no statistically significant effect on outcomes.

We also find significant dynamics of sentiments in the negotiations. We observe a more positive sentiment in the initiation phase and in the resolution phase than in the problem-solving phase. In negotiations that end in an agreement, negotiators use significantly fewer negative words in the initial relationship-building phase than in other phases. In addition, these negotiators use significantly fewer positively and negatively connoted words in the middle negotiation phase. This might indicate that they clearly distinguish between the relationship-building and the problem-solving and use a more rational and goal-oriented language during the core negotiation phase.

The remainder of this paper is structured as follows: Sect. 2 reviews related work on the role of sentiment in negotiations and states our research hypotheses. Section 3 presents the research design, data, and the methodological approach to sentiment analysis. Section 4 presents our results, which we discuss in Sect. 5. Section 6 concludes with a summary and suggestions for future research.

2 Related Work and Research Hypotheses

Businesses increasingly transfer social interactions, such as negotiations, to electronic platforms. Performed electronically, these negotiations do not require a face-to-face discussion. In such electronic settings, the role of emotions or sentiment in written, non-verbal communication, gains a lot of importance (Moore et al. 1999), since personal or in-person interaction diminishes. Recent research investigates how feelings and sentiment influence decision-making in negotiations (e.g., Hine et al. 2009).

2.1 The Role of Language in Negotiations

Research on the role of language and emotions expressed through language in the decision-making in negotiations has emerged during the last decade (e.g., Drolet and Morris 2000; Griessmair and Koeszegi 2009; Hine et al. 2009; Kopelman et al. 2006; Kopelman and Rosette 2008; Sokolova and Szpakowicz 2007; Sokolova and Lapalme 2012; Twitchell et al. 2013), as the following research review shows.

Kopelman et al. (2006) observe in an experimental setting that displaying positive emotions increases the likelihood of achieving concessions and committing to a prospective business relationship. At the same time, they report higher demands by negotiators when faced with a counterpart demonstrating negative emotions. Further, Kopelman and Rosette (2008) reveal cultural differences in the effect that emotions have on negotiation outcomes. For instance, Asian participants tend to decline an offer if a counterparty expresses negativism.

Sokolova et al. (2008) study whether language plays a similar role in face-to-face and electronic negotiations. According to their findings, language characteristics in face-to-face negotiations are particularly important and predictive of outcomes in the initial phase of the negotiator communication. In contrast, language characteristics, such as the use of negotiation- or strategy-related words, play a more important role towards the end of the communication in electronic communication. The authors, following Simons (1993), stipulate that the rationale for this different influence of language is that in a face-to-face negotiation, negotiators build trust and a relationship in the initial phase of the negotiation.

Sokolova and Lapalme (2012) apply machine learning approaches to negotiation messages. They find a positive relationship between information content of negotiations and likelihood of achieving an agreement.

Twitchell et al. (2013) rely on a machine learning-driven negotiation outcome classification model, which classifies negotiation transcripts as successful or not. Their accuracy is above 80% when accounting for all utterances. As they assume a higher importance for more recent utterances, they assign a higher weight to text occurring towards the end of the negotiation.

Hine et al. (2009) study the role of language sentiment in e-negotiations by computing linguistic variables with the Linguistic Inquiry and Word Count (LIWC) software. In their analysis, less negative language and more positive emotions lead to a greater propensity to arrive at an agreement. However, they do not study the effect of language on the final outcomes. In addition, successful e-negotiations are found to be of shorter duration than unsuccessful ones.

Griessmair and Koeszegi (2009) assess negotiators' embed emotions implicitly into their electronic communication. For that purpose, 39 business students manually classify 76 messages from six selected negotiations based on the similarity of the emotional content. They find a strong interrelationship between factual statements and a related emotional connotation. Successful negotiations spend more time on relationship building and cooperative behavior, which seeks to use factual statements to explore outcomes in the interest of both negotiators.

2.2 Language Sentiment Analysis

Information Systems (IS) research is well-equipped to study how agents process information conveyed by the language used in textual messages (Chen et al. 2012). Textual content provides relevant insights into the position of the message originator through the tone of the language. The subjective tone of text documents can be assessed using so-called *sentiment analysis*, which includes methods that measure how positive or negative the language of textual information is (Loughran and McDonald 2016).

Sentiment measures enable investigation of the role of language in the context in which it is used. Academic research deploys sentiment analysis frequently to study the role of language in financial markets (e.g., Antweiler and Frank 2004;

Karapandza 2016; Loughran and McDonald 2016; Tetlock 2007; Tetlock et al. 2008). Sentiment-related research on the role of language sentiment in negotiations also emerges.

IS research has developed various approaches to measure sentiment, since sentiment analysis is deployed across various domains and for different textual sources (Pang and Lee 2008). Overall, sentiment analysis approaches fall into two categories: dictionary-based approaches and machine learning methods.

Dictionary-based approaches produce reliable results by counting the frequency of pre-defined negative and positive words from a given dictionary, which classifies words as either positive or negative (Loughran and McDonald 2016). Commonly, psychologists develop such domain-independent dictionaries. The psychological Harvard-IV dictionary represents a widely used psychological dictionary. In addition, domain-specific dictionaries reflect the different classification of a word in a specific domain. Words such as tax or cost, which are negative in the Harvard-IV dictionary, may convey a positive meaning in a financial context. Therefore, dictionaries for the finance domain take into consideration its different perception of language. The Loughran and McDonald (2011) dictionary and the Henry (2008) dictionary represent two widely referenced sources for the finance domain. Table 1 provides an overview of popular dictionaries, their scope and examples of negatively and positively labeled words.

There are a variety of machine learning methods (e.g., Antweiler and Frank 2004; Li 2010; Mittermayer and Knolmayer 2006; Schumaker and Chen 2009; Sokolova et al. 2008), which are particularly suited for predictive studies. One caveat is that they may be subject to overfitting (Sharma and Dey 2012).

In the following, we pursue a dictionary-based approach for various reasons. First, such approaches are popular due to the replicability of their results, which is particularly beneficial when studying not merely the predictive capabilities, but also the effect of a cause, in this case language, on an effect, in this case negotiation outcomes. Second, dictionary-based approaches allow us to aggregate language into one sentiment measure, which is more suitable for interpreting the results than machine learning methods.

2.3 Hypotheses Development

We study the influence of language sentiment of the negotiation messages on the negotiation outcomes using the following hypotheses.

Van Kleef et al. (2004) and Kopelman et al. (2006) find negotiators to be more cooperative and willing to make concessions when they expose positive affect. A qualitative assessment of Griessmair and Koeszegi (2009) also indicates that friendlier language leads to a higher likelihood of achieving an agreement. Also, empirical evidence from financial markets highlights a statistically significant positive relationship between the sentiment of financial news and stock market movements (e.g., Antweiler and Frank 2004; Loughran and McDonald 2016; Tetlock

Table 1 Comparison of different dictionaries

Reference	Name	Negative Words	Positive Words	Datasets	Benchmark	Negative words, examples	Positive words, examples
Harvard-IV Dictionary	Harvard	1793	1316	Harvard's General Inquirer Dictionary	Expert categorization by psychologists	Abandon, abnormal, tax	Admit, adjust, confident, consolidate, plain
Loughran-McDonald	LM	883	146	70,925 10-K forms	Stock market return	Caution, deny, fraud	Win, succeed, enable
Henry	Henry	44	53	1,366 annual press releases	Stock market return	Decline, low, obstacle	Best, deliver, grow

2007). Therefore, we hypothesize that sentiment plays a similar role and has a positive effect on achieving an agreement.

> ***Hypothesis 1a (H1a):*** Language sentiment has a positive effect on the likelihood of achieving an agreement.
> ***Hypothesis 1b (H1b):*** Language sentiment has a positive effect on the payoff achieved in a negotiation.

According to the negativity bias, negative information outweighs positive information in human decision-making (Kanouse 1984). As an extension, Lewicka et al. (1992) differentiate between affective and informational negativity effects. The affective negativity influences human decision-making more strongly than positive affect, e.g. emotions or feelings. The informational negativity effect suggests a higher cognitive relevance of negative than positive information, e.g. a more pronounced curiosity for negative than for positive information (Fiske 1980). In economics, Kahneman and Tversky (2013) formalize this in the prospect theory according to which human agents assign a larger probability weight to negative than to positive outcomes.

We pursue two alternative approaches to study whether negotiators treat language sentiment asymmetrically. First, we decompose language sentiment into its negative and positive components, respectively, by only considering negatively or positively connoted words for either component. Consequently, we expect negative sentiment (as measured by negative language only) to play a more important role than positive sentiment (as measured by positive language only). Second, we perform quantile regressions on different payoff quantiles to understand whether the language reception intensity differs depending on whether the outcomes are poor or good.

> ***Hypothesis 2a (H2a):*** Negative sentiment is a stronger driver of negotiation outcomes than positive sentiment.
> ***Hypothesis 2b (H2b):*** Sentiment is a stronger driver of poor negotiation outcomes than of good negotiation outcomes.

Research has identified and classified negotiations into several phases and marked the progress as the process advanced from one phase to another (Weingart and Olekalns 2004). Holmes (1992) formalizes such negotiation stage models into three phases: an initiation phase, a problem-solving phase and a resolution phase. During the initiation phase, negotiators introduce themselves, establish a relationship, and clarify any preliminary questions, e.g. related to the agenda. During the middle problem-solving phase, the core negotiation takes place: negotiators exchange their positions, explore mutual opportunities, and bargain. Finally, the resolution phase serves to fix the settlement and conclude the negotiation. Similarly, Pesendorfer et al. (2007) classify negotiations into four phases, whereby the first phase (relationship building) and fourth phase (agreement reached) are less prone to conflict than the two middle phases (problem identification and solution generation). In the beginning, the negotiators focus on establishing a relationship, e.g. by exploring commonalities. Moving forward in the negotiation, they state their position and try to achieve a best-possible result, which may involve more conflictual language. Once an agreement is made (or not), the conversation turns less heated again and is, e.g., about the administrative cornerstones of realizing the concluded agreement.

Prior research also suggests that the importance of sentiment varies across the different phases in a negotiation. Twitchell et al. (2013)and Sokolova et al. (2008) report an increase in the importance of language towards the end of negotiations. At the same time, Simons (1993)emphasizes that negotiators focus on relationship-building at the beginning of a negotiation because trust needs to be established (Elkins and Derrick 2013).

Hypothesis 3 (H3): Language sentiment is more positive during the initiation phase and the resolution phase and less positive in the problem-solving phase.

Bronstein et al. (2012) and Nadler (2004)emphasize the importance of rapport in the search for common ground. Sheehy and Palanovics (2006) report a positive and statistically significant effect between rapport building and achieving an agreement. According to Jap et al. (2011), high levels of rapport induce even unethical behavior, such as lying, if this is a prerequisite for mutual success in a negotiation. They also observe that high levels of rapport increase the likelihood of agreement and outcome satisfaction of negotiators.

Kurtzberg et al. (2005) study the respective roles of similarity and familiarity between negotiators in a setting with peer negotiators (agent-agent) and a hierarchical separation (principal-agent). Their results indicate a higher success rate of achieving an agreement and a higher level of both similarity and familiarity for a peer agent-agent setting than for a hierarchical setting.

In contrast, low levels of reciprocity in communication lead to inferior outcomes (Brett et al. 1998). Similarly, displayed emotions tend to evoke reciprocal emotions (Van Kleef et al. 2004). Weingart et al. (1999) report reciprocation in negotiation techniques among negotiators.

Overall, the evidence on reciprocation and rapports leads us to hypothesize that reciprocally more positive sentiment leads to a higher reciprocal willingness to make concessions and thus a higher overall payoff.

Hypothesis 4 (H4): Similar sentiment levels as a proxy for rapport, measured as the difference between the individual sentiment levels of negotiators, are positively related to negotiation outcomes.

3 Materials and Methods

3.1 Measuring Language Sentiment

Loughran and McDonald (2016) report various ways to measure sentiment on the basis of dictionaries. Such dictionaries classify words as either positive or negative. The most widely used are the Harvard-IV dictionary, the finance-specific Henry (2008) dictionary and the Loughran and McDonald (2011) dictionary. While psychologists developed the Harvard-IV dictionary based on the likely positive or negative cognitive perception of words, the latter two dictionaries adapt to the perception of language in financial markets. Finance literature recently has increasingly adopted Loughran-McDonald's dictionary, but the psychological Harvard-IV dictionary is also widely used across different domains of business and finance research. Since the communication between the negotiators occurs in a broader business context than just financial news, we rely on the psychological Harvard-IV dictionary in the following.

Before computing a language sentiment variable of the sentiment in the negotiation messages, we need to prepare our text messages corpus (Manning and Schutze 1999): first, tokenization splits running text into single words named tokens. Then, we adjust for negations using a rule-based approach to detect negation scopes and invert the meaning accordingly (Dadvar et al. 2011). In a next step, we remove so-called stop words, which are words without relevance, such as articles (e.g. the, a, an) and pronouns (Lewis et al. 2004). Finally, we perform stemming in order to truncate all inflected words to their stem, using the so-called Porter stemming algorithm, e.g. stemming truncates announces and announced to *announc*.

Upon completion of the pre-processing, we can compute our sentiment variable. The Net-Optimism metric of Demers and Vega (2008), combined with the Harvard-IV dictionary is one such sentiment approach that yields a robust relationship. The Net-Optimism metric $S(m)$ of message m processes all words contained in message m, where $S(m)$ is the difference between the count of positive W_{pos} and negative W_{neg} words divided by the total count of words W_{tot} in message m with

$$S(m) = \frac{W_{pos} - W_{neg}}{W_{tot}} \in [-1, +1]. \tag{1}$$

We standardize the sentiment variable to zero mean and with a standard deviation of one. Henceforth, we refer to the standardized Net-Optimism language sentiment as *sentiment*.

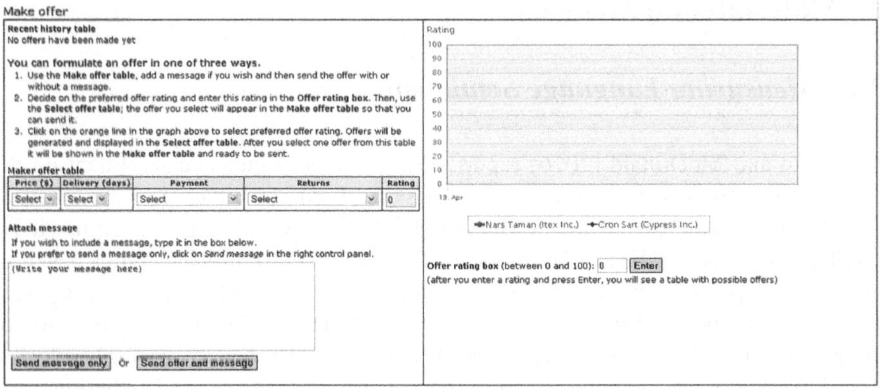

Fig. 1 Screenshot of the Inspire e-negotiation platform for the Yowl-Pop scenario

3.2 Inspire Negotiation Experiment

Inspire is an e-negotiation system supporting bi-lateral multi-issue negotiations with enhanced negotiation analytic methods and communication. The system provides dynamic and user-controlled graphical tools. The system runs on the Invite platform at the InterNeg Research Centre[1], which has regularly conducted online negotiation experiments involving students from different countries for more than a decade.

The Inspire system allows for the establishment of business scenarios to provide participants with meaningful information to enhance their engagement. In this study, we rely on the "Yowl-Pop" scenario[2], which is frequently used at InterNeg Research Centre. The scenario requires two parties to negotiate on a contract between a music artist and an entertainment company. There are four issues, including the number of promotional concerts, the number of new songs, royalties for the CDs, and a signing bonus for the contract. Each issue has several options. An agreement is reached when the two parties agree on a package that contains a chosen option for each of the issues. The process of each negotiation instance has three stages: the negotiation preparation, the negotiation itself, and a post-settlement stage. In the negotiation preparation stage, participants need to read materials on the business scenario. According to the information contained in the business scenario, they need to specify their preferences with the issues and options (see also the screenshot in Fig. 1). The system uses a hybrid conjoint analysis approach to elicit users' preferences. In the negotiation and post-settlement stage, negotiators can use the elicited preferences as a decision support. The elicited preferences also provide a payoff value in case of an agreement. Negotiators could jointly decide whether they would like to enter this stage if they had a chance of improving their agreement.

---

[1] http://www.interneg.concordia.ca.

[2] http://invite.concordia.ca/cases/inspireYowlPop.html.

3.3 Data

The negotiation instances involved in the current study are bi-lateral. In other words, each negotiation instance is a dyad of participants in the Yowl-Pop scenario, each representing either the musician (i.e., the seller) or the entertainment company (i.e., the buyer). The current study does not use messages or negotiation outcome data in the post-settlement stage because all negotiations in this stage already had an agreement. The interaction between negotiators in this stage would be integrative by default. In total, we obtain 10,929 messages generated by 2,184 individual negotiators involved in 1,092 negotiations. The reported ages of the majority sample (i.e., 93%) are between 20 and 30 years. We collected the data from 11 online experiments, conducted between 2010 and 2016. The participants were students enrolled from education institutions in 11 countries. Table 2 reports the number of participants, including a split by gender, for each experiment instance.

Although the settings for batches could vary slightly, they were largely kept being consistent. The negotiations usually took 10 days. Participants would receive their account for their negotiations three or four days before negotiations started. Upon receiving their account, participants could start to read the information regarding their negotiations.

Table 2 Participants per online experiment

Date	Participants	Gender		
		Female	Male	Missing
Dec. 2010	246	120	118	8
May 2011	232	62	86	84
Oct. 2011	184	80	104	0
Apr. 2012	134	78	56	0
Apr. 2013	248	119	129	0
Nov. 2013	156	78	78	0
Apr. 2014	340	194	146	0
Nov. 2014	80	42	38	0
Apr. 2015	298	173	125	0
Nov. 2015	66	29	37	0
Apr. 2016	200	118	82	0
Total	2184	1093	999	92

4 Results

In this section, we investigate how the sentiment in the messages exchanged between negotiators interferes with the likelihood of achieving an agreement and with the payoffs for the negotiators in case of an agreement. We run logit regressions to evaluate the effect of sentiment on whether an agreement is achieved and ordinary least squares (OLS) regressions to evaluate the effect on the negotiators' values produced by their negotiations.

4.1 Sentiment Aggregates and Reaching Agreement

We assess the impact of language sentiment on the propensity to achieve an agreement at various levels. We assess the effect of language sentiment on (i) a message level, (ii) a negotiator level, and (iii) a negotiation level. As our results in support of Hypothesis H1a reveal, the importance of sentiment is more accentuated when looking at aggregated sentiment measures instead of evaluating it on a message level.

On a message level, the language sentiment regressed on the binary agreement variable does not produce a statistically significant effect on whether negotiators achieve an agreement or not (see Table 3). Next, we calculate the overall language sentiment across all messages for each of the two negotiators in the negotiation. This provides us with one sentiment variable for each negotiator. The impact of language sentiment gains statistical significance (P-value < 0.01). As the regression coefficient indicates, more negative language sentiment diminishes the likelihood of an agreement.

4.2 The Effect of Sentiment on Negotiation Outcome

Next, we evaluate the effect of the message sentiment on negotiation outcomes. The descriptive statistics of negotiation outcomes are reported in Table 4. Negotiation outcome at individual level is measured by using value of agreement yield to either the buyer or the seller. Negotiation outcome at dyadic level is measured by using the joint agreement value of the buyer or seller. Negotiation outcome would have zero value at both individual and dyadic level if no agreement were reached.

Consistent with our above findings and in support of Hypothesis H1b, the effect of sentiment intensifies when aggregating the messages from a negotiator- to a negotiation-level. In addition, negative sentiment plays a dominant role as it is exerts a statistically and economically significant influence on the payoffs, while the positive sentiment is not statistically significant, in line with Hypothesis H2. For instance, a one-standard-deviation increase in language sentiment links to a 2.363 units (P-value smaller than 0.05) higher payoff for a negotiator, which is a 4.51% payoff

Table 3 Probit regression results on the likelihood to achieve an agreement in negotiations

	(1) Message level	(2) Negotiator level	(3) Negotiator level	(4) Negotiator level	(5) Negotiation level	(6) Negotiation level	(7) Negotiation level	(8) Negotiation level
Intercept	1.006*** (0.014)	0.463*** (0.052)	0.489*** (0.054)	0.461*** (0.000)	0.404*** (0.079)	0.430*** (0.080)	0.394*** (0.079)	0.470*** (0.092)
Sentiment (Harvard)	-0.001 (0.014)	0.061 (0.046)	0.082 (0.047)		0.116*** (0.085)	0.146 (0.085)		0.130 (0.084)
Squared sentiment (Harvard)			-0.049 (0.026)			-0.095 (0.068)		
Negative sentiment (Harvard)				0.156** (0.050)			0.309** (0.096)	
Positive sentiment (Harvard)				-0.006 (0.046)			-0.002 (0.084)	
Delta sentiment Btw. negotiators								-0.103 (0.072)
Number of messages		0.114*** (0.013)	0.112*** (0.013)	-0.115 (0.060)	0.065*** (0.001)	0.065*** (0.001)	0.066*** (0.010)	0.064*** (0.010)
AIC	9498.5	2456.6	2455.1	2450.5	1224.2	2455.1	2547.2	1224.2

Stated: Coefficients, t-statistics in parentheses. Significance is indicated at 10% (^), 5% (*), 1% (**) and 0.1% (***).

Table 4 Descriptive statistics of negotiations

Date	Number of negotiations	Number of agreements	Average of seller value	Average of buyer value	Average of joint value
Dec. 2010	123	110	71.6	73.6	153.7
May 2011	116	91	62.8	72.1	152.1
Oct. 2011	92	74	57.8	77.6	149.4
Apr. 2012	67	31	37.6	68.7	149.9
Apr. 2013	124	113	72.5	71.4	151.0
Nov. 2013	78	66	67.8	67.2	147.3
Apr. 2014	170	144	66.0	81.5	159.4
Nov. 2014	40	33	63.0	76.2	152.5
Apr. 2015	149	126	66.6	76.4	155.2
Nov. 2015	33	26	60.0	73.2	149.4
Apr. 2016	100	80	64.9	74.7	155.8
Aggregation	1092 [s]	894 [s]	62.8[a]	73.9[a]	152.4[a]

s: sum; a: average

increase over the mean payoff (regression model 1 in Table 5). We also split our sentiment variable into its positive and negative language components. The negative language sentiment component exerts a stronger effect on negotiation outcomes than positive language. Using less negatively connoted words increases the probability of achieving an agreement and a higher payoff, while using more positively connoted words has no statistically significant effect on the payoff. For instance, a one-standard-deviation reduction in negative sentiment in a negotiation leads to a 7.703 units higher payoff. This is about a 14.73% of the mean payoff per player across all negotiations and thus of high economic relevance (regression model 4 in Table 5).

Furthermore, we cannot reject Hypothesis 4 that reciprocity in the sentiment plays a crucial role in achieving a higher payoff. The less coherent the sentiment of the two parties, the lower the payoff. As such, the more the sentiment of the two players diverges, the lower the mean payoff for both players, however only at a weakly statistically significant level (*P*-value smaller than 0.1).

Table 5 OLS regressions on the achieved payoffs in negotiations

	(1) Negotiator level	(2) Negotiator level	(3) Negotiator level	(4) Negotiation level	(5) Negotiation level	(6) Negotiation level	(7) Negotiation level
Intercept	52.392***	53.155***	52.310***	52.511***	53.27***	52.291***	54.22***
	(1.148)	(1.188)	(1.146)	(1.596)	(1.642)	(1.591)	(1.900)
Sentiment	2.363*	2.990**		3.141^	3.993*		3.424^
(Harvard)	(1.060)	(1.090)		(1.810)	(1.861)		(1.816)
Squared sentiment		-1.615*			-3.045^		
(Harvard)		(0.655)			(1.582)		
Negative sentiment			4.070***			7.703***	
(Harvard)			(1.161)			(2.088)	
Positive sentiment			0.786			0.402	
(Harvard)			(1.045)			(1.756)	
Delta sentiment							-2.688^
Btw. Negotiators							(1.624)
Number of messages	2.380***	2.331***	2.383***	1.171***	1.159***	1.180***	1.132***
	(0.245)	(0.245)	(0.244)	(0.173)	(0.173)	(0.172)	(0.174)
Multiple-R^2	0.036	0.038	0.039	0.034	0.037	0.042	0.036
Adjusted-R^2	0.035	0.037	0.038	0.033	0.035	0.040	0.034

Stated: Coefficients, t-statistics in parentheses. Significance is indicated at 10% (^), 5% (*), 1% (**) and 0.1% (***).

4.3 Dynamics of the Sentiment During Negotiations

Next, we want to understand whether the message sentiment differs across successful and failed negotiations (i.e., negotiation without an agreement). We thus evaluate how our sentiment measure develops throughout the negotiation process. To examine this sentiment evolution throughout the communication process, we decompose our negotiations into three phases following Holmes (1992): an initiation (relationship building) phase, a problem-solving (negotiation) phase and a resolution phase.

We assign the first two and last two messages, respectively, to the first and third phase. We assume that in a standard negotiation protocol, one of the two parties initiates the negotiation with a first (welcoming or introductory) message, while the other negotiator replies with a similar message. Then, the negotiation takes place. Finally, the last two messages serve to close the negotiation with one party summarizing the (possibly) made agreement and the negotiating party confirming the made agreement.

As the results in Table 6 reveal, sentiment is more positive at the beginning and end of the negotiation, similar to Simons (1993) and Elkins and Derrick (2013), with

Table 6 Regression output

	(1) Sentiment (GI)	(2) Sentiment (GI)	(3) Pos. Sentiment (GI)	(4) Pos. Sentiment (GI)	(5) Neg. Sentiment (GI)	(6) Neg. Sentiment (GI)
Intercept	0.107***	0.219***	0.133***	0.157***	-0.024	0.178***
	(0.020)	(0.039)	(0.020)	(0.039)	(0.020)	(0.040)
Agreement		-0.188**		-0.099		-0.221***
		(0.065)		(0.065)		(0.065)
Negotiation phase	0.170***	0.074^	0.096***	0.094*	0.187***	-0.020
	(0.027)	(0.044)	(0.027)	(0.044)	(0.028)	(0.045)
Nego. phase × Agreement		-0.420***		-0.283***		-0.374***
		(0.053)		(0.053)		(0.054)
Closing phase	-0.287***	0.015	-0.300***	-0.005	-0.046	0.042
	(0.024)	(0.071)	(0.024)	(0.072)	(0.024)	(0.072)
Closing phase × Agreement		-0.051		-0.141*		0.164**
		(0.060)		(0.059)		(0.060)
Multiple-R^2	0.039	0.039	0.032	0.033	0.010	0.011
Adjusted-R^2	0.039	0.039	0.032	0.032	0.010	0.010

Stated: Coefficients, t-statistics in parentheses. Significance is indicated at 10% (^), 5% (*), 1% (**) and 0.1% (***).

the difference across the three periods being statistically significant. Thus, we cannot reject Hypothesis 3. Whereas the initiation phase serves to establish a relationship based on a positive sentiment, the sentiment in the middle problem-solving stage is more negative, driven by a decreased use of positively connoted words and an increased use of negatively connoted words. Towards the end, once an agreement has been reached (or abandoned), the outcome becomes clear, and the tone of messages rebounds and gets more positive again.

A closer look at the comparative sentiment dynamics, depending on whether an agreement is made or not reveals intriguing insights into how successful and abortive negotiations differ. While the differences in sentiment across the phases are all significant, the sentiment in negotiations leading to an agreement is more positive in both the relationship building and closing phase and lower during the core stage compared to futile negotiations. While this interaction effect is only statistically significant for the building phase, the directional effect size reveals that successful negotiations are characterized by a more diligent (social) opening and closing phase, but also a more intense negotiation period.

Interestingly, successful negotiations differ from aborted negotiations in the way negotiators use of both positive and negative language. In negotiations that achieve an agreement, the language is more positive in the opening phase (P-value smaller than 0.05), but less positive in the central negotiation period. Meanwhile, the degree of positivity in the final closing period does not reach statistically relevant levels. At the same time, successful negotiators (in terms of coming to an agreement) also reduce their amount of negatively connoted words during the negotiation phase as compared to abortive negotiators (P-value smaller than 0.01).

5 Discussion

Our results contain some very relevant implications for the negotiation research domain as well as practitioners.

From a research perspective, the importance of sentiment across different negotiation phases is of particular interest. The fact the negotiations are more likely to succeed if they rely on more positive language during the opening phase and reduce the overall amount of positively or negatively charged words during the middle, negotiation phase, is of high relevance for both academics and researchers. This means that investing in initial trust- and relationship-building pays off and at the same time allows the negotiators to focus during the core negotiation phase on the negotiation itself in a matter-of-facts communication or language style.

For companies and negotiation practitioners, our findings imply that it is important to train their employees with negotiation responsibilities in written negotiation communication skills. The language in negotiations still matters in a digitized or electronic context and thus can be a significant contributor to better negotiation results.

Thereby, practitioners can for instance leverage decision-support systems empowered with sentiment analysis features to steer the tone in written communication (Alfano et al. 2018).

6 Conclusion and Outlook

Throughout human history, negotiations have been a central element in the exchange of goods and services. With continuing digitization, e-negotiations become increasingly important and are thus highly relevant for academic research. Yet, knowledge about the role of language sentiment in exchanged communication is scant.

As this research shows, language plays a crucial role in e-negotiations. Negotiations are more likely to succeed when negotiators use more positive language. In that context, it is particularly advantageous if negotiators elicit a high level of reciprocity in their language, since mutually high levels of sentiment lead to higher payoffs. Intriguingly, the language in negotiations that reach an agreement is more positive in the opening, trust-building phase, but also more negative in the core negotiation phase, compared to failed negotiations.

Going forward, research should focus on better understanding the underlying drivers of the communication process, i.e. whether certain personal characteristics have an effect on negotiation outcomes. In addition, research on this area may evaluate how different cultural backgrounds shape the way negotiators communicate and select their language sentiment.

References

Alfano, S., Pröllochs, N., Feuerriegel, S., Neumann, D.: Say it right: IS prototype to enable evidence-based communication using big data. In Deokar, A., Gupta, A., Iyer, L., Jones, M. (eds.) Analytics and Data Science, pp. 217–221. Springer, Heidelberg (2018)

Antweiler, W., Frank, M.Z.: Is all that talk just noise? The information content of internet stock message boards. J. Financ. **59**(3), 1259–1294 (2004)

Brett, J.M., Shapiro, D.L., Lytle, A.L.: Breaking the bonds of reciprocity in negotiations. Acad. Manag. J. **41**(4), 410–424 (1998)

Bronstein, I., Nelson, N., Livnat, Z., Ben-Ari, R.: Rapport in negotiation: the contribution of the verbal channel. J. Conflict Resolut. **56**(6), 1089–1115 (2012)

Chen, H., Chiang, R.H., Storey, V.C.: Business intelligence and analytics: from big data to big impact. MIS Quarterly. **36**(4), 1165–1188 (2012)

Dadvar, M., Hauff, C., de Jong, F.: Scope of negation detection in sentiment analysis. In: Proceedings of the Dutch-Belgian Information Retrieval Workshop. DIR 2011, (2011)

Demers, E., Vega, C.: Soft information in earnings announcements: news or noise? Board of Governors of the Federal Reserve System International Finance Discussion Papers, Number 951, (2008)

Drolet, A.L., Morris, M.W.: Rapport in conflict resolution: accounting for how face-to-face contact fosters mutual cooperation in mixed-motive conflicts. J. Exp. Soc. Psychol. **36**(1), 26–50 (2000)

Elkins, A.C., Derrick, D.C.: The sound of trust: voice as a measurement of trust during interactions with embodied conversational agents. Group Decis. Negot. **22**(5), 897–913 (2013)

Fiske, S.T.: Attention and weight in person perception: the impact of negative and extreme behavior. J. Pers. Soc. Psychol. **38**(6), 889 (1980)

Griessmair, M., Hippmann, P., Gettinger, J.: Emotions in e-negotiations. In: Martinovsky, B. (ed.) Emotion in Group Decision and Negotiation, pp. 101–135. Springer, Netherlands, Dordrecht (2015)

Griessmair, M., Koeszegi, S.T.: Exploring the cognitive-emotional fugue in electronic negotiations. Group Decis. Negot. **18**(3), 213–234 (2009)

Henry, E.: Are investors influenced by how earnings press releases are written? J. Bus. Commun. **45**(4), 363–407 (1973)

Hine, M.J., Murphy, S.A., Weber, M., Kersten, G.: The role of emotion and language in dyadic e-negotiations. Group Decis. Negot. **18**(3), 193–211 (2009)

Holmes, M.E.: Phase structures in negotiation. In Putnam, L.L., Roloff, M.E. (eds.) Communication and Negotiation, pp. 83–105. Thousand Oaks, CA: Sage Publications (1992)

Jap, S., Robertson, D.C., Hamilton, R.: The dark side of rapport: agent misbehavior face-to-face and online. Manage. Sci. **57**(9), 1610–1622 (2011)

Kahneman, D., Tversky, A.: Prospect theory: an analysis of decision under risk. In: MacLean, L.C., Ziemba, W.T. (eds.) Handbook of the Fundamentals of Financial Decision Making: Part I, pp. 99–127. World Scientific (2013)

Kanouse, D.E.: Explaining negativity biases in evaluation and choice behavior: theory and research. In: Kinnear, T.C. (ed.) ACR North American Advances in Consumer Research, vol. 11, pp. 703–708. Provo, UT: Association for Consumer Research (1984)

Karapandza, R.: Stock returns and future tense language in 10-K reports. J. Bank. Financ. **71**, 50–61 (2016)

Kersten, G.E., Noronha, S.J.: WWW-based negotiation support: design, implementation, and use. Decis. Support Syst. **25**(2), 135–154 (1999)

Kopelman, S., Rosette, A.S.: Cultural variation in response to strategic emotions in negotiations. Group Decis. Negot. **17**(1), 65–77 (2008)

Kopelman, S., Rosette, A.S., Thompson, L.: The three faces of Eve: strategic displays of positive, negative, and neutral emotions in negotiations. Organ. Behav. Hum. Decis. Process. **99**(1), 81–101 (2006)

Kurtzberg, T., Dunn-Jensen, L., Matsibekker, C.L.: Multi-party e-negotiations: agents, alliances, and negotiation success. Int. J. Conflict Manag. **16**(3), 245–264 (2005)

Lax, D.A., Sebenius, J.K.: The Manager as Negotiator: Bargaining for Cooperation and Competitive Gain. The Free Press, New York, London, Toronto, Sydney (1986)

Lewicka, M., Czapinski, J., Peeters, G.: Positive-negative asymmetry or when the heart needs a reason. Eur. J. Soc. Psychol. **22**(5), 425–434 (1992)

Lewis, D.D., Yang, Y., Rose, T.G., Li, F.: Rcv1: a new benchmark collection for text categorization research. J. Mach. Learn. Res. **5**, 361–397 (2004)

Li, F.: Textual analysis of corporate disclosures: a survey of the literature. J. Account. Lit. **29**, 143–165 (2010)

Loughran, T., McDonald, B.: When is a liability not a liability? Textual analysis, dictionaries, and 10-Ks. J. Financ. **66**(1), 35–65 (2011)

Loughran, T., McDonald, B.: Textual analysis in accounting and finance: a survey. J. Account. Res. **54**(4), 1187–1230 (2016)

Manning, C., Schutze, H.: Foundations of Statistical Natural Language Processing. MIT Press, Cambridge (1999)

Mittermayer, M.-A., Knolmayer, G.: Text mining systems for market response to news: a survey, Institut für Wirtschaftsinformatik der Universität Bern (2006)

Moore, D.A., Kurtzberg, T.R., Thompson, L.L., Morris, M.W.: Long and short routes to success in electronically mediated negotiations: group affiliations and good vibrations. Organ. Behav. Hum. Decis. Process. **77**(1), 22–43 (1999)

Nadler, J.: Rapport: rapport in negotiation and conflict resolution. Marquette Law Rev. **87**(4), 25 (2004)
Pang, B., Lee, L.: Opinion mining and sentiment analysis. Found. Trends® Inf. Retrieval **2**(1–2), 1–135 (2008)
Parlamis, J.D., Geiger, I.: Mind the medium: a qualitative analysis of email negotiation. Group Decis. Negot. **24**(2), 359–381 (2015)
Pesendorfer, E.-M., Graf, A., Koeszegi, S.T.: Relationship in electronic negotiations: Tracking behavior over time. J. Bus. Econ. **77**(12), 1315–1338 (2007)
Schumaker, R.P., Chen, H.: A quantitative stock prediction system based on financial news. Inf. Process. Manage. **45**(5), 571–583 (2009)
Sharma, A., Dey, S.A.: Comparative study of feature selection and machine learning techniques for sentiment analysis. In: Proceedings of the 2012 ACM Research in Applied Computation Symposium, pp. 1–7 (2012)
Sheehy, B., Palanovics, N.: E-negotiations: rapport building, anonymity and attribution. In: Proceedings of the International Conference of Group Decison and Negotiation. GDN 2006, Karlsruhe, Germany. June 25–28, (2006)
Simons, T.: Speech patterns and the concept of utility in cognitive maps: the case of integrative bargaining. Acad. Manag. J. **36**(1), 139–156 (1993)
Sokolova, M., Lapalme, G.: How much do we say? Using informativeness of negotiation text records for early prediction of negotiation outcomes. Group Decis. Negot. **21**(3), 363–379 (2012)
Sokolova, M., Nastase, V., Szpakowicz, S.: The telling tail: signals of success in electronic negotiation texts. In: Proceedings of the Third International Joint Conference on Natural Language Processing: Volume-I (2008)
Sokolova, M., Szpakowicz, S.: Strategies and language trends in learning success and failure of negotiation. Group Decis. Negot. **16**(5), 469–484 (2007)
Tetlock, P.C.: Giving content to investor sentiment: the role of media in the stock market. J. Financ. **62**(3), 1139–1168 (2007)
Tetlock, P.C., Saar-Tsechansky, M., Macskassy, S.: More than words: quantifying language to measure firms' fundamentals. J. Financ. **63**(3), 1437–1467 (2008)
Twitchell, D.P., Jensen, M.L., Derrick, D.C., Burgoon, J.K., Nunamaker, J.F.: Negotiation outcome classification using language features. Group Decis. Negot. **22**(1), 135–151 (2013)
Van Kleef, G.A., De Dreu, C.K., Manstead, A.S.: The interpersonal effects of emotions in negotiations: a motivated information processing approach. J. Pers. Soc. Psychol. **87**(4), 510 (2004)
Weingart, L.R., Olekalns, M.: Communication processes in negotiation: frequencies, sequences, and phases. In: Gelfand, M.J., Brett, J.M. (ed.) The Handbook of Negotiation and Culture, pp. 143–157 (2004)
Weingart, L.R., Prietula, M.J., Hyder, E.B., Genovese, C.R.: Knowledge and the sequential processes of negotiation: a Markov chain analysis of response-in-kind. J. Exp. Soc. Psychol. **35**(4), 366–393 (1999)

Applications

Mathematical Based Models for Group Decision Support in Telecommunication Network Design and Management—Challenges and Trends

José Craveirinha, João Clímaco, and Rita Girão-Silva

In memoriam of Professor Gregory Kersten – a great scientist and scholar
José Craveirinha, João Clímaco, Rita Girão-Silva

To the faithful and generous Human Being
To the irreverent and freethinker Citizen
To my dearest friend Gregory
João

Abstract The extremely rapid evolutions of telecommunication network technologies and services and their interactions with complex socio-economic environments, justify the increasing importance in applying, in certain areas of network planning, design and management, group decision approaches. In fact, there is a significant number of decision problems focused on issues of network planning and design, of multiple natures, where more than one decision maker intervenes or where it is possible to develop mathematical formulations of the problems considering multiple DMs, representing entities of the network itself. Moreover, the evolution of these networks and related industries leads to a great variety of multifaceted and complex problems, usually involving multiple dimensions, very frequently of conflicting nature. These factors justify the interest in addressing the applications of mathematical based models for group decision support in telecommunications. Although

J. Craveirinha (✉) · J. Clímaco · R. Girão-Silva
Institute for Systems Engineering and Computers at Coimbra, University of Coimbra, Coimbra, Portugal
e-mail: jcrav@deec.uc.pt

J. Clímaco
e-mail: jclimaco@fe.uc.pt

R. Girão-Silva
e-mail: rita@deec.uc.pt

R. Girão-Silva
Department of Electrical and Computer Engineering, University of Coimbra, Coimbra, Portugal

T. Szapiro and J. Kacprzyk (eds.), *Collective Decisions: Theory, Algorithms And Decision Support Systems*, Studies in Systems, Decision and Control 392,
https://doi.org/10.1007/978-3-030-84997-9_10

mathematical based models have been developed in the framework of operations research, systems science, game theory, etc., and they are an essential part of group decision and negotiation support systems, the scope of this study is limited to multi-criteria models and game theory models. In fact, we believe these methodologies are adequate to discuss the challenges and trends of formal models for telecommunication applications. An outline of more relevant evolutions of telecommunication network technologies and services will be presented, followed by a brief overview of major concepts concerning multicriteria group decision (MCGD) and game theory (GT) approaches and methods, relevant to these areas. After identifying the major domains of application of MCGD and GT approaches in telecommunications, an overview of representative contributions in these areas, based on MCGD and GT methodologies, will be put forward. Finally, an analysis and outline discussion of current and future research trends and challenges concerning the use of MCGD and GT approaches in this broad area of decision support, also focusing on some relevant methodological issues, will be presented.

Keywords Multicriteria group decision · Game theory · Multicriteria decision analysis · Telecommunication networks · Network design · Telecommunication policies

1 Introduction and Motivation

Telecommunication networks and technologies, as well as the services they support, have been subject and are expected to continue in a process of very fast evolution. This has been fostered by an exponential increase in offered traffics, of multiple types, in parallel with a drastic increase in the demand for more advanced services, with better QoS (Quality of Service). These mega-trends constitute a process of the outmost importance, not only concerning technological advances, but also with respect to their great impacts on the economy and on the society as a whole. The evolution of these networks and related industries leads to a great variety of multifaceted and complex problems, usually involving multiple dimensions, very frequently of conflicting nature. This is related to strong interactions between the complex socio-economic environment of societies and the extremely fast evolution of telecommunication networks and services. Moreover, there is a great number of decision problems focused on issues of network planning and design, of multiple natures, where more than one decision maker (DM) intervenes, such as managers, regulators, network designer experts, for which it is possible to develop mathematical formulations of the problems considering multiple DMs, representing entities of the network itself.

These factors clearly justify the interest in using, in this broad problematic area, group decision evaluation approaches concerning multiple problems of network planning and design. These issues and also some reflections raised by the past research experience of the authors in some of these problems, laid the *leitmotiv* for this work, namely an overview on mathematical based models for group decision support, let it be multicriteria group decision (MCGD) or game theory (GT) models.

Here, we would like to draw attention of the reader to two texts that, in some sense, are the memory of the participation of Gregory Kersten in two conferences—MCDM 94 and GDN 2008—organized by our research team in Coimbra, Portugal. Namely, (Kersten 1997) and (Clímaco et al. 2015a). We believe that the first is useful for introducing the reader to the roots of this research area, and the second one gives a broad idea of its evolution.

We will also discuss major trends and challenges of MCGD and GT in relation to current and expected evolutions of telecommunication networks in a near future.

This contribution is organized as follows. In the next section we will outline most relevant evolutions of telecommunication network technologies and services, emphasizing recent developments in relation to group decision problems. Also in this section we will include a brief overview of major concepts concerning MCGD and GT approaches and methods, relevant in the considered application areas. In Sect. 3 we present an overview of applications of MCGD approaches to telecommunication network planning and design problems, including strategic planning issues. Also an identification of the main areas of application of GT approaches, where a very large number of papers has been presented, will be put forward. An overview of some representative papers will also be presented, in order to illustrate typical applications and formulations of GT in each of these areas. Furthermore, in these overviews we will seek to briefly discuss main modelling and methodological aspects of MCGD and GT applications in these areas. Finally, a discussion of current and future trends and challenges in these areas, will be outlined. Particular attention will be paid to modelling and methodological issues concerning MCGD and GT approaches.

2 Telecommunication Networks and Group Decision Analysis

2.1 Highlights on Telecommunication Networks Evolution

The first integrated service broadband networks based on the ATM (Asynchronous Transfer Mode) technology, developed in the 90’s, were rapidly abandoned after 2000. This basically resulted from the emergence of cost effective multiservice Internet based technologies, supporting the implementation of connection oriented services and advanced QoS routing and network management mechanisms. In subsequent years MPLS (Multiprotocol Label Switching) and GMPLS (Generalized MPLS), based on optical networks, emerged as more advanced technologies for use in IP (Internet Protocol) based networks. These evolutions were supported, at the level of the transport telecommunication infrastructure—*wired transmission networks*—by the development of advanced optical networks with extremely large bandwidths associated with a great number of very low wavelengths that may be carried by the optical fibers. Another area where there have been extremely rapid evolutions concerns *cellular mobile wireless networks,* driven by the exponential increase in the demand for mobile data services, namely Internet access. The widespread of 4G (4th generation) networks that are interoperable with existing wireless standards, enabled significant improvements in QoS performance and provided an extensive range of services, including HD (High Definition) broadcast, video calls and mobile TV, as well as a multitude of applications for entertainment, business, social networking, education, and so on—see (Tran et al. 2017). Other important aspect should be mentioned, concerning new types of service demand, namely the unprecedented increase in *cloud computing* and the emergence of *IoT* (Internet of Things) in which a plethora of devices are equipped with electronic systems, sensors and software, enabling to exchange data through the Internet. The current step in this technology evolution, the 5G (5th generation) mobile networks is expected to provide important quantitative and qualitative advances regarding increased bandwidth and transmission latency. Also, multiple wireless technologies for data/computer networks of various sorts have been developed such as *Cognitive Radio, Multi-Radio, Wireless LANs, Fiber-wireless (FiWi) Access, Ad hoc Wireless Networks* including *Ad hoc Sensor Networks,* that pose new problems of network design typically involving multiple technic-economic dimensions. A new technological paradigm that was devised for overcoming important limitations in the working and management of current network structures is Software Defined Networking (SDN) (the basic concept underlying SDN is the separation between the control logic of the network and the devices that implement the forwarding of traffic flows).

We also would like to draw attention to the increasing relevance of multidimensional QoS/QoE (QoE-Quality of Experience, i.e. the multiplicity of performance measures as perceived by the end users, e.g. service availability or communication latency, in a given service) in relation to the various technological platforms. Note that the possible interplay between various technologies, both in wired and wireless networks, in the different functional layers of communication networks, enables the use of a great variety of network architectures and technical solutions in terms of network design.

These trends lead to many new decision problems in multiple areas of network planning, design and management where, frequently, multiple criteria, possibly conflicting and of incommensurate nature, are at stake. This led to an increasing interest in using multicriteria approaches. Furthermore, in various of such problems, namely involving market competition, selection of vendors or evaluation and selection of techno-economic alternatives (or of similar nature) there are several DMs intervening in the decision process. This opens clearly a field for the development of MCGD and Game theoretical approaches for tackling several of these problems, as shown in the overview in Sect. 3 of this work.

2.2 *Formal Mathematical Models—Brief Overview*

Although mathematical based models have been developed in the framework of operations research, systems science, game theory, etc., and they are an essential part of group decision and negotiation support systems, the scope of this study is limited to multicriteria models and game theory models. In fact, we believe these are adequate for discussing the challenges and trends of formal models for telecommunication applications.

2.2.1 Multicriteria Analysis

In recent years, multicriteria decision analysis (MCDA) models have been used in some telecommunication group decision problems—see Sect. 3.1.

Considering that different and conflicting criteria are involved in the model, the concept of optimal solution should be replaced by the concept of non-dominated (also designated as Pareto optimal or efficient) solutions set.

In general, we can state that MCDA approaches seek to obtain one or more non-dominated (or at least approximately non-dominated), solution(s) which may be considered as satisfactory by the DM(s). An important issue in developing these types of models is to evaluate *a priori* if there is the possibility of using interactive procedures, especially taking into account the required speeds of calculation, for a given application. This means that, in certain areas of telecommunication network operational design, an interactive procedure cannot be used because of the time consumed in an interaction, such as in the case of dynamic routing or real-time bandwidth assignment models. The above considerations refer mostly to multicriteria/multiobjective mathematical programming based models, that may be linear, non-linear and, in many cases, may have a specific structure, resulting from the particular features of the addressed problem and of the used mathematical formulation. For instance, Granat and Wierzbicki (2004) argue the adequacy of reference point methods for decision aiding in some telecommunication problems.

There is another type of MCDA models, normally designated as *multiattribute decision models*, that have also suffered significant developments and have important application in certain telecommunication decision problems (including group decision models), for example in the area of market competition and techno-economic evaluations. It should be noted that, while in multicriteria mathematical programming formulations it is assumed that the set of feasible solutions (or alternatives) results implicitly from the constraints, in multiattribute models a discrete and small set of alternatives is specified explicitly. The alternatives in this set have to be analyzed by the DM(s) with respect to the considered criteria (or attributes). It is important to note that in multiattribute decision models it is possible for the DM(s) to carry out a more detailed evaluation of the alternatives and consider a larger consistent family of criteria. Moreover, this can be done without having to pay the price of a computational explosion. Nevertheless, in a large number of problems of network design (such as in typical routing and facility capacity related problems) this may lead to a reductive point of view, which may be unrealistic since it does not enable an adequate exploration of the decision space.

Concerning multiattribute models, we can identify, in the so-called American School, the construction of completely compensatory approaches, where a linear or non-linear multiattribute utility function (defined in the framework of multiattribute utility theory) is used (Keeney and Raiffa 1993). As for the Analytical Hierarchy Process (AHP) method, it can be viewed as a specific branch of the American School, involving the identification of a hierarchy of interrelated decision levels (Saaty 1980; 1994a,1994b). AHP models have been used in several telecommunication multicriteria group decision models as it will be reviewed in Sect. 3.1. Moreover, it must be referred to a ranking approach, TOPSIS—Technique for Order Preference by Similarity to an Ideal Solution (Hwang and Yoon 1981), also used by many authors in the works reviewed in this paper. The use of TOPSIS in these problems is probably due to the fact that it is very simple, comprehensible and computationally efficient. In a few words: ideal and anti-ideal solutions are calculated; a metric is used to calculate the distance of each alternative to these points. Finally, alternatives are ranked according to a coefficient representing the relative closeness to the ideal solution. Note that this coefficient is obtained for each alternative, from the distances to the ideal and anti-ideal solutions.

Furthermore, there are non-compensatory approaches (of the so-called French School), not allowing a complete ranking of alternatives, hence not guaranteeing the principle of optimality, that is neither transitivity nor full comparability of alternatives are verified. Therefore, we can say that their results are less conclusive with respect to the aggregation of the preferences of the DM in comparison. A most relevant methodology of the French School, is the ELECTRE family of methods (Roy and Bouyssou 1993; Figueira et al. 2016). Resulting from the features of the problem, the purpose of ELECTRE is the classification of alternatives, the ranking of alternatives, or selection of the most preferred alternative.

More recently, multiattribute and mathematical programming approaches such that the preference aggregation is based on inductive rules, were developed. In particular, the approaches which are rooted in an adaptation of rough sets concepts must be mentioned (Slowinski et al. 2012).

Although other multiattribute methods, namely outranking approaches, should be tested in the addressed application areas, till now, as far as we know, only compensatory based approaches were applied to group decision frameworks dealing with telecommunication problems. In the literature we can find MAUT, AHP, TOPSIS and VIP-Analysis. VIP-Analysis (Variable Interdependent Parameters Analysis), which is structured according to MAUT, is an interactive software package dedicated to the choice problematic of the evaluation of a discrete set of alternatives according to a multiattribute additive value function (Dias and Clímaco 2000, 2005). The principal characteristic of this tool is that no precise values, for the scaling constants/weights, are required. Instead, it can accept imprecise information (i.e. intervals and/or linear constraints) on these values, usually identified by indirect ways, as for example by comparing swings, by ordering scaling constants, etc. The major objectives are the identification of robust conclusions (holding for every feasible combination of the scaling constants), and secondly identifying what is the variability of the results due to the imprecision in the parameter values.

Interactive tools, and in particular learning oriented tools, as VIP-Analysis, seem to be the most interesting in many future developments. In fact, as it is emphasized in (Clímaco 2015), in most of the situations "… the intended help does not consist of showing the various actors involved in the course to follow, but rather of constructing a set of coherent recommendations that contribute to the clarification of the process. Thus, the models' goals and values do not run the risk of being replaced by any calculated rationale".

It must be emphasized that MCGD Support is mostly concerned with cooperative group decision, rather than with negotiation processes, though the frontier between these two decision settings is often fuzzy. In fact, MCGD deals with common sets of alternatives and objectives, while in negotiations the proposals are sequentially presented by parties, which involves making concessions. This peculiar interdependence among actors, "rather than conflict, distinguishes negotiation from other forms of decision making" (Kersten and Cray 1996). Nevertheless, as it is explored in (Granat and Wierzbicki 2004), multicriteria interactive decision tools can be useful in a preparation phase of negotiation processes. Furthermore, as analyzed in (Dias and Clímaco 2005) specific group decision problems where the DMs involved may have significantly different and divergent priorities cannot be excluded from consistent and adequate treatment by MCGD approaches. The architecture of the group decision support tool is a key point regarding this issue. See, e.g., in (Dias and Clímaco 2005) the proposed architecture that makes VIP-Analysis adequate for group settings. Extensions of the methodology of the VIP analysis to address explicitly the differences among the DMs in terms of the weight space are in (Clímaco and Dias 2006).

2.2.2 Game Theory

As it is stated in (Clímaco 2015), "Game Theory is dedicated to the choice of an optimal strategic behavior of two or more rational players (decision agents) interacting strategically. Cost and benefits of each option for one player depend on the choices of the other players" (so decision agents take into account the interdependence of their decisions) and constitute the so-called payoffs of each player. "It is clearly a rigorous approach for dealing with conflicts". To foresee the result of a game, the analyst must focus his attention on possible combinations of the strategies, as the interaction among the strategies of the players determines the outcomes.

Remembering that game theory is particularly appropriate to deal with the economics of imperfect economy and that telecommunications live a stormy period immersed in deep and sharp alterations of technology, it seems an adequate approach to address some of its fundamental issues. So, it must be emphasized that this type of mathematically based models is a potentially suitable tool for many telecommunication group decision problems, where, for instance, the analysis of the stability of outcomes is one of its key issues. Many researchers have exploited a great number of game theory approaches, some of them considered in the papers reviewed in Sect. 3.2.

Nevertheless, many situations of complex real applications are not adequately tractable by GT due to the need of oversimplifying the models. Otherwise, it would be impractical to put them into operation. Plus, as it is emphasized in (Rapoport 1964): "The weakness of game-theoretic approaches includes the treatment of the process and its impact on the game itself, and strict rationality assumptions which, for numerous reasons, rarely hold (e.g. imperfect information, parties cognitive limitations, and deception) ... Thus, while game-theoretic methods have a significant role to play in the prior or posterior analysis of the group decision or negotiation problems, their usefulness as a support tool during the process is limited."

Next, we make a very short summary of some Game Theory key concepts relevant in the applications overviewed in Sect. 3.2:

- We can classify the games as cooperative, when there is cooperation among the players, and non-cooperative otherwise. Note that in many situations cooperation and conflict can coexist.
- A game is static when all players make their moves simultaneously and independently. On the other hand, in dynamic games the moves of the players may happen successively, the most common case in the application to network design models (such as in models of channel and bandwidth assignment, routing and congestion control).
- A zero-sum game is a game where the sum of payoffs of the players is zero, independently of the chosen strategies.
- Pure strategies of a player are the strategies belonging to his space of strategies.
- A mix strategy of a player consists of selecting it randomly, assigning probabilities to the strategies belonging to the strategies space of the player. It is chosen when the player is indifferent regarding pure strategies.
- An equilibrium is a combination of strategies formed by the "best" strategy for each one of the players.
- A strategy of a player is dominant if all the others are strictly dominated; and a strategy is strictly dominated by another one if it is strictly worse than the other one, independently of the choices of the other players.
- An equilibrium of dominant strategies is a combination of dominant strategies including a dominant strategy of each player.
- A Nash equilibrium (NE) is a combination of strategies such that no player takes advantage of dodging from his strategy assuming all the others maintain their strategies. So, in equilibrium, the strategy of each player is the best answer to the strategies of the others.
- A combination of strategies is Pareto optimal if it is not possible to improve the payoff of a player without worsening the payoff of at least another one.
- A game is a perfect information game if a player knows the full history of the game when he/she makes a move. Otherwise it is an imperfect information game.
- A game is a complete information game if the rules of the game (structure and payoffs) are common knowledge of the players.
- Evolutionary games assume the players learn about the process progressively. Their strategies are obtained by trial and error. So, it is assumed that complete rationality of the players does not occur in practice many times.

- Repeated game: a modelling interaction in a repeated manner, in cases where the players interact several times.
- Mean-field game theory considers strategic decision making in very large populations involving small interactions among the decision agents.

3 Overview on Applications of Formal Mathematical Models to Group Decision in Telecommunication Networks

3.1 Applications of Multicriteria Group Decision Approaches to Telecommunication Networks

Herein, we will present an overview of works focusing on the application of Multicriteria Group Decision (MCGD) to telecommunication network planning and design.

An area where there has been a number of proposals on applications of MCGD has to do with problems related to *competition in telecommunication network markets* including the role of regulatory entities, or management related decision problems. This includes as typical decision problems: vendor selection by a telecommunication company, the role and influence of regulatory entities or the selection of a manager for a telecommunication company. Also, there are works concerning the evaluation and selection of technological or architectural network solutions, usually in association with cost and/or investment analysis, in a given market context, which we may designate as *techno-economic evaluation* problems. These two types of issues may be aggregated in a broad area which we will identify as a *competition and evaluation decision problem* domain. In this area we may include also the contributions on applications of MCGD approaches to specific operational design problems in *wireless networks* and the evaluation and selection of *routing methods* in wired networks.

Firstly, we will refer to some papers that, although not addressing *per se* a group decision making problem (that is how a given group decision problem could be supported with a given modelling and decision support method approach), seem very useful in tackling preliminary and conditioning issues of problem modelling and decision support method implementation.

Granat and Wierzbicki (2004) present an overview of multicriteria analysis methods that may be applied to the planning and design of telecommunication networks, focusing on important methodological issues. The authors identify design and management problems where MCDA techniques can be used and show why they are particularly adequate for dealing with such problems, involving multiple criteria, giving particular attention to the possible application of reference point methods. This paper also discusses, in realistic *problems of strategic management*, involving several DMs, how MCDA methods could be used by the DMs as an adequate methodological tool in a preparation phase of a negotiation process. In (González and Verhoest 2020) a model of analysis on *de facto* interaction between regulatory actors (within a given country) involved in making regulatory decisions about the telecommunications market, is presented. It uses inferential Social Network Analysis (SNA) techniques to analyze the dynamics of the relationships of the DMs/actors (including regulators, operators and user organizations) and the underlying factors. The second part of the proposed data analysis method involves hypothesis testing, using a technique of inferential network analysis with Exponential Random Graph Models (ERGM) by Cranmer and Desmarais (2011)—a statistical technique of inferential analysis with relational data that uses the characteristics of nodes (corresponding to DMs) and links (types of relations between DMs, in the form a network). Ahmad and Shahid (2015) analyze factors influencing the process of decision making in the telecommunication sector concerning the capabilities of the DMs and of the organizations, namely company management. A statistical software of regression analysis is used to implement the analytical model. Al-Shehri et al. (2017) provide a systematic analysis of technical and economic frameworks for the metrics involved in the characterization and performance evaluation of broadband networks in the context of the fast growing digital economy, based on OECD recommendations (OECD 2012, 2014). A comprehensive list of the metrics of different classes and their features as well as of the associated measurements, is presented.

Another important issue related to the practical implementation of MCGD, is the characterization, from a decision science point of view, of communication technologies and techniques which can be used for supporting MCGD tools in a geographical distributed environment. This issue is thoroughly analyzed in (Bordetsky and Iz 1996) where it is proposed that four sets of dimensions for evaluating the features of communication systems/technologies adequate for implementation of a MCGD friendly electronic environment, be considered.

We will now address concrete contributions of MCGD models in the wide area concerning *competition and evaluation decision problems*.

In (Tam and Tummala 2001) a very important decision problem in the telecommunication industry market is addressed: the selection of a vendor of a telecommunication system by a telecom company, considering a MCGD approach. A key issue of the decision aid model is centered on the identification and prioritization of the criteria and sub-criteria for vendor selection through a systematic approach based on the inputs of the DMs, originated from different functional departments of the company, taking into account the existence of conflicting criteria. The authors then develop a AHP model for tackling this problem, considering a four phases approach. Sahin and Pehlivan (2018) present a MCGD approach for selecting a manager of a telecommunication company by using a fuzzy set based methodology. The authors use Type-2 fuzzy sets (cf. (Zadeh 1975)) and consider three kinds of fuzzy ranking methods, proposed in (Qin and Liu 2015), based on arithmetic average, geometric average and harmonic average operators, for computing the ranking of the fuzzy intervals. The DMs inputs to the model are made through the specification of fuzzy numbers (concerning the relations between alternatives and the criteria values) and criteria weights. The results of the model are compared with those of a TOPSIS method based on Type-1 fuzzy sets (originally defined in (Zadeh 1965)) for a numerical example.

Let us now consider a network design area where evaluation/selection methods based on MCGD approaches have been proposed: wireless networks.

Falowo and Chan (2012) propose a MCGD model for dynamic selection of a RAT (Radio Access Technology) by a call or multiple calls in a Heterogeneous Wireless Network (HWN). These are particular types of wireless networks where an end-to-end connection may use different technological platforms such that a service demand (call) or a group of calls (from an end user), originated at a Multimode Terminal (MT) has to select one of the available RAT telecommunication systems/networks. Note that both the calls/groups of calls (corresponding to the DMs/groups of DMs in the MCGD model) and the available RATs may vary in time. Each user is supposed to specify his/her preference information for choosing a particular RAT for each class of calls, in terms of weights associated with the RAT selection criteria, concerning different technic-economic attributes. A dynamic RAT selection algorithm using the fuzzy TOPSIS MCGD method in (Wang and Lee 2007) is then described and its performance analyzed through simulation. The same type of decision problem as in (Falowo and Chan 2012) is addressed in (Luo et al. 2017) where a network, in a set of available networks with different technologies, has to be selected (in an automated manner) at a MT terminal, by considering that each service profile demand corresponds to a 'DM'. The MCGD model uses the AHP method for obtaining the weight vector of all network attributes for each service, then enabling to synthetize these weight vectors. A utility function is used to calculate the utility value of each network attribute, that is it represents the degree to which a certain 'DM' (corresponding to a service class) is satisfied with that attribute for each available network with a certain technology. The aggregation of the attribute utility values, results from a weighted sum, constructed from synthetic performance matrices previously obtained from network design experts, using the method in (Wang 2000).

The performance *comparison of routing models* in telecommunication networks usually implies the necessity of an evaluation in terms of multiple, potentially conflicting, frequently incommensurate criteria, often involving imprecise information regarding the relative importance of the various network performance criteria. This is particularly relevant for flow-oriented, decentralized routing optimization methods, having in mind their inherent limitations (see e.g. the analysis in (Craveirinha et al. 2008)). In the short paper (Clímaco et al. 2015b) a multicriteria decision problem concerning the evaluation and selection of flow-oriented, decentralized, optimization routing models in telecommunication wired networks, is formulated. In the proposed model various measurable network performance attributes are considered that enable the evaluation of the global effect in the network of using the various routing methods. Moreover, it is assumed that the additive value function to be used in this interactive decision model should be inherently prepared to deal with imprecise information, associated with the scaling constant values (also designated as importance parameters or weights), ascribed to a DM for each attribute. The features of a MCDA method for tackling this problem, based on the VIP-Analysis software in (Dias and Clímaco 2000), also considering its possible extension to a cooperative group decision setting, are outlined. An overview on this issue, concerning *comparison and selection of routing models* and the complete development and application, in a given network setting, of a MCDA method with the above methodological features, are described in (Clímaco et al. 2019). The adequacy of the features of this interactive multiattribute decision analysis model, based on the VIP software, prepared for coping with imprecise preference parameters, are analyzed. The proposed method is applied to a case study in a reference network setting in the context of Carrier-Ethernet and MPLS-TP (Multiprotocol Label Switching-Transport Profile) technologies, as described in (Martins et al. 2013). Also the extension of this approach to a face-to-face cooperative group decision (with a facilitator) situation is carried out. The interplay between a tolerance parameter defining quasi-dominance relations between two alternatives and α-majority relations concerning preferences elicited by the DMs is analyzed, following the concepts in (Dias and Clímaco 2005), and applied to the case study.

3.2 Applications of Game Theory Approaches to Telecommunication Networks

There has been a very large number of papers concerning applications of game theoretical approaches to communication network planning and design for the last two decades. We would like to refer that a complete overview of papers on game theory applications in these areas is out of the scope of our contribution. Instead, we will discuss main modelling and methodological aspects of these approaches and key issues vis a vis other Operation Research based approaches of different nature, namely MCGD (Multicriteria Group Decision). We will also present an overview of some representative papers, illustrating typical applications and formulations in the main areas where GT has been applied.

We will begin by identifying the main application areas and sub-areas of GT in this context. Firstly, we can consider decision problems concerning *competition in telecommunication network markets*. Also a problematic area of decision problems focused on *techno-economic evaluation* and selection of technologies/technical network solutions and/or strategic investments, can be identified, similarly to what happens in MCGD applications. It is important to note that the players (or decision makers, DMs) of the game models in these two broad areas—which we have identified in the previous subsection as 'competition and evaluation problem domain'—are essentially human agents external to the networks, namely company managers, service directors, network designers, customers or a mix of similar agents. Therefore, from this point of view, *unlike what happens in the other areas of application of GT in telecommunications*, as identified hereafter, alternative approaches to GT in these two areas are precisely multicriteria group decision approaches (MCGD).

The other main areas of GT applications where there has been an explosion of publications refer to problems of network design—classically encompassed in the category of *network operational planning*—have been focused on a great variety of technic or technic-economic issues, namely: *Wireless network design* (usually data packet networks with multiple technological and architectural types including ad-hoc structures); *Medium access control mechanisms* (in different types of networks); *congestion control in Internet type networks* and *routing methods* (in various types of networks). An overview of some representative contributions in these areas, will be presented next.

We would like to draw attention to the fact that, in these network design areas, the 'players' of the GT approaches are typically technical entities internal to the system/network upon which the decision falls on, such as routers, switches, base stations, moving transceivers, satellites or centralized network management entities, typically represented by nodes (or particular sets of nodes) of the network representation. The strategies of each player of the formulated game are technical actions (for example channel assignment, bandwidth allocation in a link or choice of an end-to-end route for given traffic or service demand). The payoffs are expressed through utility functions encompassing QoS and/or economic related measures. This means that the definition of 'players', in this context, is a conceptual artifice so that these decision problems may be formulated mathematically in GT terms, thence having an inherent nature different from the DMs in MCGD, assumed as agents external to the system, usually human. Therefore, we can state that, in these particular areas, unlike in the *competition and techno-economic evaluation problem domain*, MCGD should not be considered as the alternative methodological choice to GT. Instead we can state that the alternative OR approaches to GT, in these areas, are single or multicriteria decision approaches (with one DM alone), mainly based on mathematical programming (and, in particular, in network flow programming), combinatorial optimization methods or on optimal control algorithms.

Regarding the GT approaches to Internet design problems, very common in the literature, these have three fundamental elements: a set of players (congestion-sensitive data flow sources, typically the routers or a subset of routers), a set of possible actions/strategies for each player (congestion control and/or channel/bandwidth assignment strategies or, in routing design, a concatenation of transmission links), and a set of utility functions (such as throughput, packet delay, other quality of service parameters or pricing related parameters).

Altman et al. (2006) present a comprehensive survey of applications of game theoretical approaches to telecommunication network design problems, which can be considered as a particular application domain of 'networking games'. The authors identify some of the mathematical challenges and methodologies that are involved in these problems and classify, under different telecom network design topics, a quite significantly large number of publications in this area. Also particular attention is paid to the application of NE solutions or its variants, namely by applying equilibrium concepts used in transportation networks. Also in (Shah et al. 2012) it is presented a review of basic concepts of GT and a summary overview of possible applications to the design of communication networks.

Next, we will refer to papers representative of each of the identified application areas.

Market Competition and Techno-Economic Evaluations. The early reference (Turpin 1998) reviews concepts and game theory models related to the 'struggle for the first move' and the origins of cooperation in the context of the modelling of the economics of imperfect competition (namely in duopoly or oligopoly scenarios) in the *telecom network markets*. Some broad considerations concerning the game theoretical models that could be used in different conditions and an analysis of advantages/disadvantages of cooperation versus confrontation of the incumbent operator with new competitors when facing the expansion and exploration of optical networks, are put forward. Note that although the application context of this publication is much outdated some of its broad conclusions may be useful in similar telecommunication oligopolistic contexts. One of the first publications concerning the application of game theory approaches to the *analysis of competitive markets in telecommunications industry* is in (Adjali and Olafsson 1998). The authors describe a general analytical model for this purpose, using an *evolutionary game theory* approach based on the theoretical framework in (Smith 1982; Zeeman 1981). Different scenarios of possible strategies of the incumbent/primary operator and of the newcomers are constructed, considering mixed strategies. A work in (Katsianis et al. 2007) presents a methodology, coupled with the detailed description of a game theory model, for *techno-economic evaluation* of the market competition for 3G mobile networks, involving two operators: a dominant operator and a newcomer competitor. The payoff functions of the players, defined in terms of NPVs (Net Present Values), represent the techno-economic evaluations of each competitor and are described in terms of NPVs concerning investments, costs and revenues. The authors apply the resulting non-cooperative game model to a case study developed in European Projects, to obtain NE solutions. Maillé et al. (2008) present a full game-theoretic analysis model for another type of problem in a telecom market, focused on competing telecom service providers and concerning the *migration of customers* from one operator to another. Assumptions are made regarding the customer behavior (a 4-state Markov model, for two providers) and the strategies are defined through the retention times imposed by the losing provider; the utility functions also depend on the net revenues of retaining the customer and the sanction cost resulting from the possible suing in court of the operator by the customer. The application of the mathematical model to a simplified setting shows that both stable and unstable NE solutions do exist.

Wireless Networks In (Min 2008) a review on GT approaches in wireless networks, mainly focusing on network design problems concerning power control, radio channel access control, cooperation between mobile terminals and security, is presented. Lin et al. (2018) present an updated review on papers with applications of game theory approaches to wireless networks, namely *wireless local area networks* (WLAN), *wireless sensor networks* (WSNs), *ad hoc wireless networks* (AWNs) and *satellite communication networks*. These approaches are based on the fact that the network nodes (the 'players') compete for network resources, mainly data transmission wireless channels and associated bandwidths. Furthermore, in many situations, a node needs the other nodes' collaboration to relay the data message, situations which can be associated with price coefficients. The review is particularly focused on problems concerning the choice of resource allocation strategies taking into account QoS objectives and relay pricing and in problems involving energy efficiency objectives.

A game theoretic model for the problem of spectrum pricing in a *cognitive radio network* where multiple primary service providers compete to offer spectrum access to the secondary users, is described in (Niyato and Hossain 2008a). Cognitive radio networks are a special type of the software defined radio networks concept where it is possible to operate in multiple frequency bands by using multiple transmission protocols and to estimate the communication parameters so that the users can adapt to a changing communication environment. This approach is formulated in terms of an oligopoly market problem with a few firms and a consumer. A dynamic game model for price competition is formulated to analyze the impacts of several technical parameters on the NE. The players in this game are the primary services and the strategies of the players concern the prices per unit of spectrum. An interesting conclusion of the case study, with two primary services, is that the NE solution is inefficient in the sense that is not Pareto optimal with respect to the total profit of the primary services and that a collusion among primary services could be obtained in order to maximize such profits. A similar problem, focused on the spectrum sharing among a primary user and multiple secondary users is tackled in (Niyato and Hossain 2008b), also using a dynamic game approach.

Another problem, concerning adaptive channel allocation, in cognitive radio networks, was tackled in (Nie and Comaniciu 2006) using a GT approach. The authors consider possible utility functions for selfish and for cooperative nodes (the players), depending on several technical parameters related to transmission quality and discuss the advantages and limitations of such functions in relation to the game resolution algorithms and their theoretical properties.

Another problem of *channel assignment* in a special type of wireless networks—MRMC (*Multi-Radio Multi-Channel Wireless Networks*) with mesh topology—MRMC Mesh Networks (these are broadband wireless access networks in the user premises, such that the routers are mutually interconnected and use multiple transceivers that can be tuned to multiple non-overlapping channels) —is tackled in (Shah et al. 2010) using a game theoretic model. The model is a non-cooperative game with incomplete information where the players correspond to flows originated by end nodes and considers a network structure with multiple transmission collision-domains. This model is shown to converge, in certain conditions, to a stable NE solution and a distributed algorithm is obtained for its resolution in finite time. This work extends the analysis of the pioneering work in this area in (Felegyhazi et al. 2007) and particularly the results in (Felegyhazi and Hubaux 2006) which addressed the same problem, formulated in a single collision domain network and where it was proved that this type of game converges to a stable NE solution such that each node gets an equal share of the channel resources. Shah et al. (2012) address the same problem of channel assignment in MRMC networks with multiple collision-domains, now considering a non-cooperative bargaining mechanism among end users. The motivation for this variant of the game theoretic model in (Shah et al. 2010) was the fact that, in these networks, the NE solution of the pure non-cooperative formulation does not always lead to maximal data rate (throughput) for end users; experimental results of the case study confirm that non-cooperative games with bargaining can, in many instances, perform better than NE, in terms of end-to-end throughput. An extension of this non-cooperative game model study for the same type of MRMC Mesh Networks, now considering network topologies with explicit channel interference constraints, is presented in (Shah et al. 2012). These works are closely related to (Chen and Zhong 2009) which also addressed the same problem and extends the initial result of (Felegyhazi and Hubaux 2006) by showing that it is possible to obtain NE solutions by requiring a payment from each player, that enables that optimal fairness (defined as the minimum of the max–min throughput difference, for all users) be attained. Gao and Wang (2008) also studied a channel assignment problem in *multi-hop ad hoc wireless networks* modelled as a static cooperative game, in which some players collaborate in order to achieve high data rates across end-to-end paths. The authors derive the necessary conditions of a 'min–max coalition-proof Nash equilibrium' allocation scheme which aims at maximizing the throughput of the transmission links. Mkiramweni et al. (2018) present an overview of applications of game theoretical approaches to another particular type of *ad hoc* wireless network that recently has gain increasing interest: *wireless communications with Unmanned Aerial Vehicles* (*UAVs*). In these *ad hoc* networks UAVs equipped with high performance transceivers, are used as data relay transmitters, working as aerial base stations to provide services in geographical areas without network infrastructure and also enabling interconnections to ground stations. The addressed design problems are specially focused on the optimization of energy consumption, enhancement of network coverage and connectivity improvement. For example, in (Giagkos et al. 2016) a non-cooperative game theory model with perfect information for locating the UAVs, is proposed. The UAVs act as players that seek selecting the best actions (in a set of flying instructions) to maximize the number of mobile users for which the UAVs provide coverage. The game theoretical solution is compared with evolutionary algorithm solutions according to various technical criteria. A power optimization problem in this context is tackled by Koulali et al. (2016), considering the use of periodic beaconing for UAVs that act as substitutes of aerial base stations. The UAVs are the players of a non-cooperative game model where each payoff function is defined as the difference between the successful encounter rate and energy consumption during the beaconing period and the strategies correspond to the beaconing periods scheduling. The conditions for the existence and uniqueness of the NE are checked and the scheduling procedure is tested via simulations.

Medium Access Control Mechanisms. Another important issue in some wireless networks is the design of *medium access control mechanisms*, that is the mechanisms that determine the access and share of a wireless channel by contending wireless nodes. This is in fact a complex optimal control problem (see (Kelly et al. 1998)) and several techniques may be used with the aim of obtaining high throughput, low packet collision frequency and improved fairness. Chen et al. (2010) put forward a GT model for tackling this issue, called 'random access game', a game in which the wireless node's (player) strategy is its channel access probability and its payoff function includes the utility gain from channel access and the cost associated with packet collision. A general mathematical framework for this game and its application to single-cell wireless LANs (Local Access Networks), and multicell wireless LANs is presented and the existence and convergence properties for NE solutions, in certain conditions, are derived. Also, a practical medium access procedure for distributed control is proposed.

A particular problem of medium access control in a specific type of wireless network architecture—*Fiber-wireless* (*FiWi*) *access mesh networks* (where the fibers go as far as possible from the central office and then the network becomes wireless at the front, up to the end users)—is tackled in (Coimbra et al. 2010), using a repetitive game approach. The players are certain nodes of the network, corresponding to the forward routers, and the strategies of a player indicate the cooperation levels that it uses in every stage of the game, in terms of how much bandwidth that wireless router needs to share for 'foreign' traffic forwarding; the aim is that services of local users and services of foreign users can have access to bandwidth in the available channels in a balanced way. Preliminary results indicate that NE solutions are possible, depending on the parameterization of a scheduling algorithm.

Another important technical problem in wireless networks is the design of *power control mechanisms*. This problem, in conjunction with the pricing of a single resource among several users, is addressed in (Alpcan et al. 2002) in the context of *CDMA* (*Code Division Multiple Access*) *mobile wireless networks*, via a non-cooperative game theoretic approach, using a cost function depending on power levels (pricing component) and signal-interference ratios (transmission utility component). The underlying optimization problem for each user is to minimize its cost, given the sum of powers of other users as received at the base station. The existence and mathematical properties of the NE solutions of the game are analyzed and two practical control algorithms, based on this formulation, are presented.

Congestion Control in Generic Internet-Type Networks. A critical problem in Internet-type networks is the design of the *congestion control mechanism*, since the congestion control algorithm of the TCP (Transmission Connection Protocol), was introduced in 1988. An in-depth mathematical analysis and modelling study for this type of algorithms was addressed in the pioneering work of Kelly et al. (1998) where the resource allocation problem underlying the congestion control design, was addressed as an optimal control problem involving a non-linear mathematical programming formulation (maximization of the sum of user utilities within the bandwidth constraints of the links). Primal and dual algorithms were proposed and their equilibrium and dynamic properties, focusing on fairness, delay instability and stochastic instability, were thoroughly analyzed. An extensive study based on this mathematical framework is in (Kelly 2003). Inspired on some of these results, a non-cooperative GT approach was developed in (Alpcan and Basar 2005) for tackling the control mechanism design problem. An objective function (to be minimized) is defined for each user (player), which includes a pricing function proportional to the experienced delay and a general utility function of the type in (Kelly et al. 1998) expressing the user demand for bandwidth; the strategies of each user correspond to flow rates assigned to the available communication channel. A unique approximation to the NE solution is shown to exist and its stability conditions are analyzed. A discretization of the user cost function enables the formulation of a distributed control procedure (a possible improvement to the TCP/IP protocol) and tested in simple network topologies using a packet simulator.

A problem of *congestion control in satellite networks* is addressed in (Huang and Jiao 2008) seeking to achieve an allocation of bandwidth (of an outgoing link) by the input flows (more or less congestion sensitive) that be fair and eliminate unnecessary bandwidth waste. The authors propose a non-cooperative game model where routers implement certain queueing scheduling mechanism (associated with possible bandwidth allocations strategies) to the input flows (selfish agents) and show that there exists a fair NE solution. The resulting scheduling mechanism is tested in a very simple network and the authors draw attention to the importance of achieving a tradeoff between efficiency and fairness in this context.

Routing Methods. Ergün et al. (2016) present a summary overview and a classification, from a game theory point of view, of a list of papers on game theory applications in telecommunication network routing.

Many papers in this area are based on the application to telecommunication networks of the concept of *selfish routing models*, a type of non-cooperative '*congestion game*' originally formulated by Wardrop (1952) for road transport routing problems. In this game the origin–destination pairs correspond to the players, the arcs of the network represent the resources and the strategies available to a particular player type are the paths in the network while the cost of an arc is the delay experienced by traffic in that arc. A 'social optimum' is defined (it corresponds to an optimal multicommodity flow solution with minimum total delay), while a NE solution corresponds to an equilibrium flow solution, where every player is traveling on a shortest path under certain conditions. This modelling approach to routing can be applied straightforwardly to telecommunication routing problems with adequate adaptations concerning cost/payoff functions or traffic engineering constraints, depending on the particular nature of the network or the features of the specific design problem being addressed. It is important to note, as analyzed in (Papadimitriou 2001), that NE are quite often inefficient solutions since in general these solutions may not minimize the social cost (the global network optimum) leading the author to introduce the so-called *price of anarchy* defined as he ratio of the worst social cost of a NE solution to the cost of an optimal routing solution. Theoretical results on this 'price of anarchy' are analyzed in (Correa et al. 2005) in terms of the mathematical properties of the cost functions. A closely related paper with theoretical results on selfish routing game models in capacitated networks (where there are upper bounds on arc flows, which is the case of telecommunication networks) is (Correa et al. 2004), also including results for nonconvex and non-differentiable arc cost functions (having in mind that convexity and differentiability properties for the cost/utility functions are usually assumed for obtaining NE).

Concerning the mathematical properties of 'selfish routing models' based on GT in relation to network capacity design it should be referred to that some standard approaches of network capacity expansion conjugated with that type of non-cooperative routing methods, may lead to severe degradation of network performance. This is the case when the expansion strategy is based on bottleneck analysis, that is when the network operator adds capacity to identified bottleneck links so that they cease to be a bottleneck for the expected traffic offered under a certain routing method. It was originally shown in (Braess 1968) that in road transport networks (these are analog to pure delay queueing communication networks), in a non-cooperative routing approach, the bottleneck based capacity expansion may lead to a drastic increase in the delays of all users thence to a much worse network design solution—the so-called *Braess paradox*. An extension of this theoretical result to other networking contexts was carried out by other authors, such as Cohen and Kelly (1990); Calvert et al. (1997), and, for loss networks, in (Bean et al. 1997; Altman et al. 2002). Altman et al. (2003) review this problematic and propose forms of avoiding the Braess paradoxical situation when upgrading a communication network for a general payoff function for each user (utility or cost function). Note that some models of channel assignment in wireless networks with multiple domains such as in (Shah et al. 2010), also enable the joint calculation of end-to-end transmission routes. This is the case in (Xiao et al. 2008) where a routing and channel assignment problem in wireless mesh networks (WMNs) is tackled. This is formulated as a non-cooperative game the players of which are all the sink-source node pairs and the aim is to choose a path (one of the possible strategies) between them with feasible channel assignment, satisfying certain radio transmission constraints—a game model designated as 'Strong Transmission Game'. The utility function of each player encompasses transmission costs and QoS parameters. A proof of the existence of a pure strategy NE solution is shown, and a heuristic for obtaining a feasible computational solution, is proposed and tested.

Problems focused on *security issues* are clearly important in communication networks. The general problem of detecting intruding packets in TCP/IP based communication networks, based on packet sampling (of the network links), was tackled in (Kodialam and Lakshman 2003) using a GT approach with two players: the service provider and the intruder. The payoff functions of the players are min–max functions expressed in terms of the sampling rates and the probability of a path being used by the intruder and lead to a classical two persons zero-sum game. The authors show that the optimal strategy for the service provider is obtained from the resolution of a max-flow non-convex optimization problem and devise heuristic procedures for solving it. Chang et al. (2012) propose a dynamic routing method for Internet type networks, modeled as a non-cooperative routing game, where the players are the routers. The model assumes traffic splitting among feasible paths and that frequent traffic measurements are performed by the routers in the network arcs so that the possible paths (player strategies) are associated with a cost function depending on the arc sampling rate and on traffic engineering parameters; thence routers compete in order to minimize their own costs for the downstream paths. The authors derive the existence of NE solutions and analyze conditions for their stability;

a distributed routing procedure is obtained and tested through packet simulations. In (Yang et al. 2013) a routing problem in Internet type networks, assuming a certain congestion control mechanism (max–min fair congestion), is addressed considering a non-cooperative game model. This formulation is a type of the 'bottleneck routing game' model described in (Banner and Orda 2007). Each player corresponds to a source–destination node pair of the network and seeks to choose a route which maximizes its available bandwidth but it is subject to a max–min fairness congestion scheme such that all paths using an arc are assigned an equal share of bandwidth unless a path receives less bandwidth at another link. A distributed routing procedure is devised, enabling a user to find a path with maximal bandwidth under max–min fair congestion control in polynomial time, supposing the paths of other users to be fixed. The authors prove the existence of NE solutions and obtain the corresponding theoretical convergence speed as well bounds for the price of anarchy. In (Bianzino et al. 2011) a very specific network design problem involving a routing problem in Internet type networks, aiming at selecting the nodes which can be switched off in order to *optimize energy saving*, while guaranteeing that the resulting routes (set of available paths for each node pair, after switching off those nodes) satisfy traffic carrying objectives. This complex problem is tackled through a coalition game designated as 'Green Game' (where the nodes are the 'players' and the strategies correspond to associated feasible routes) and a heuristic is used to obtain the solutions. The solutions are tested in a reference Internet access/metropolitan network segment and puts in evidence that there is a tradeoff between QoS and energy saving. Related papers on this problematic area focused on the so-called 'green Internet', using different modelling approaches are for example (Chiaraviglio et al. 2009; Fisher et al. 2010; Bianzino et al. 2010).

Concerning problems on *security issues* these are most relevant in wireless networks, namely in AWNs, particularly vulnerable to malicious attacks. In the report (Michiardi and Molva 2002) a problem of this type, aimed at enforcing node cooperation in AWNs and detect possible intruders, is addressed. For this purpose, a GT approach using a particular formulation of the 'prisoner's dilemma' game, is developed. The players are the mobile nodes of the network and can choose to defect or to cooperate. The security mechanism is modeled through the payoff structure of this game (where the energy cost has a central role), using as inputs data monitoring and based on a model of preferences, designated as 'equity, reciprocity, and competition' due to Bolton and Ockenfels (2000). The resulting routing procedure enables the calculation and storage in every node of the network of the reputation ratings associated with other nodes and so detect passive attacks.

4 Trends and Challenges

We will put forward an outline of future research trends and challenges in various areas of network planning, design, and management focusing topics where it seems likely that more opportunities and challenges for MCGD and GT approaches may arise. For facilitating this exercise, in the following these trends will be organized around three great areas: decision problems associated with the interactions between *telecommunication networks evolution and socio-economic issues*; problems *involving market competition and techno-economic evaluation and selection*; *new problems of network planning and design* in wired and wireless networks. Note that these types of decision issues have no strict boundaries and are often mixed, explicitly or implicitly, in the modelling of the problem.

Telecommunication Networks Evolution and Socio-Economic Issues. The widespread social penetration of communications technologies and services and the resulting socio-economic implications are in the agenda nowadays. In fact, their present remarkable relevance is indisputable, however the future trends are still of non-expectable dimensions stemming from the very rapid evolution of the underlying technological and socio-economic factors. Of course, the associated problems are multidimensional and it seems that multicriteria group decision models can be a very helpful tools of analysis. However, as these issues are relatively new, evolving quickly and requiring also very fast options, the number of studies testing the usefulness of new models is still very limited. Some key issues deserve attention, such as those associated with the evaluation of interactive multimedia and various types of distributional services (such as video streaming or cloud computing) based on broadband network technologies and the explosion of wireless networks in most regions/countries. In fact, these revolutions in telecommunication networks are making it available many new services with great socio-economic impacts. Among them we emphasize those related to healthcare, education, e-commerce, financial services, administrative services, etc. Of course, there are very positive socio-economic impacts, but also potential risks and negative consequences. We believe that, in the future, MCGD analysis can be used in order to help policy makers in a reality evolving very fast, requiring timely decisions. Besides the consequences of broadband capabilities and the availability of new societal/personal services, it must be emphasized some structural consequences of the communications revolution, namely related to the growth of classical and virtual economy, the innovation, the equity and employment issues, the mass media evolution, the local governance vs global governance, as well as regulatory and policy issues. Concerning the issues briefly addressed in this paragraph, we would refer to the contributions in the following papers: (Ogunsola 2005; BSR/CTIA 2012; Bauer 2014,2018; Desruelle and Stančík 2014; Napoli 2015; Castronova et al. 2015; Jorgenson and Vu 2016; Cave and Nicholls 2017; Stocker and Whalley 2018). It must be remarked that in applications of this type it is particularly relevant the complexity of the problems, namely "involving conflicting/collaborations, tactics/strategies, cognitive/emotional and social/cultural issues, but also the cross-fertilization of large number of disciplinary areas, such as theory of organizations, political science, sociology, psychology, telecommunications/internet, systems science, operations research, information systems, decision support systems, etc." (Clímaco 2015). So, it is particularly relevant building new learning oriented multicriteria group decision support systems, open to the combination of inputs from diversified areas. Note that, in the end of Sect. 2.2.1, some considerations related to this issue, were made.

Market Competition and Techno-Economic Evaluation Issues. Having in mind the extremely rapid evolution of telecommunication technologies and associated markets, ranging from those concerning international and national networks to a very wide variety of smaller range networks serving regions, companies, public institutions or private premises, it is expected that new problems (or new forms of similar problems already tackled in the literature) will appear. Needless to say that market competition in these areas is usually very intense regarding all levels of demand supply let it be the choice of network operator(s), service operator(s), technology for given network/service requirements, equipment vendors or any market contests involving simultaneously various of these issues. As it is clear from the literature overview these problems may typically involve various DMs. Also new problems of competition, among service operators or service providers, for customers in different market scenarios, will appear, both in the context of wired and wireless networks of all types. Furthermore, as already stressed, these problems are inherently multicriteria in nature since multiple, often conflicting, incommensurate aspects are at stake in the decision process. Therefore, this is clearly an area where MCGD or GT approaches and models, adequate to each new problem setting, are needed. We could present, as an example, the public tenders for installing and operating the 5G mobile wireless networks, under the auspices of public regulators. In terms of modelling issues raised by this type of problems one could put forward the representation of the role of the regulator(s) and of possible oligopolistic situations, a theme already treated in similar market contexts by classical GT frameworks—see e.g. (Turpin 1998). The main involved challenges are, in our view, the form in which the representation and aggregation of the DM's preferences is considered in the model and the way of tackling the interactivity of the DMs with the decision support system as well as the interaction among the DMs, having in mind that a learning process should be enabled. Similar considerations apply to problems of vendor selection for a particular network technological context with the necessary adaptations. Concerning the use of MCGD approaches in these new market competition problems it should be noted, as pointed out in Sect. 3.1, that the scope of applications of MCGD is essentially concerned with cooperative group decision, not negotiation processes involving non-cooperative/antagonistic decision agents, although the frontier between the two settings be often fuzzy. Nevertheless, as exemplified in (Granat and Wierzbicki 2004) in a strategic management problem involving conflicting actors, MCGD can be very useful for the DMs in a preparation phase of the negotiation process. Of course, this may be a relevant issue to explore in the future, concerning the application of MCGD methods in some telecommunication decision problems of similar nature.

Regarding problems of *technic-economic evaluation/selection* a wide range of new problems is arising, both in wired and wireless networks, having in mind the increasingly great variety of technologies, service types and technical alternatives that may be used in a given networking context. An example was presented in Sect. 3.1 for wired networks concerning the choice of a routing method for a transport network, using a MCGD method. Similar decision problems arise concerning the choice of fault protection mechanisms for various types of transmission networks, taking also into account the multilayer functional nature of modern telecommunication networks. The aspects involved, namely resiliency objectives in the event of equipment or software failures or abnormal working conditions (for example, disaster situations or malicious attacks), and the economic evaluation of the network functioning in failure conditions in multiple and uncertain scenarios, clearly open challenging issues for the development of GT or MCGD approaches. An overview of research trends in the area of *robust network design*, including MCDA applications in this area, is in (Rak et al. 2015). Problems of techno-economic evaluation also will arise in wireless networks having in mind the great variety of technological and architectural solutions already available and the foreseeable technological evolution. An example of decision problem in this area concerns LANs, when alternative networking solutions are considered such as fibre/cable LAN, Hybrid Fiber-Wireless LAN or pure wireless LAN (WLAN). Note that for these types of problems, after selecting a networking techno-economic solution the network operator may have to address the problem of vendor selection for particular types of needed new equipment, again a MCGD problem since usually more than one DM is involved. Ideally, the two problems should be integrated in one bi-level, more complex, challenging group decision problem involving, simultaneously, the evaluation/selection of the networking solution and the equipment vendor. A specific type of problems in this area where MCGD and GT approaches could be used is *modernization planning* of the access networks for residential or institutional customers, having in mind the generalized introduction of broadband services and the existence of different technical solutions, so that a preliminary level of decision analysis for evaluating the alternatives, is worth considering. Again MCGD (this is clearly a multiattribute decision problem where more than one DM may be involved) and GT models could be used in this context. Note that, in many situations in this context, complex real applications are not adequately tractable in GT due to the necessity of oversimplifying the model, otherwise its operationalization would become impractical.

New Problems of Network Planning and Design. The rapid evolution of communication technologies and network architectures has raised and will continue to raise new problems of network planning and design which may be formulated as GT problems, as illustrated in the overview section. Also some of these problems, typically involving multiple criteria, may also be tackled with MCGD approaches, when more than one DM external to the networks, is involved.

This is particular true concerning the use of GT approaches for network design problems in *wireless networks* and IP based networks. Here, the development of new technologies/architectures will naturally foster the appearance of many contributions of GT to new problems the types of which were already identified, namely focused on network resource allocation issues (such as *channel assignment*, *bandwidth allocation*, *route selection* and *congestion control*). These problems take multiple forms depending on the types of wireless networks, the QoS/economic performance objectives and the technical–economic constraints. Concerning the objectives/constraints, great importance has to be given to the energy issues. This kind of issues may naturally continue to be tackled by GT approaches, raising significant challenges concerning the development of cooperative or non-cooperative networking games capable of leading to feasible resolution algorithms in terms of computational requirements. Lin et al. (2018) present a review on applications of GT to wireless networks and briefly outline future research directions. Altman et al. (2006) discuss theoretical challenges in the framework of GT, namely the difficult application of hierarchical optimization/Stackelberg equilibrium in telecommunications and the need to develop appropriate models and solution procedures for the new networking game problems.

It should be added that the developments in SDN technologies, both in wired and wireless networks, enabling the *softwarization* of various control functions opens field for new capabilities in network control related design, thence leading to new design problems in these areas, where GT and multicriteria approaches can give relevant contributions.

New forms of *routing problems*, associated with technological evolutions and the rapid increase in the demand for new and more bandwidth greedy services in wired and wireless networks, are expected to come forward. There is great advantage (or, in many cases, a methodological necessity) of tackling these problems through multicriteria approaches taking into account the multidimensional nature of QoS/cost metrics and the need to address tradeoffs between often conflicting and incommensurate metrics. This is particularly important having in mind the trend for the increasing importance of certain forms of multipath routing, namely: calculation of two or more node-to-node maximally risk disjoint paths in *resilient routing* methods; *multicast routing* (when a set of paths has to be calculated from an originating node to a given set of destination nodes, as in distributional services or for interconnecting two given sub-sets of network nodes, as in teleconferencing; if all nodes have to be connected it is designated as *broadcast routing*) and *anycast routing* (involving the calculation of paths from one originating node to one of many possible destination nodes, as in cloud computing), see e.g. (Contreras et al. 2012). Naturally, appropriate modelling formulations will have to be considered, capable of taking into account the technic-economic specificities of each particular decision environment both in terms of optimization objectives or constraints. This is clearly an area where a quite significant number of multicriteria models, assuming only one DM (the network routing designer) have been proposed as analyzed in the state-of-art review (Clímaco et al. 2016) and in the overviews in (Clímaco et al. 2007; Clímaco and Craveirinha 2019). Wierzbicki and Burakowski (2011) analyze, in depth, the features required by routing approaches which may be considered as consistently multicriteria; a conceptual analysis of various forms of understanding "hierarchies" in routing and in multicriteria routing optimization, is also presented. Note that these MCDA-based routing methods can be fully automated, since that is usually required by the application, namely in dynamic routing, by imbedding in the routing method a system of preferences previously determined by the network designer/DM. This can be done in various ways, namely: by defining dynamic preference regions in the objective function space (see e.g. in (Craveirinha et al. 2016) for broadcast routing), by using reference point methods, a combination of reference points and priority regions (see e.g. in (Clímaco et al. 2006) for unicast routing) or by using a method based on the concept of objective ranking and recurring to achievement functions, in a reference point approach context, as proposed in (Wierzbicki and Burakowski 2011).

Of course, these routing problems can also be formulated in terms of GT approaches by aggregating in the utility functions assigned to the originating nodes (the players) the various optimization objectives. This was the form, already used in many of the GT models (see examples in Sect. 3.2), for tackling routing issues and other design problems when more than one objective is to be explicitly included in the payoff representation. This type of procedure uses, in general, a weighted sum of the measures of each of the considered objectives (i.e. the criteria to be optimized by each user) and we think that there are various issues concerning possible limitations of this type of approach that deserve a discussion in this context. Firstly, the problem of setting of weights which although being devised as importance coefficients, are in most situations associated with tradeoffs between the criteria, so leading to fully compensatory parameters concerning the considered dimensions. Note that the fully compensatory nature of such parameters is questionable in many circumstances, namely when we are dealing with incommensurate dimensions, for example when comparing the loss in a QoS measure with an increase in revenue. Also, beyond the problem of the construction and normalization of scales for each criterion there is the underlying assumption of the independence of the dimensions, essential when an additive model is used, something that is often unrealistic for some pairs of criteria. In this respect we think that MCDA based approaches seem more adequate to deal with the exploration of the tradeoffs. This is clear in interactive models in MCGD approaches as it is typically the case in the applications to socio-economic and technical–economic evaluation problems, as far as the models and the associated decision support tools are adequately prepared to deal with the inherently imprecise nature of the importance parameters and enabling the exploration of the interactivity with the DM(s). If these methodological requirements are met the multicriteria approach may lead, desirably, to a learning process by the DM(s), developed around the choice of a final solution in the Pareto optimal set. On the other hand, it can be argued that GT approaches are inherently better prepared to deal with a direct representation of the conflict between users/players, a feature that may be particularly relevant in market competition problems.

These issues are also relevant in the case of models requiring an automated selection of a solution, for given input information in terms of the service demand requirements and network status information. This is the case in most problems of routing (the only exception is static/quasi static routing or dynamic routing methods with large updating periods) and in network operational design problems as those analyzed in the overview section. As noted above, in these types of problems the alternative approaches to GT, are single or multicriteria decision approaches (with one DM alone), mainly based on mathematical programming (and, in particular cases, in network flow programming), combinatorial optimization methods or on optimal control algorithms. Concerning routing problems, multicriteria shortest paths, muticriteria spanning trees and multicriteria Steiner tree algorithms are most relevant for application in unicast, broadcast and multicast/anycast problems, respectively. Note that for specific problems in this area there are procedures for calculating Pareto optimal solutions, computationally very efficient, namely based on shortest path, k-shortest paths and minimal cost spanning tree algorithms. A review on multicriteria shortest path and tree problems can be seen in (Clímaco and Pascoal 2012). Note that in these areas multicriteria approaches can be used (with some intrinsic advantages) but require that the system of preferences of the DM be imbedded in the resolution method, through an appropriate computational procedure, for example the choice of the first non-dominated solution obtained in a highest priority region in the objective function space, after calculating the convex hull of the Pareto set, or using a weighted Chebyshev distance to an ideal reference point or a combination of both techniques.

We also would like to mention some methodological and theoretical challenges. Firstly, an important issue has to do with the treatment of the uncertainties in various types of MCGD or GT models. In particular, in many network design models, the uncertainty associated with service demands or traffic flows offered to the network is of great relevance but it is an issue the importance of which is frequently underestimated. The representation of this uncertainty is a task with two major aspects: the use of adequate stochastic models (even as mere approximations) in the context of the model, and the determination of estimates of the statistical parameters of these stochastic sub-models. Uncertainties and/or imprecisions inherent to other quantities involved in the MCGD models, that may be of different natures, for instance data collection or modelling of preference aggregation (see (Bouyssou 1990)) are also relevant issues in this regard. This type of concerns and challenges also apply to GT approaches having also in mind the key role of the utility functions representing the player payoffs and the importance, in this context, of the associated parameters. Secondly, several of the more complex of the addressed decision problems, in particular those involving several interrelated levels of decision/optimization, lead to clear modelling and methodological challenges. A known problem of hierarchical optimization in GT in this context has to do with the hierarchical relationship between the objectives of the network manager (the player who defines the network parameters so as to optimize some objective), and other, lower level players (the users) who respond to the values of such parameters (for example related to pricing) by seeking (through their strategies) to achieve some equilibrium solution. This type of problem can be modeled trough a bilevel optimization program (see (Bard 1998)) also known as the Stackelberg leader–follower problem (due to von Stackelberg (1934)). As analyzed in (Altman et al. 2006) this is clearly a relevant problem in some applications to telecommunication networks but also very difficult to solve and it is a challenging issue for GT. Thirdly, the possible choice between GT and MCGD approaches for a given network application—beyond the discussion of possible advantages or limitations associated with methodological differences, in a given decision environment context—, should take into consideration several other aspects. A first issue is the adequacy of the model and of the mathematical formulation developed for the addressed decision problem (involving the evaluation of the inevitable underlying assumptions and of the simplifications on the network representation) to the particular technical features and the available information on the network under analysis. A second and relevant issue is the form in which the interaction of the DMs with the developed computational system should be made in association with the representation in the model of the DM's preferences. A third issue is to obtain sets of data, relevant to the addressed problem, as realistic as possible, namely concerning the network features and the DMs preferences, related to technical and economic aspects. In this respect, the involvement of multidisciplinary contributing teams it is clearly very advantageous, namely by including experts in the used methodology and in the relevant network design aspects, agents capable of understanding each other in the essential aspects of the application of their expertise to the addressed problem. Finally, the development of a resolution procedure computationally feasible and as efficient as possible for the desired application environment, that can vary significantly, is a key issue. For example, in dynamic routing methods we may have CPU time limitations that, at transport network level, may vary from a few minutes to few tenths of milliseconds, depending on the application setting. Although studies of a more prospective or theoretical nature do not impose necessarily these types of concerns, these are important aspects that raise many challenges in the forthcoming research in the new problematic areas of application of GT and MCDA in telecommunications.

Acknowledgements This work was funded by ERDF Funds through the Centre's Regional Operational Program and by National Funds through the FCT—Fundação para a Ciência e a Tecnologia, I.P. under the project CENTRO-01-0145-FEDER-029312. This work was also partially supported by FCT under project UIDB/00308/2020.

References

Adjali, I., Olafsson, S.: Evolutionary game theory: applications to the telecommunications industry. In: Economics of Technological and Institutional Change (ETIC) Conference, Strasbourg, France, vol. 2 (1998)

Ahmad, S.F., Shahid, M.K.: Factors influencing the process of decision making in telecommunication sector. Eur. J. Bus. Manage. **7**(1) (2015)

Alpcan, T., Basar, T.: A globally stable adaptive congestion control scheme for Internet-style networks with delay. IEEE/ACM Trans. Networking **13**(6), 1261–1274 (2005)

Alpcan, T., Başar, T., Srikant, R., Altman, E.: CDMA uplink power control as a noncooperative game. Wireless Netw. **8**(6), 659–670 (2002)

Al-Shehri, S.M., Loskot, P., Numanoğlu, T., Mert, M.: Metrics for broadband networks in the context of the digital economies. In: Haidine, A., Aqqal, A. (eds.) Broadband Communications Networks – Recent Advances and Lessons from Practice, IntechOpen (2017)

Altman, E., El-Azouzi, R., Abramov, V.: Non-cooperative routing in loss networks. Perf. Eval. **49**(14), 257–272 (2002)

Altman, E., El-Azouzi, R., Pourtallier, O.: Avoiding paradoxes in multi-agent competitive routing. Comput. Netw. **43**(2), 133–146 (2003)

Altman, E., Boulogne, T., ElAzouzi, R., Jiménez, T., Wynter, L.: A survey on networking games in telecommunications. Comput. Oper. Res. **33**(2), 286–311 (2006)

Banner, R., Orda, A.: Bottleneck routing games in communication networks. IEEE J. Sel. Areas Commun. **25**(6), 1173–1179 (2007)

Bard, J.F.: Practical Bilevel Optimization – Algorithms and Applications, Nonconvex Optimization and Its Applications, vol. 30. Springer, New York (1998)

Bauer, J.M.: Platforms, systems competition, and innovation: reassessing the foundations of communications policy. Telecommun. Policy **38**(89), 662–673 (2014)

Bauer, J.M.: The Internet and income inequality: socio-economic challenges in a hyperconnected society. Telecommun. Policy **42**(4), 333–343 (2018)

Bean, N.G., Kelly, F.P., Taylor, P.G.: Braess's paradox in a loss network. J. Appl. Prob. **34**(1), 155–159 (1997)

Bianzino, A.P., Chaudet, C., Larroca, F., Rossi, D., Rougier, J.L.: Energy-aware routing: a reality check. In: 2010 IEEE Globecom Workshops, pp. 1422–1427. IEEE, Miami (2010)

Bianzino, A.P., Chaudet, C., Rossi, D., Rougier, J.L., Moretti, S.: The green-game: Striking a balance between QoS and energy saving. In: 2011 23rd International Teletraffic Congress (ITC), pp. 262–269. IEEE, San Francisco (2011)

Bolton, G.E., Ockenfels, A.: ERC: a theory of equity, reciprocity, and competition. Am. Econ. Rev. **90**(1), 166–193 (2000)

Bordetsky, A.B., Iz, P.H.: Dimensions for incorporating multiple criteria group decision support and negotiation support techniques into a geographically distributed environment. In: Proceedings of HICSS-29: 29th Hawaii International Conference on System Sciences, vol. 3, pp. 251–260. IEEE, Wailea (1996)

Bouyssou, D.: Building criteria: a prerequisite for MCDA. In: Bana e Costa, C.A. (ed.) Readings in Multiple Criteria Decision Aid, chap. 4, pp. 58–80. Springer, Heidelberg (1990)

Braess, D.: Über ein paradoxon aus der verkehrsplanung. Unternehmensforschung/Math. Methods Oper. Res. **12**(1), 258–268 (1968)

BSR/CTIA: Socioeconomic impacts of wireless technology: a review of opportunities and challenges in health care, finance, education and community empowerment. Business for Social Responsibility (BSR), CTIA - The Wireless Association (2012). https://www.academia.edu/28795561/Socioeconomic_Impacts_of_Wireless_Technology

Calvert, B., Solomon, W., Ziedins, I.: Braess's paradox in a queueing network with state-dependent routing. J. Appl. Prob. **34**(1), 134–154 (1997)

Castronova, E., Knowles, I., Ross, T.L.: Policy questions raised by virtual economies. Telecommun. Policy **39**(9), 787–795 (2015)

Cave, M., Nicholls, R.: The use of spectrum auctions to attain multiple objectives: policy implications. Telecommun. Policy **41**(5–6), 367–378 (2017)

Chang, C.W., Liu, H., Huang, G., Lin, B., Chuah, C.N.P.: Distributed measurement-aware routing: Striking a balance between measurement and traffic engineering. In: 2012 Proceedings IEEE INFOCOM, pp. 2516–2520. IEEE, Orlando (2012)

Chen, T., Zhong, S.: Perfectly fair channel assignment in non-cooperative multiradiomulti-channel wireless networks. Comput. Commun. **32**(6), 1058–1061 (2009)

Chen, L., Low, S.H., Doyle, J.C.: Random access game and medium access control design. IEEE/ACM Trans. Networking **18**(4), 1303–1316 (2010)

Chiaraviglio, L., Mellia, M., Neri, F.: Reducing power consumption in backbone networks. In: 2009 IEEE International Conference on Communications, pp. 1–6. IEEE, Dresden (2009)

Clímaco, J.: Formal models. In: Kamiński, B., Kersten, G., Szapiro, T. (eds.) Outlooks and Insights on Group Decision and Negotiation – 15th International Conference, GDN 2015, Warsaw, Poland, 22–26 June 2015, Proceedings, Lecture Notes in Business Information Processing, vol. 218, pp. XXIV–XXVIII. Springer (2015)

Clímaco, J., Craveirinha, J.: MCDA/M in telecommunication networks - challenges and trends. In: Amor, S.B., de Almeida, A.T., de Miranda, J.L., Aktas, E. (eds.) Advanced Studies in Multi-Criteria Decision Making, chap. 2, Chapman and Hall/CRC, CRC Press, Taylor and Francis, New York (2019)

Clímaco, J.N., Dias, L.C.: An approach to support negotiation processes with imprecise information multicriteria additive models. Group Decis. Negot. **15**(2), 171–184 (2006)

Clímaco, J.C.N., Pascoal, M.M.B.: Multicriteria path and tree problems: discussion on exact algorithms and applications. Int. Trans. Oper. Res. **19**(1–2), 63–98 (2012)

Clímaco, J.C.N., Craveirinha, J.M.F., Pascoal, M.M.B.: An automated reference point-like approach for multicriteria shortest path problems. J. Syst. Sci. Syst. Eng. **15**(3), 314–329 (2006)

Clímaco, J.C.N., Craveirinha, J.M.F., Pascoal, M.M.B.: Multicriteria routing models in telecommunication networks - overview and a case study. In: Shi, Y., Olson, D.L., Stam, A. (eds.) Advances in Multiple Criteria Decision Making and Human Systems Management: Knowledge and Wisdom, pp. 17–46. IOS Press (2007)

Clímaco, J., Costa, J.P., Kersten, G.: Introduction. Group Decis. Negot. **24**(4), 563–566 (2015a)

Clímaco, J., Craveirinha, J., Martins, L.: Cooperative group multiattribute analysis of routing models for telecommunication network. In: Kamiński, B., Kersten, G., Szufel, P., Jakubczyk, M., Wachowicz, T. (eds.) The 15th International Conference on Group Decision and Negotiation Letters, pp. 177–184, Warsaw School of Economics Press (2015b)

Clímaco, J., Craveirinha, J., Girão-Silva, R.: Multicriteria analysis in telecommunication network planning and design: a survey. In: Greco, S., Ehrgott, M., Figueira, J.R. (eds.) Multiple Criteria Decision Analysis: State of the Art Surveys, International Series in Operations Research & Management Science, 2nd edn., chap. 26, vol. 233, pp. 1167–1233. Springer, New York (2016)

Clímaco, J., Craveirinha, J., Martins, L.: Comparison of routing methods in telecommunication networks – an overview and a new proposal using a multicriteria approach dealing with imprecise information. In: Doumpos, M., Figueira, J.R., Greco, S., Zopounidis, C. (eds.) New Perspectives in Multiple Criteria Decision Making, Multiple Criteria Decision Making, chap. 16, pp. 397–427. Springer, Cham (2019)

Cohen, J.E., Kelly, F.P.: A paradox of congestion in a queuing network. J. Appl. Prob. **27**(3), 730–734 (1990)

Coimbra, J., Schutz, G., Correia, N.: Forwarding repeated game for end-to-end QoS support in fiber-wireless access networks. In: 2010 IEEE Global Telecommunications Conference (GLOBECOM 2010), pp. 1–6. IEEE, Miami (2010)

Contreras, L.M., Lopez, V., De Dios, O.G., Tovar, A., Munoz, F., Azanon, A., Fernandez-Palacios, J.P., Folgueira, J.: Toward cloud-ready transport networks. IEEE Commun. Mag. **50**(9), 48–55 (2012)

Correa, J.R., Schulz, A.S., Stier-Moses, N.E.: Selfish routing in capacitated networks. Math. Oper. Res. **29**(4), 961–976 (2004)

Correa, J.R., Schulz, A.S., Stier-Moses, N.E.: On the inefficiency of equilibria in congestion games. In: Jünger, M., Kaibel, V. (eds.) Integer Programming and Combinatorial Optimization, IPCO 2005. Lecture Notes in Computer Science, vol. 3509, pp. 167–181. Springer, Heidelberg (2005)

Cranmer, S.J., Desmarais, B.A.: Inferential network analysis with exponential random graph models. Polit. Anal. **19**(1), 66–86 (2011)

Craveirinha, J., Girão-Silva, R., Clímaco, J.: A meta-model for multiobjective routing in MPLS networks. CEJOR **16**(1), 79–105 (2008)

Craveirinha, J.M.F., Clímaco, J.C.N., Martins, L.M.R.A., Pascoal, M.M.B.: An exact method for constructing minimal cost/minimal SRLG spanning trees over optical networks. Telecomm. Syst. **62**(2), 327–346 (2016)

Desruelle, P., Stančík, J.: Characterizing and comparing the evolution of the major global economies in information and communication technologies. Telecomm. Policy **38**(89), 812–826 (2014)

Dias, L.C., Clímaco, J.N.: Additive aggregation with variable interdependent parameters: the VIP analysis software. J. Oper. Res. Soc. **51**(9), 1070–1082 (2000)

Dias, L.C., Clímaco, J.N.: Dealing with imprecise information in group multicriteria decisions: a methodology and a GDSS architecture. Eur. J. Oper. Res. **160**(2), 291–307 (2005)

Ergün, S., Aydoğan, T., Alparslan Gök, S.Z.: An overview of game theory and its applications in network routing. Afyon Kocatepe Univ. J. Sci. Eng. **16**, 196–204 (2016)

Falowo, O.E., Chan, H.A.: Dynamic RAT selection for multiple calls in heterogeneous wireless networks using group decision-making technique. Comput. Netw. **56**(4), 1390–1401 (2012)
Felegyhazi, M., Hubaux, J.P.: Game theory in wireless networks: a tutorial. Technical Report LCA-REPORT-2006-002. EPFL, Switzerland (2006). https://infoscience.epfl.ch/record/79715
Felegyhazi, M., Cagalj, M., Bidokhti, S.S., Hubaux, J.P.: Non-cooperative multiradio channel allocation in wireless networks. In: IEEE INFOCOM 2007 - 26th IEEE International Conference on Computer Communications, pp. 1442–1450. IEEE, Barcelona (2007)
Figueira, J.R., Mousseau, V., Roy, B.: ELECTRE methods. In: Greco, S., Ehrgott, M., Figueira, J.R. (eds.) Multiple Criteria Decision Analysis: State of the Art Surveys, International Series in Operations Research & Management Science, 2nd edn., chap. 5, vol. 233, pp. 155–185. Springer, New York (2016)
Fisher, W., Suchara, M., Rexford, J.: Greening backbone networks: reducing energy consumption by shutting off cables in bundled links. In: Green Networking 2010: Proceedings of the First ACM SIGCOMM Workshop on Green Networking, pp. 29–34. ACM Press, New Delhi (2010)
Gao, L., Wang, X.: A game approach for multi-channel allocation in multi-hop wireless networks. In: MobiHoc 2008: Proceedings of the 9th ACM International Symposium on Mobile Ad Hoc Networking and Computing, pp. 303–312. ACM Press, Hong Kong (2008)
Giagkos, A., Wilson, M.S., Tuci, E., Charlesworth, P.B.: Comparing approaches for coordination of autonomous communications UAVs. In: 2016 International Conference on Unmanned Aircraft Systems (ICUAS), pp. 1131–1139. IEEE, Arlington (2016)
González, C.I., Verhoest, K.: De facto regulatory decision-making processes in telecommunications regulation: explaining influence relationships with exponential random graph models. J. Public Policy **40**(1), 144–170 (2020)
Granat, J., Wierzbicki, A.P.: Multicriteria analysis in telecommunications. In: Proceedings of the 37th Annual Hawaii International Conference on System Sciences. IEEE, Big Island (2004)
Huang, X., Jiao, C.: Congestion control in satellite networks. In: 2008 4th International Conference on Wireless Communications, Networking and Mobile Computing (WiCOM 2008), pp. 1–4. IEEE, Dalian (2008)
Hwang, C.L., Yoon, K.: Multiple Attribute Decision Making – Methods and Applications A State-of-the-Art Survey. Lecture Notes in Economics and Mathematical Systems, vol. 186. Springer, Heidelberg (1981)
Jorgenson, D.W., Vu, K.M.: The ICT revolution, world economic growth, and policy issues. Telecommun. Policy **40**(5), 383–397 (2016)
Katsianis, D., Gyürke, A., Konkoly, R., Varoutas, D., Sphicopoulos, T.: A game theory modeling approach for 3G operators. NETNOMICS **8**(1–2), 71–90 (2007)
Keeney, R.L., Raiffa, H.: Decisions with Multiple Objectives: Preferences and Value Tradeoffs. Cambridge University Press (1993)
Kelly, F.: Fairness and stability of end-to-end congestion control. Eur. J. Control **9**(23), 159–176 (2003)
Kelly, F.P., Maulloo, A.K., Tan, D.K.H.: Rate control for communication networks: shadow prices, proportional fairness and stability. J. Oper. Res. Soc. **49**(3), 237–252 (1998)
Kersten, G.E.: Support for group decisions and negotiations an overview. In: Clímaco, J. (ed.) Multicriteria Analysis, chap. 32, pp. 332–346. Springer, Heidelberg (1997)
Kersten, G.E., Cray, D.: Perspectives on representation and analysis of negotiation: towards cognitive support systems. Group Decis. Negot. **5**, 433–467 (1996)
Kodialam, M., Lakshman, T.V.: Detecting network intrusions via sampling: a game theoretic approach. In: IEEE INFOCOM 2003. Twenty-second Annual Joint Conference of the IEEE Computer and Communications Societies, vol. 3, pp. 1880–1889. IEEE, San Francisco (2003)
Koulali, S., Sabir, E., Taleb, T., Azizi, M.: A green strategic activity scheduling for UAV networks: a sub-modular game perspective. IEEE Commun. Mag. **54**(5), 58–64 (2016)
Lin, D., Wang, Q., Yang, P.: The game theory: applications in the wireless networks. In: Tuljak-Suban, D. (ed.) Game Theory – Applications in Logistics and Economy, IntechOpen (2018)

Luo, R., Zhao, S., Zhu, Q.: Network selection algorithm based on group decision making for heterogeneous wireless networks. In: 2017 IEEE 9th International Conference on Communication Software and Networks (ICCSN), pp. 397–402. IEEE, Guangzhou (2017)

Maillé, P., Naldi, M., Tuffin, B.: Competition for migrating customers: a gametheoretic analysis in a regulated regime. In: IEEE GLOBECOM 2008 – 2008 IEEE Global Telecommunications Conference, pp. 1–5. IEEE, New Orleans (2008)

Martins, L., Lopes, J., Craveirinha, J., Clímaco, J., Cadime, R., Mónica, C.: Network performance improvement through evaluation of bicriteria routing methods in transport networks. In: Proceedings of the 9th International Conference on Network and Service Management (CNSM 2013), pp. 34–41. IEEE, Zurich (2013)

Michiardi, P., Molva, R.: Game theoretic analysis of security in mobile ad hoc networks. Research Report RR-02-070. Institut Eurécom, Sophia-Antipolis, France (2002)

Min, M.: Game theoretical approaches in wireless networks. In: Chinchuluun, A., Pardalos, P.M., Migdalas, A., Pitsoulis, L. (eds.) Pareto Optimality, Game Theory And Equilibria, Springer Optimization and Its Applications, vol. 17, pp. 645–663. Springer, New York (2008)

Mkiramweni, M.E., Yang, C., Li, J., Han, Z.: Game-theoretic approaches for wireless communications with unmanned aerial vehicles. IEEE Wirel. Commun. **25**(6), 104–112 (2018)

Napoli, P.M.: Social media and the public interest: governance of news platforms in the realm of individual and algorithmic gatekeepers. Telecommun. Policy **39**(9), 751–760 (2015)

Nie, N., Comaniciu, C.: Adaptive channel allocation spectrum etiquette for cognitive radio networks. Mobile Netw. Appl. **11**(6), 779–797 (2006)

Niyato, D., Hossain, E.: Competitive pricing for spectrum sharing in cognitive radio networks: dynamic game, inefficiency of Nash equilibrium, and collusion. IEEE J. Sel. Areas Commun. **26**(1), 192–202 (2008a)

Niyato, D., Hossain, E.: Competitive spectrum sharing in cognitive radio networks: a dynamic game approach. IEEE Trans. Wirel. Commun. **7**(7), 2651–2660 (2008b)

OECD: Measuring the digital economy: A new perspective (2014)

OECD: OECD Workshop on broadband metrics (London, June 2012): Summary of recommendations (2012). https://one.oecd.org/document/DSTI/ICCP/CISP(2012)13/FINAL/en/

Ogunsola, L.A.: Information and communication technologies and the effects of globalization: twenty-first century "digital slavery" for developing countries – Myth or reality? Electr. J. Acad. Spec. Librariansh. **6**(1–2) (2005)

Papadimitriou, C.H.: Algorithms, games, and the Internet. In: STOC 2001: Proceedings of the Thirty-Third Annual ACM Symposium on Theory of Computing, pp. 749–753. ACM Press, Hersonissos (2001)

Qin, J., Liu, X.: Multi-attribute group decision making using combined ranking value under interval type-2 fuzzy environment. Inf. Sci. **297**, 293–315 (2015)

Rak, J., Pickavet, M., Trivedi, K.S., Lopez, J.A., Koster, A.M.C.A., Sterbenz, J.P.G., Çetinkaya, E.K., Gomes, T., Gunkel, M., Walkowiak, K., Staessens, D.: Future research directions in design of reliable communication systems. Telecommun. Syst. **60**(4), 423–450 (2015)

Rapoport, A.: Strategy and Conscience. Harper & Row, New York (1964)

Roy, B., Bouyssou, D.: Aide multicritère à la décision: méthodes et cas. Economica (1993)

Saaty, T.L.: The Analytic Hierarchy Process: Planning, Priority Setting, Resource Allocation. Decision Making, McGraw-Hill (1980)

Saaty, T.L.: Highlights and critical points in the theory and application of the Analytic Hierarchy Process. Eur. J. Oper. Res. **74**(3), 426–447 (1994a)

Saaty, T.L.: How to make a decision: the analytic hierarchy process. Interf./informs J. Appl. Anal. **24**(6), 19–43 (1994b)

Sahin, A., Pehlivan, N.Y.: A novel multi-criteria decision making method based on the ranking values of interval type-2 fuzzy sets: an application of a manager selection for a telecommunication company. Süleyman Demirel Üniversitesi Fen Bilimleri Enstitüsü Dergisi **22**, 290–296 (2018)

Shah, I., Jan, S., Loo, K.K.: Selfish flow games in non-cooperative multi-radio multi-channel wireless mesh networks with imperfect information. In: 2010 6th International Conference on Wireless and Mobile Communications (ICWMC), pp. 219–225. IEEE, Valencia (2010)
Shah, I.A., Jan, S., Khan, I., Qamar, S.: An overview of game theory and its applications in communication networks. Int. J. Multidisc. Sci. Eng. **3**(4), 5–11 (2012)
Slowinski, R., Greco, S., Matarazzo, B.: Rough set and rule-based multicriteria decision aiding. Pesquisa Operacional **32**(2) (2012)
Smith, J.M.: Evolution and the Theory of Games. Cambridge University Press, Cambridge (1982)
Stocker, V., Whalley, J.: Speed isn't everything: a multi-criteria analysis of the broadband consumer experience in the UK. Telecommun. Policy **42**(1), 1–14 (2018)
Tam, M.C.Y., Tummala, V.M.R.: An application of the AHP in vendor selection of a telecommunications system. Omega **29**(2), 171–182 (2001)
Tran, T.X., Hajisami, A., Pandey, P., Pompili, D.: Collaborative mobile edge computing in 5G networks: new paradigms, scenarios, and challenges. IEEE Commun. Mag. **55**(4), 54–61 (2017)
Turpin, E.: Lessons from game theory for telecommunications. Réseaux **6**(1), 15–39 (1998)
von Stackelberg, H.: Marktform und Gleichgewicht. J. Springer (1934)
Wang, L.F.: Compatibility and group decision making. Syst. Eng. Theory Pract. **20**(2), 93–97 (2000)
Wang, Y.J., Lee, H.S.: Generalizing TOPSIS for fuzzy multiple-criteria group decision-making. Comput. Math. Appl. **53**(11), 1762–1772 (2007)
Wardrop, J.G.: Road paper. Some theoretical aspects of road traffic research. Proc. Inst. Civil Eng. **1**(3), 325–362 (1952)
Wierzbicki, A.P., Burakowski, W.: A conceptual framework for multiple-criteria routing in QoS IP networks. Int. Trans. Oper. Res. **18**(3), 377–399 (2011)
Xiao, J., Xiong, N., Yang, L.T., He, Y.: A joint selfish routing and channel assignment game in wireless mesh networks. Comput. Commun. **31**(7), 1447–1459 (2008)
Yang, D., Xue, G., Fang, X., Misra, S., Zhang, J.: A game-theoretic approach to stable routing in max-min fair networks. IEEE/ACM Trans. Netw. **21**(6), 1947–1959 (2013)
Zadeh, L.A.: Fuzzy sets. Inf. Control **8**(3), 338–353 (1965)
Zadeh, L.A.: The concept of a linguistic variable and its application to approximate reasoning–I. Inf. Sci. **8**(3), 199–249 (1975)
Zeeman, E.C.: Dynamics of the evolution of animal conflicts. J. Theor. Biol. **89**(2), 249–270 (1981)

Group Decision Process for Evaluating a Mango Variety to Be Planted in New Agricultural Farms

Danielle C. Morais, Andre M. Araújo, Eduarda A. Frej, and Adiel T. de Almeida

Abstract Fruticulture is one of the most important sectors of Brazilian agribusiness, being a strategic segment for the country's socio-economic development, with mango having significant participation. One of the most complex problems faced by this sector is the assessment of which variety of mango to grow in new farms, given the long period of time to have the first production and only then, to verify the result of the cultivation. Furthermore, this kind of choice may consider different technical aspects and stakeholders' viewpoints. In that perspective, this paper presents a case study of an agribusiness organization, which is one of the greatest exporting company of Mango from Brazil, that needs to evaluate which variety to plant in new farms intending to double its cultivable area in the next five years. It was developed a group decision process, appropriate to the company's organizational structure, with four phases: 1) definition of the actors, criteria, and identification of alternatives, 2) individual assessment by each decision-maker, 3) application of the framework for choosing a voting procedure; 4) collective result. Based on the results achieved, besides the recommendation of the mango variety to be planted, with this new approach of group decision, it was also possible to enrich the discussion in the process of analysis of the expansion of planted areas, in addition to fostering support for strategic planning for the company's growth in a sustainable way.

Keywords Multicriteria group decision-making (MCGDM) · Voting procedure · Mango culture · FITradeoff · Promethee-ROC · Framework for choosing VP

D. C. Morais (✉) · A. M. Araújo · E. A. Frej · A. T. de Almeida
Departamento de Engenharia de Produção, Universidade Federal de Pernambuco, Av. da Arquitetura - Cidade Universitária, Recife, PE, Brazil
e-mail: dcmorais@insid.org.br

E. A. Frej
e-mail: eafrej@cdsid.org.br

A. T. de Almeida
e-mail: almeida@cdsid.org.br

T. Szapiro and J. Kacprzyk (eds.), *Collective Decisions: Theory, Algorithms And Decision Support Systems*, Studies in Systems, Decision and Control 392,
https://doi.org/10.1007/978-3-030-84997-9_11

1 Introduction

Brazil is the seventh-largest mango producer in the world, behind India, China, Thailand, Indonesia, Mexico, and Pakistan (FAO 2019). In 2018, Brazil harvested about 57 thousand hectares of mango, concentrated basically in two regions: Northeast (73%) and Southeast (27%) (CEPEA 2019).

Data from the ComexStat website (2019) shows that in 2018, Brazil exported approximately quantity of 170 thousand tons of mango and for the first quarter of 2019, Brazil expresses a 41% increase in export volume.

This scenario can be observed by increasing the productive areas, as well as reallocating market windows, whose strategies are outlined by producers, to serve increasingly demanding markets.

Nevertheless, the growing competitiveness in the international fruit trade has been causing, in a way, a segmentation among producers. On the one hand, some agricultural companies with cutting-edge technologies, following the new trends and, therefore, are looking for strategies that guarantee differentiation and exponential participation. On the other side, there is most producers, whose organizational management is based on obsolete routines and without innovations (Ramos 2013).

Even so, with the constant development of processes and imminent competition between organizations, new technological alternatives have been established with the purpose of improving production processes. Such evolution causes a range of data that require treatment so that, when in a state of information, they provide the involvement of sectors in companies and, based on this assumption, they can guarantee greater support for decision making (Slack and Lewis 2009).

This, combined with the demands of new markets and globalization, has demanded a new dimension from Brazilian agriculture, leading to the emergence of technological innovations in this area, as well as shaping existing competitive factors. Consequently, those who do not improve their production and export techniques for their products may lose space in market disputes.

It is evident, therefore, the need for methodologies that serve as input for decision-making for the expansion of new areas, which will be planted in the coming years. However, this kind of decision must contemplate the desires and needs of different stakeholders, such as lower initial costs for investors, minimal environmental degradation, protection of conservation areas, social impacts, productive management, and so on. Thus, considering the different interests of stakeholders, the occurrence of conflicts is imminent.

In general, the choice of the most suitable alternative is currently carried out through a technical analysis only or based on the feeling and know-how of the team's agronomists, which includes information on the initial investments in carrying out the project, maintenance, degree impacts, productive potential, management, among other information. This technical report already points out a preferred option, which is usually accepted by the board and presented to stakeholders, however, it demands long debates with the main parties involved to analyze these alternatives and uncertainties.

Nevertheless, considering the investments required for an expansion of productive capacity, it is extremely important that the decision made transcends the technical indication and, in addition, meets other interests as much as possible.

In view of this problem, it is necessary to elaborate support for the group decision to recommend a solution better adequate for all decision-makers and in relation to all the evaluated criteria, with a holistic view on which variety to plant in new farms.

In this context, multicriteria methods aim to support such types of decision problems, incorporating preferences of the decision-makers and leading them to a solution recommendation. To minimize the generation of an unsatisfactory solution, care related to understanding the problem, establishing the criteria, and applying the appropriate method should be observed. It should be noted that many methods can be long and require time and information that may not be available to the decision maker, making their applicability unfeasible for many contexts.

From then, it was developed a group decision process for a Brazilian exporting mango company for choosing which variety of the fruit should be planted on new farms, intending to double its cultivable area in the next five years. It is important to note that, considering the decision structure of this organization, the problematic that best meets the needs of decision-makers is ranking, i.e., to make an order of the potential alternatives of variety to be planted. Nevertheless, to deal with a group decision, it is important to observe how the group acts.

According to Leyva-López and Fernandéz-González (2003), group decision is a reduction of different individual preferences into a collective preference. There are two main approaches for aggregating these individual preferences: a) input level aggregation (Fig. 1a), or b) output level aggregation (Fig. 1b) (Leyva-López and Fernandéz-González 2003; Dias and Climaco 2005). If the group members are very homogeny and they agree with initial parameters requested by the decision model, i.e., if there is little divergence amongst the group members in their choosing of parameters, so, the input level aggregation is more appropriated. Otherwise, if the group members would like to keep their own individual result, the output level is more adequate. In this case, each member constructs his own individual result, which will be aggregated with other results into a final collective one.

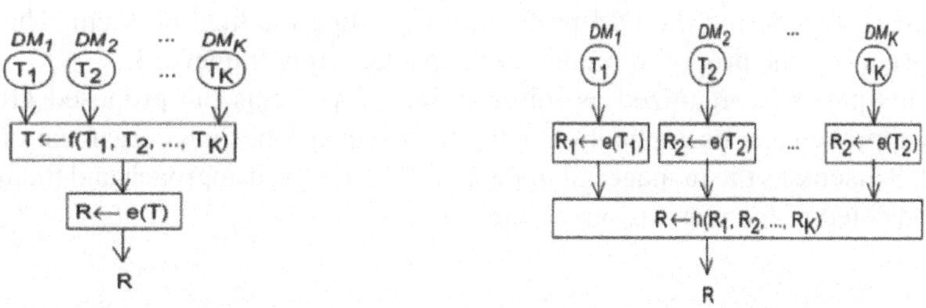

(a) Individual aggregation at input level (b) Individual aggregation at output level

Fig. 1 Main approaches for preference aggregation. *Source* Dias and Climaco (2005)

In that perspective, for the case studied, the characteristic of group members suggests that the most appropriate approach to aggregate preferences is output level aggregation, so each decision-maker will know how each one can solve the problem.

To support these individual decisions, the FITradeoff method for ranking problematic (Frej et al. 2019) is applied. This method has shown to be adequate to the analyzed problem, in view of the compensatory rationality of the decision maker. Furthermore, FITradeoff is based on the axiomatic structure of the classical tradeoff procedure, but being much more accessible and flexible, requiring only partial information regarding preferences, making the elicitation process simpler and with less effort (de Almeida et al. 2016; Frej et al. 2019). In addition, the potential of the FITradeoff method to solve practical problems is evidenced by a wide range of applications areas in which it has been applied. Some of them include supplier selection problems (Frej et al. 2019), selection of agricultural technology packages (Carrillo et al. 2018); problems in the field of energy sector (Kang et al. 2018; Fossile et al. 2020); and scheduling problems (Pergher et al. 2020). Mendes et al. (2020) conducted a simulation study that shows promising results on the capability of the FITradeoff method to find a solution for the decision process with few efforts required from the DMs. Methodological developments have also been proposed to extend the applicability of the FITradeoff method for dealing with sorting problematic (Kang et al. 2020) and portfolio problematic (Frej et al. 2021). Moreover, the use of neuroscience tools have also been applied to improve the decision process in FITradeoff method (Roselli et al. 2019). Recently, the use of holistic evaluations integrated with the classical elicitation by decomposition was proposed as a new flexibility feature of the method, to fasten the decision process (de Almeida et al. 2021).

Considering this, another issue to be evaluated in this decision process is how to aggregate the individual results. Since the results of each decision-maker will be a ranking of the alternatives, voting procedure is one of the easiest ways to deal with this problem and is acceptable by the organization. However, there are several voting procedures that can be used, and each one can give a different result. In this context, De Almeida et al. (2019) proposed a framework for choosing a voting procedure (VP) based on the context that it will be applied. The authors argued that the choice of VP can be based on technical issues as characteristics and proprieties of the VPs and also considering the DM preference regarding the final problem. Therefore, to choose a VP, the proposed model will considerer this framework.

This paper is organized as follows: Sect. 2 presents the proposed group decision approach and its application in the Brazilian agrobusiness organization context; Sect. 3 discusses the managerial impacts of the proposed approach and finally, Sect. 4 is dedicated to the conclusions of the work.

2 Group Decision Process for Choosing a Mango Variety

The group decision process for choosing a Mango variety to be planted in new farms was developed appropriate to the Brazilian company's organizational structure and consists of four phases: 1) definition of the actors, criteria, and identification of alternatives, 2) individual assessment by each decision-maker, 3) application of the framework for choosing a voting procedure; 4) collective result, as can be seen in Fig. 2.

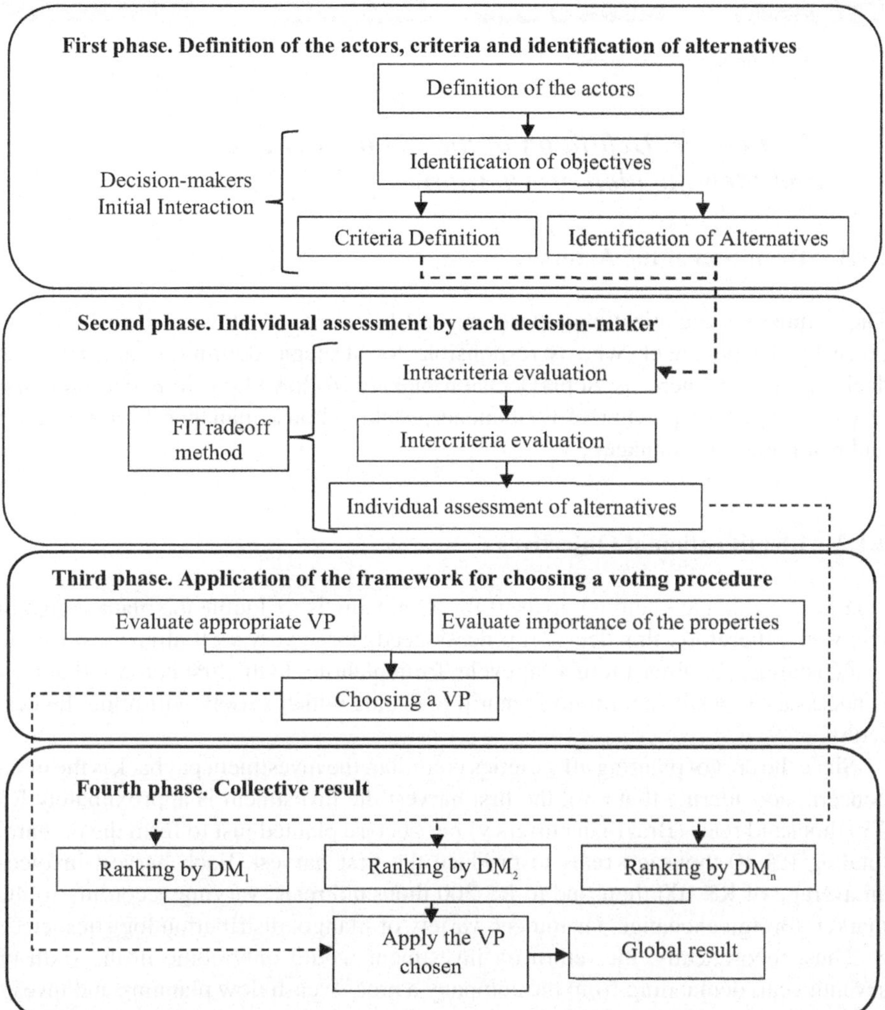

Fig. 2 Flowchart of the proposed group decision process

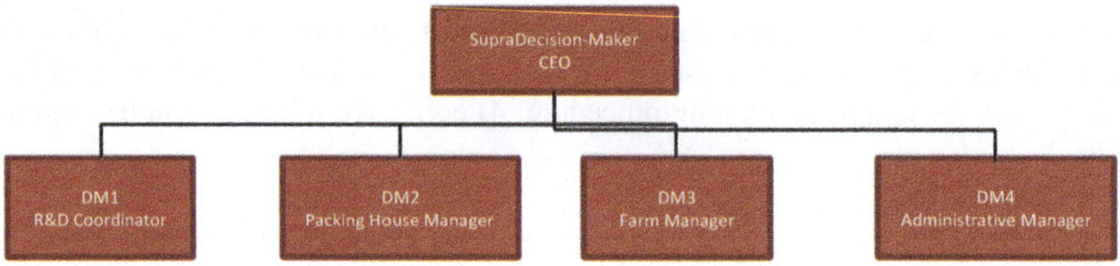

Fig. 3 Decision-makers hierarchy

2.1 First Phase. Definition of the Actors, Criteria and Identification of Alternatives

2.1.1 Definition of the Actors

The company's organizational structure and hierarchy, as illustrated in Fig. 3, is given by the president, who is responsible for strategic definitions and business decisions (called here as SupraDecision-maker); followed by four coordinators: research and development (R&D) manager; packing house manager; farm manager; and administrative manager.

2.1.2 Identification of Objectives

One of the strategic planning goals of this Company is to double the planted area in five years, therefore, this decision process needs to be very well aligned, since the mango culture requires a four-year cycle, from planting to the first harvest, that is, it is necessary to analyze a future scenario to decide which variety will bring the best return.

Since the cost of planting all varieties is similar, the investment payback is the main concern, considering that until the first harvest the investment is approximately R$ 150 thousand reais (Brazilian currency) per hectare planted just to form the orchard, totaling R$ 40 thousand reais to produce the first harvest. Each harvest invoices an average of R$ 100 thousand to R$ 200 thousand reais, varying according to the market, foreign exchange, customers, variety of mango, distribution logistics, etc.

Thus, theoretically, the return on investment would only come in the sixth or seventh year, demanding from the company a precise cash flow planning and investment prospecting. That said, the decision of which variety to plant on new farms is highly strategic, requiring models that enable decision support.

Then, each decision-maker involved in this process, has a different profile and viewpoint. Table 1 shows the relationship between these profiles and viewpoints considered.

Table 1 Decision-makers' profiles and viewpoints

Decision-makers	Concerns	Dimension	Relates to	Objectives
DM1 R&D	Innovation	Technological	Innovative technical aspects that influence good productions	Maximize agronomic performance
DM2 PH	Packaging efficiency	Industrial	Adding value to the finished product	Maximize operational performance and efficiency of the production process
DM3 Farm	Farm production	Agronomic	Produce the best fruit at the lowest cost	Maximize production with quality
DM4 Admin	Administrative	Financial	Costs related to investing in technology for generating electricity	Minimize costs of production

2.1.3 Defining Criteria

The definition of criteria is an extremely important stage for the proper functioning of the decision model. According to de Almeida (2013), we seek to develop the means criteria so that the objectives identified in the previous step can be measured in relation to the performance in relation to the objectives.

The selection of the criteria was based on four decision-makers' perspectives. In view of all the aspects presented above, the following criteria were defined for the problem: Cost; Productivity; Shelf-life; BRIX; Refuse; Industrial income; Demand; Crop and Appearance Calendar. Table 2 shows these criteria and their respective parameters, which were raised in the interviews with decision-makers, meeting their all views.

The measurement of some marketing criteria, sales prices, post-harvest, was not the focus of this research. It is important to note that the company has significant social and environmental purposes, due to its history in the region, with social projects, integration with the community, besides several socio-environmental certifications. Thus, as it has a concise culture, there would be no differences to include this kind of criterion in the evaluation of alternatives.

The "cost" criterion considers its value for monetary amounts to plant the alternative. The same behavior can be seen in the quantitative criteria: "productivity", "shelf life", "°BRIX", "refuse", "industrial" and "calendar".

Table 2 Set of criteria

Code	Criteria	Definition	Unit
C1	Cost	The cost variable can fluctuate according to the cultivar, whether due to the use of labor for specific activities or the use of agricultural inputs. Value considering Cost/kg (Labor + Inputs + Mechanization)	R$/kg
C2	Productivity	Production refers to the quantity of tons that a variety can produce per hectare, using the same spacing between plants	Ton/ha
C3	Shelf-life	It makes reference to the number of days the fruit can take after processing	Days
C4	BRIX	It consists of the proportion of "sugars" that the fruit has	°BRIX
C5	Refuge	Inference by the % of Refuge whose cause was nutritional aspects or pest attack	%
C6	Industrial Yield	It deals with the processing itself, machine capacity, flow within the line and performance	Ton/h
C7	Demand	Commercial evaluation of the company based on contact with customers, in an unstructured interview, to estimate the ranking	–
C8	Harvest Calendar	It denotes the number of months that the variety is produced throughout the year	Month
C9	Appearance	It makes reference to the appearance of the fruit, physical characteristics, shape, consistency, color, aesthetics, being a ranking in consensus	–

Table 3 Scale of evaluation for criterion Demand

Description	Level
Normal demand in foreign market	1
Moderate demand in foreign market	2
High demand in foreign market	3
Very high demand in foreign market	4

Table 4 Scale of evaluation for criterion Appearance

Description	Level
Appetizing	1
Moderate appetizing	2
Very appetizing	3
Extremely appetizing	4

The other criteria, "demand" and "appearance" are subjective, being necessary to create a scale of values to allow comparison. In these criteria, the differences between the alternatives can be small and difficult to measure. Table 3 and Table 4 show the levels defined for the consequences for these criteria, based on their impact, respectively, the demand and appearance scales.

2.1.4 Identification of Alternatives

The identification of alternatives is guided by the universe of everyday knowledge of the decision-makers. Thus, the set of alternatives for the problem is formed by the mango varieties that the company has quick access to:

Tommy and Keitt: varieties of greater dominance, with extensive knowledge of management, however, they are varieties that are falling in the market.

Kent: a variety that has greater demand and higher price, however, only produces in the second half with greater intensity, mango harvest period in Brazil and competitors, that is, a lot of mangoes in general.

Palmer: a variety that has been growing a lot on the market, however, with little knowledge of the cycle on the part of the company that has many mechanical problems to process it due to its shape.

Table 5 shows the varieties of mango considered as alternatives and their characteristics.

Table 5 Set of alternatives

Variety	Characteristics	Appearance
Kent	Large fruits (600g to 750g), with a skin between light green and yellow, acquire a reddish tone with ripening. The pulp is yellow-orange, has a 19 ° Brix, without fiber, aromatic and juicy. Maturation is late. It is prone to internal disorders.	
Keitt	Large fruits (600g to 800g), greenish-yellow peel, pulp of intense yellow tone, without fibers, firm, juicy, sweet and with a Brix of 21 °. Sensitive to internal disorders and prone to browning of pulp	
Palmer	The fruits are large (500g to 900g), orange in color and bright red. The pulp is yellowish, with little fiber, firm and soft aroma, 19 ° Brix. Maturation is late. Growing acceptance in the consumer market. Strong blooms, however, sensitive to internal disorders.	
Tommy	Medium to large fruits (400g to 600g), with predominantly red skin color. Pulp with firm texture, dark yellow color, 17 ° Brix and few fibers. It presents medium resistance to anthracnose, being, however, one of the most sensitive to the internal collapse of the fruits.	

2.2 *Phase 2. Individual Evaluation by Each Decision-Maker*

2.2.1 Intracriteria Evaluation

The intracriteria assessment, according to de Almeida (2013), is interconnected with the evaluation of each alternative for each criterion previously established by the decision-makers, which allows obtaining the consequences matrix. Table 6 is composed of the consequence's matrix of the alternatives, in which the survey of realistic data (use of adjustment factor), occurred by opening the case study company in providing the parameters to identify the value that each alternative has in each criterion.

For the criteria Demand and Appearance, a meeting was held to reach consensus among decision-makers, to define the evaluation of the alternatives, based on the scales presented in Tables 4 and 5.

2.2.2 Intercriteria Evaluation

Different preferences structures were assumed by the decision-makers when ranking the criteria weights and expressing preferences. Then, the FITradeoff elicitation process was performed with each decision-maker based on data from the consequence matrix (Table 6), which obviously, led to different results.

Using the FITradeoff support tool (available upon request at: http://fitradeoff.org/), each decision-maker could experience to solve alone the problem. As an example of the application into the decision support system, let us consider DM1 R&D. After the input file containing the consequences matrix and the criteria (Table 6), the DM ranks the criteria weights according to his preferences, which can be made either by a holistic assessment considering all criteria or by pairwise comparisons, which is one of the flexibility features of the method. Figure 4 shows the screen of the decision support system for the step of ranking of weights by holistic assessment.

The output of this process is the criteria weights order for each DM. Table 7 shows the final criteria rankings per decision-maker, where k_j corresponds to the weight of a criterion c_j.

After ranking the criteria weights, the system starts asking questions for the DM, in which he/she has to declare his/her preference between two consequences (A and B), considering tradeoffs amongst different criteria. Figure 5 shows a screen of the FITradeoff DSS at this step, in which the DM makes a comparison between these consequences: consequence A has an intermediate outcome for criterion "Demand" and the worst outcome for all other criteria; and consequence has the best outcome for criterion "BRIX" and the worst outcome for the others.

Table 6 Consequences matrix

Alternatives	Criteria								
	C1 Cost	C2 Productivity	C3 Shelf Life	C4 BRIX	C5 Refuge	C6 Industrial	C7 Demand	C8 Calendar	C9 Appearance
	Min	Max	Max	Max	Min	Max	Max	Max	Max
Tommy	R$ 1,15	29,3 ton/ha	40 days	11,9	3,6%	17 ton/h	2	10 months	4
Keitt	R$ 1,09	34,6 ton/ha	25 days	12,9	2,6%	23 ton/h	1	12 months	1
Kent	R$ 1,13	33,5 ton/ha	35 days	14,9	3,0%	25 ton/h	4	4 months	2
Palmer	R$ 0,98	38,1 ton/ha	30 days	11,6	5,2%	12 ton/h	3	12 months	3

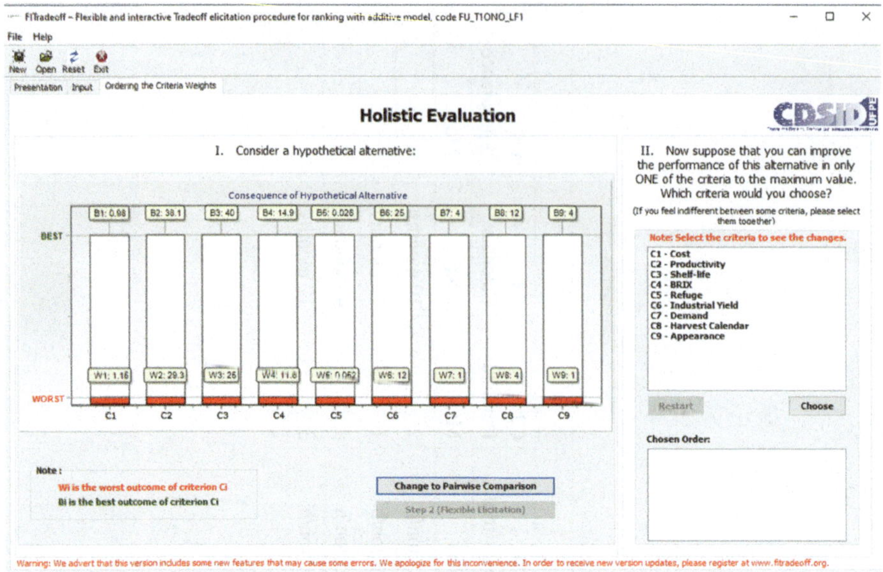

Fig. 4 Ranking of criteria weights by DM1(FITradeoff DSS)

Table 7 Ranking of criteria weights per decision-maker

DMs	Ranking of criteria weights
DM1 R&D	$k_{(\text{Demand})} = k_{(\text{Appearance})} > k_{(\text{Cost})} = k_{(\text{Productivity})} = k_{(\text{Harvest Calendar})} > k_{(\text{Shelf-life})} = k_{(\text{BRIX})} > k_{(\text{Refuge})} = k_{(\text{Industrial Yield})}$
DM2 PH	$k_{(\text{Industrial Yield})} > k_{(\text{Refuge})} = k_{(\text{Harvest Calendar})} > k_{(\text{Shelf-life})} > k_{(\text{Productivity})} = k_{(\text{Demand})} = k_{(\text{Appearance})} > k_{(\text{Cost})} = k_{(\text{BRIX})}$
DM3 Farm	$k_{(\text{Cost})} = k_{(\text{Productivity})} > k_{(\text{Shelf-life})} = k_{(\text{Refuge})} > k_{(\text{BRIX})} = k_{(\text{Industrial Yield})} = k_{(\text{Harvest Calendar})} = k_{(\text{Demand})} = k_{(\text{Appearance})}$
DM4 Adm	$k_{(\text{Cost})} = k_{(\text{Productivity})} = k_{(\text{Demand})} > k_{(\text{Refuge})} > k_{(\text{Industrial Yield})} = k_{(\text{Harvest Calendar})} > k_{(\text{Shelf-life})} = k_{(\text{Appearance})} > k_{(\text{BRIX})}$

The process is interactive, and after each question has been answered, a linear programming model is applied to compute dominance relations between alternatives, in order to build a ranking of them (Frej et al. 2019). This question-answering process goes on until a complete order (or preorder) of the alternatives has been achieved. Alternatively, the elicitation process may be interrupted before that, in case the partial order obtained at some step during the process is sufficient for the DMs purposes. The next section shows the results obtained for each DM with the application of the FITradeoff method.

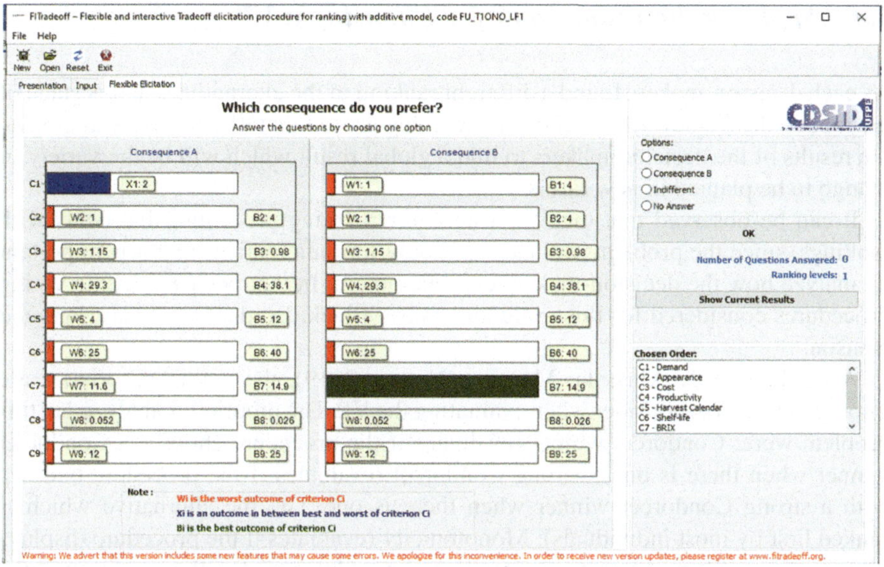

Fig. 5 Elicitation question put for DM1 (FITradeoff DSS)

2.3 *Individual Results*

Table 8 presents the results found by FITradeoff per each decision-maker, also it presents how many questions were answered by each one to get the final ranking.

When analyzing the results for the four decision-makers, it can be observed that the decision-makers had different priorities of varieties, as expected for a decision whose interests are different. The varieties Kent and Palmer are considered good possibilities as the best option for two decision-makers each (Kent for DM2 PH and DM3 Farm and Palmer for DM1 R&D and DM4 Adm). While the Tommy option is the worst for two decision-makers (DM3 and DM4).

This result evidenced the necessity of a voting procedure (VP) to evaluate these rankings. So, in the next phase, the framework for choosing a VP is applied to indicate which VP is better appropriate to analyze this problem.

Table 8 Results per decision-maker

Ranking	DM1	DM2	DM3	DM4
1	Palmer	Keitt, Kent	Kent	Palmer
2	Kent	Tommy	Palmer	Kent
3	Tommy	Palmer	Keitt	Keitt
4	Keitt		Tommy	Tommy
Number of questions answered	9	22	6	2

2.4 Applying the Framework for Choosing a VP

As each decision-makers found a different ranking of the alternatives, at this stage of the model, the framework for choosing a voting procedure (VP) is used to aggregate the results of the decision-makers to find a global result which will be the variety of mango to be planted in new farms.

It can be observed that there is a need for a voting procedure that deals with rankings, since the problem evaluated has only four alternatives and it is important to analyze how the decision-makers classified them. In that perspective, the voting procedures considered for this evaluation were: Black, Borda, Copeland, Hare and Nanson.

According to de Almeida, Morais e Nurmi (2019), the voting proprieties are important aspect to consider when evaluating the VP. The proprieties analyzed in this problem were: Condorcet winner (evaluates if the procedure chooses a Condorcet winner when there is one); Strong Condorcet (evaluates if the procedure ends up with a strong Condorcet winner when there is one, i.e., the alternative which is ranked first by most individuals); Monotonicity (evaluates if the procedure displays monotonicity); Consistency (evaluates if the procedure satisfies the condition of the invariance of the set chosen when different decision-making groups are gathered together to make social choices), and Invulnerability to the no-show paradox (evaluates if a DM may achieve a better result by not voting, thus prompting him/her to manipulate the voting result by abstaining).

The proprieties of Condorcet loser (evaluates if the procedure does not choose a Condorcet loser when there is one) and Pareto (evaluates if the procedure has a collective rationality, i.e., whenever all individuals strictly prefer x to y, then y is not chosen) were not considered since all VPs analyzed satisfy these conditions. Similarly, the Chernoff (evaluates if an alternative is a winner in a set of alternatives, it must be the winner in every subset of these alternatives) and Independence of irrelevant alternatives (evaluates if the procedure satisfies this property) were not considered since none of the VPs analyzed satisfies these conditions.

Table 9 shows the consequences matrix of the VPs and their proprieties based on a discrete binary outcome. For this multicriteria analysis, it was used the PROMETHEE-ROC method (Morais et al. 2015), since this method does not require a value for the criteria weights to be a-priori established. It uses surrogate weights defined based on the ranking of the criteria, applying the Rank Order Centroid (ROC). Then, in Table 8 is also presented the order of priority of these voting proprieties that were given by the SupraDecision-maker.

Table 10 presents the results after applying the PROMETHEE-ROC method to evaluate the decision matrix.

Thus, the Borda voting procedure was identified as the most appropriate to aggregate the decision-makers rankings to find a variety of mango to be planted in new farms.

Table 9 Matrix of consequence of the VP considered

	Criteria/ weights				
Voting system	Condorcet winner	Strong Condorcet	Mono-tonicity	Consis-tency	Invulnerability to the no-show paradox
	4	5	3	1	2
Black	1	1	1	0	0
Borda	0	0	1	1	1
Copeland	1	1	1	0	0
Hare	0	1	0	0	0
Nanson	1	1	0	0	0

Table 10 Results after applying the PROMETHEE-ROC method

Rank	VP	Phi+	Phi−	Phi
1	Borda	0,81	0,09	0,72
2	Copeland	0,16	0,14	0,02
3	Black	0,16	0,14	0,02
4	Nanson	0,07	0,24	-0,17
5	Hare	0,02	0,29	-0,27

Table 11 Results for the group decision-making

Alternatives	Points				
	DM1	DM2	DM3	DM4	Results
Kent	3	4	4	3	14
Palmer	4	2	3	4	13
Tommy	2	3	1	1	7
Keitt	1	4	2	2	9

2.5 Global Result

Since the Borda voting procedure was identified as the most appropriate for aggregating the decision-makers rankings, Table 11 presents the global result with the application of the Borda count to the data presented in Table 8.

Based on the ranking obtained by the Borda count, the Kent variety of mango was the first alternative, followed by Palmer, Keitt and finally Tommy.

3 Managerial Implications of the Model

In view of the company's current top-down strategic decision-making model, the group decision approach has made it possible to open a communication channel between the company's various tactical actors. Initially, it was believed that consensus would only exist if unanimity were reached. However, with the development of debates and points of view, the objective was not unanimous, but an agreement, in which all participants believe they are getting the best solution for the group.

When decision-makers have converging opinions, that is, close responses, the overall result is fluid and comes close to a consensus, however, in cases of divergence, the result has little value for the group's opinion. Thus, it was necessary to propose mechanisms that minimize the discomfort generated by discordant scenarios, as well as from the proposed model, the mathematical complexity was low, enabling consensus among decision-makers.

After outlining the individual preferences of the decision-makers, the same model could be reformulated to address other issues of the company, diluting the responsibility and bias of the SupraDecision-maker, in addition to fostering analysis and support for a structured decision, thus generating a significant impact on the outline of strategic planning.

4 Conclusions

This study presented a group decision model, which meets the problem of choosing which mango variety to plant in new farms in a Brazilian Exporting Organization. The model was developed consisting of four phases: definition of the actors and parameters; individual assessment by each decision-maker; application of the framework for choosing a voting procedure; and finally, the collective result. For that, four multicriteria decision models (representing the four decision-makers) were developed, considering nine criteria and four alternatives. The FITradeoff method was applied for eliciting DMs' individual preferences for build a ranking of the alternatives for each one.

Afterward, the ranking of the decision-makers was aggregated, using the VP chosen using the framework proposed by De Almeida et al. (2019).

Thus, the model developed covered a large part of the company's strategic organization chart, in which, by requiring and generating significant information, the model proved to be an important organizational support tool, bringing to the surface points of convergence and divergence of opinion, in addition to offering support decision making.

In this work, all stages were built with interaction with the organization, the object of the case study, however, it can be applied and replicated in any problem involving choices of fruit cultivation or adapting to the context of any other culture.

Acknowledgements The authors acknowledge the CNPq (Conselho Nacional de Desenvolvimento Científico e Tecnológico) that partially supported this study and the Brazilian Agrobusiness Organization that served as case study and partially support it.

References

Araújo, A.M., Da Silva, A.L.C.L., Morais, D.C., Costa, A.P.C.S.: Escolha de cultura de manga com apoio do método fitradeoff. Anais... LI SBPO – Simpósio Brasileiro de Pesquisa Operacional, Limeira, SP, Brasil (2019)

Carrillo, P.A.A., Roselli, L.R.P., Frej, E.A., de Almeida, A.T. . Selecting an agricultural technology package based on the flexible and interactive tradeoff method. Ann. Oper. Res. (2018). https://doi.org/10.1007/s10479-018-3020-y

Cavalcante, C.A.V., Almeida, A.T.: Modelo multicritério de apoio a decisão para o planejamento de manutenção preventiva utilizando PROMETHEE II em situações de incerteza. Revista Pesquisa Operacional. Rio De Janeiro **25**(2), 279–296 (2005)

CEPEA. Manga/CEPEA: Com bons resultados, produtores devem aumentar investimentos. Disponível em: https://www.cepea.esalq.usp.br/br/diarias-de-mercado/manga-cepea-com-bons-resultados-produtores-devem-aumentar-investimentos.aspx. Acesso em 28 abril de 2019

COMEX STAT. Estatísticas do Comércio Exterior. Disponível em: http://comexstat.mdic.gov.br/. Acesso em: 29 de abril de 2019

De Almeida, A.T.: (2013). *Processo de Decisão nas Organizações: Construindo Modelos de Decisão Multicritério*, 1a Edição. Editora Atlas, São Paulo.

de Almeida, A.T., de Almeida, J.A., Costa, A.P.C.S., de Almeida-Filho, A.T.: A new method for elicitation of criteria weights in additive models: flexible and interactive tradeoff. Eur. J. Oper. Res. **250**(1), 179–191 (2016)

de Almeida, A.T., Frej, E.A., Roselli, L.R.P.: Combining holistic and decomposition paradigms in preference modeling with the flexibility of FITradeoff. Cent. Eur. J. Oper. Res. (2021). https://doi.org/10.1007/s10100-020-00728-z.

De Almeida, A.T., Morais, D.C., Nurmi, H.: Systems, Procedures and Voting Rules in Context: A Primer for Voting Rule Selection. Springer (2019). https://doi.org/10.1007/978-3-030-30955-8

Dias, L.C., Clímaco, J.N.: Dealing with imprecise information in group multicriteria decisions: a methodology and a GDSS architecture. Eur. J. Oper. Res. **160**, 291–307 (2005)

Food and agriculture organization of the united nations (FAO). FAOTSTAT. Disponível em: http://www.fao.org/faostat/en/#data/QC. Acesso em 29 de abril de 2019

Fossile, D.K., Frej, E.A., da Costa, S.E.G., de Lima, E.P., de Almeida, A.T.: Selecting the most viable renewable energy source for Brazilian ports using the FITradeoff method. J. Cleaner Prod. 121107 (2020)

Frej, E.A., de Almeida, A.T., Costa, A.P.C.S.: Using data visualization for ranking alternatives with partial information and interactive tradeoff elicitation. Oper. Res. **19**, 909–931 (2019). https://doi.org/10.1007/s12351-018-00444-2

Frej, E.A., Ekel, P., de Almeida, A.T.: A benefit-to-cost ratio based approach for portfolio selection under multiple criteria with incomplete preference information. Inf. Sci. **545**, 487–498 (2021)

Frej, E.A., Roselli, L.R.P., Araújo de Almeida, J., de Almeida, A.T.: A multicriteria decision model for supplier selection in a food industry based on FITradeoff method. Math. Probl. Eng. 1–9 (2017)

Kang, T.H.A., Frej, E.A., de Almeida, A.T.: Flexible and interactive tradeoff elicitation for multicriteria sorting problems. Asia Pac. J. Oper. Res. **37**, 2050020 (2020)

Kang, T.H.A., Soares Jr, A.M.C., De Almeida, A.T.: Evaluating electric power generation technologies: a multicriteria analysis based on the FITradeoff method. Energy **165**, 10–20 (2018)

Leyva-López, J.C., Fernandéz-González, E.: A new method for group decision support based on ELECTRE III methodology. Eur. J. Oper. Res. **148**, 14–27 (2003)
Marechal, B., Brans, J.P.: Geometrical representation for MCDM, the GAIA procedure. Eur. J. Oper. Res. **34**, 69–77 (1988)
Mendes, J.A.J., Frej, E.A., Almeida, A.T.D., Almeida, J.A.D.: Evaluation of flexible and interactive tradeoff method based on numerical simulation experiments. Pesquisa Operacional **40** (2020)
Morais, D.C., De Almeida, A.T., Alencar, L.H., Clemente, T.R.N., Cavalcanti, C.Z.B.: PROMETHEE-ROC model for assessing the readiness of technology for generating energy. Math. Probl. Eng. (online) **2015**, 1–11 (2015)
Pergher, I., Frej, E.A., Roselli, L.R.P., de Almeida, A.T.: Integrating simulation and FITradeoff method for scheduling rules selection in job-shop production systems. Int. J. Prod. Econ. **227**, 107669 (2020)
Ramos, S.F.: Uso do território no Vale do São Francisco: sistema técnico agrícola da fruticultura irrigada. GEOUSP - Espaço e Tempo **34**, 62–81 (2013)
Roy, B.: Multicriteria Methodology for Decision Aiding. Kluwer Academic Publishers, Netherlands. University Press, Cambridge (1996)
Roselli, L.R.P., De Almeida, A.T., Frej, E.A.: Decision neuroscience for improving data visualization of decision support in the FITradeoff method. Oper. Res. Int. J. **19**, 1–21 (2019)
Slack, N., Lewis, M.: Estratégia de operações. Tradução Sandra de Oliveira. 2nd edn. Bookman, Porto Alegre (2009)
Souza, M.A., Dowsley, B.S., De Almeida, A.T.: Um problema multicritério de seleção de fornecedores com modelo de agregação aditivo. In: Encontro Nacional de Engenharia de Produção, Salvador (2013)

The manufacturer's authorised representative in the EU is Springer Nature Customer Service Centre GmbH, Europaplatz 3, 69115 Heidelberg, Germany. If you have any concerns regarding our products, please contact ProductSafety@springernature.com

Printed and bound by CPI Group (UK) Ltd, Croydon, CR0 4YY
30/09/2026
02225329-0009